D1130485

State Increment
Dynamic Programming

MODERN ANALYTIC AND COMPUTATIONAL METHODS IN SCIENCE AND MATHEMATICS

A Group of Monographs and Advanced Textbooks

Editor: Richard Bellman, University of Southern California

Already Published:

1. R. E. Bellman, R. E. Kalaba, and Marcia C. Prestrud, Invariant Imbedding and Radiative Transfer in Slabs of Finite Thickness, 1963
2. R. E. Bellman, Harriet H. Kagiwada, and Marcia C. Prestrud, Invariant Imbedding and Time-Dependent Transport Processes, 1964
3. R. E. Bellman and R. E. Kalaba, Quasilinearization and Nonlinear Boundary-Value Problems, 1965
4. R. E. Bellman, R. E. Kalaba, and Jo Ann Lockett, Numerical Inversion of the Laplace Transform: Applications to Biology, Economics, Engineering, and Physics, 1966
5. S. G. Mikhlin and K. L. Smolitskiy, Approximate Methods for Solution of Differential and Integral Equations, 1967
6. R. N. Adams and E. D. Denman, Wave Propagation and Turbulent Media, 1966
7. R. L. Stratonovich, Conditional Markov Processes and Their Application to the Theory of Optimal Control, 1968
8. A. G. Ivakhnenko and V. G. Lapa, Cybernetics and Forecasting Techniques, 1967
9. G. A. Chebotarev, Analytical and Numerical Methods of Celestial Mechanics, 1967
10. S. F. Feshchenko, N. I. Shkil', and L. D. Nikolenko, Asymptotic Methods in the Theory of Linear Differential Equations, 1967
12. R. E. Larson, State Increment Dynamic Programming, 1968

In Preparation:

11. A. G. Butkovskii, Optimal Control Theory for Distributed Parameters
13. J. Kowalik and M. R. Osborne, Methods for Unconstrained Optimization Problems
14. S. J. Yakowitz, Mathematics of Adaptive Control Processes
15. S. K. Srinivasan, Stochastic Theory and Cascade Processes
16. D. U. von Rosenberg, Methods for the Numerical Solution of Partial Differential Equations

R. B. Banerji, Theory of Problem Solving: An Approach to Artificial Intelligence

R. Lattès and J.-L. Lions, The Method of Quasi-Reversibility: Applications to Partial Differential Equations. Translated from the French edition and edited by Richard Bellman

T. Parthasarathy and T. E. S. Raghavan, Some Topics in Two-Person Games

J. R. Radbill and G. A. McCue, Quasilinearization and Nonlinear Problems in Fluid and Orbital Mechanics

MODERN ANALYTIC AND COMPUTATIONAL METHODS IN SCIENCE AND MATHEMATICS

MÉTHODES MODERNES D'ANALYSE ET DE COMPUTATION EN SCIENCE ET MATHÉMATIQUE

NEUE ANALYTISCHE UND NUMERISCHE METHODEN IN DER WISSENSCHAFT UND DER MATHEMATIK

НОВЫЕ АНАЛИТИЧЕСКИЕ И ВЫЧИСЛИТЕЛЬНЫЕ МЕТОДЫ В НАУКЕ И МАТЕМАТИКЕ

Editor
RICHARD BELLMAN, UNIVERSITY OF SOUTHERN CALIFORNIA

State Increment Dynamic Programming

by

Robert E. Larson

Stanford Research Institute, Menlo Park, California

AMERICAN ELSEVIER PUBLISHING COMPANY, INC.
NEW YORK 1968

AMERICAN ELSEVIER PUBLISHING COMPANY, INC.
52 Vanderbilt Avenue
New York, N.Y. 10017

ELSEVIER PUBLISHING COMPANY
Barking, Essex, England

ELSEVIER PUBLISHING COMPANY
335 Jan Van Galenstraat, P.O. Box 211
Amsterdam, The Netherlands

Standard Book Number 444-00042-9

Library of Congress Catalog Card Number 68-15006

PRINTED IN THE UNITED STATES OF AMERICA

To Susan R. Larson

PREFACE

Dynamic programming, originally developed by Richard Bellman, has long been recognized as an extremely powerful approach to solving optimization problems. It can be applied to fundamental problems from many different fields, including aerospace engineering, electrical engineering, chemical engineering, physics, economics, and operations research. The basic approach is ideally suited for implementation on a digital computer. The standard computational algorithm based on dynamic programming is very desirable from a number of points of view, including the generality of problems to which it can be applied, the nature of the solution that is obtained, and the ease with which it can be programmed.

Despite the attractive features of the standard algorithm, its applicability thus far has been limited to relatively simple cases. This is due to the large computational requirements of this algorithm. The most severe restriction is generally the amount of high-speed storage required to implement the basic calculations. Another difficulty is the amount of computing time required to obtain the complete solution. Thus, while dynamic programming is frequently used as an analytical and conceptual tool, the computational difficulties associated with the standard algorithm have severely limited its application to large-scale optimization problems.

As a result of this situation a number of researchers, including Bellman and his colleagues, have been motivated to develop new computational procedures that retain the desirable properties of the standard algorithm but that have reduced computational requirements. The primary purpose of this book is to present a detailed exposition of one of the most important of these new algorithms, state increment dynamic programming. This procedure always reduces the high-speed storage requirement, often by orders of magnitude. In a number of cases a substantial reduction in computing time can be achieved as well. These savings are obtained without sacrificing the desirable properties of the standard algorithm. Thus, state increment dynamic programming represents a significant step in increasing the range of optimization problems that can be solved with state-of-the-art computer facilities.

Although this book emphasizes state increment dynamic programming, a number of other new algorithms are also discussed. For example, an entire chapter is devoted to a successive approximation technique due to Bellman. A comprehensive survey of the new algorithms will be the subject of a later Number in this Series.

This book serves two other purposes as well. First, it presents a complete treatment of the basic theory of dynamic programming and the standard computational algorithm. This discussion is valuable not only for its own sake but also as a convenient summary of the prerequisite material for the remainder of the book. The other purpose is to discuss the application of dynamic programming algorithms to optimization problems from a number of different fields. This material is of interest not only to readers who are concerned with the particular applications discussed but also as illustrations of a methodology for formulating problems in terms such that modern optimization techniques can be applied.

This additional material makes the book self-contained and greatly broadens its appeal. The book is thus valuable not only to specialists doing research in computational methods for system optimization but also to readers whose orientation ranges from a general interest in optimization theory to a strong concern with practical applications.

The treatment of the basic theory of dynamic programming is on an introductory level. This presentation is oriented toward the viewpoint of the control engineer; however, it is general enough for use by any reader who has been exposed to the elementary concepts of optimization theory. It differs from the usual expositions of dynamic programming in that the computational aspects are heavily emphasized. Throughout the volume the important concepts are carefully explained and illustrated by detailed examples.

This presentation begins with a thorough discussion of the class of problems to which dynamic programming applies. Next, the basic concepts of dynamic programming are discussed. Then, the standard computational algorithm is described in detail; particular attention is given to the computational difficulties that arise. Finally, the extension of this approach to problems in which stochastic effects are present is described.

Following this discussion the state increment dynamic programming computational procedure is described in detail. A number of variations of the basic approach are discussed. The procedures are illustrated by detailed examples, including several problems having four state variables. This material can be understood with no background other than the introductory material already mentioned.

The discussions of other new algorithms, while not as detailed, are on a similar level with the treatment of state increment dynamic programming. In all cases the explanations are complete, and illustrative examples are provided.

The material covering applications of these algorithms assumes that the reader has grasped the basic concepts used in these procedures. The cases treated are intended to be representative of problems encountered in practice, not just additional illustrative examples. The emphasis of these discussions is on the formulation of problems in terms such that dynamic programming algorithms can be applied and on the presentation of actual results. In many cases the formulation has not been previously published, and a number of the problems for which results are given are the largest problems of their type ever solved by dynamic programming techniques.

The applications come from many of the fields. The problems include: (1) minimum-fuel trajectories for a supersonic transport or other advanced aircraft, (2) minimum-time-to-intercept trajectories for an anti-missile missile, (3) optimum attitude control for aerospace vehicles, (4) optimum operation and planning for multipurpose, multireservoir water resource systems, (5) optimum scheduling of aircraft for an airline, (6) optimum operation of a natural gas pipeline network, (7) optimum adaptive control of a robot equipped with sensors operating in an unknown environment. In many cases it is found that not just one of the new algorithms, but a combination of them, is required to obtain an effective solution method; this is characteristic of real-world problems.

While much of the work reported in this book is indicative of the limits of achievement with the known algorithms using currently available computers, it is recognized that advances in both areas will in the not-too-distant future make possible the solution of problems that dwarf any of the cases considered here. It is the sincere hope of the author that this volume in some way may stimulate these advances.

I would like to take this opportunity to acknowledge some of the many people who have contributed to the completion of this book. I would particularly like to express my gratitude to Professor Richard Bellman of the University of Southern California; not only is there the obvious technical debt owed by anyone writing on the subject of dynamic programming, but in my case his personal encouragement was instrumental in the initial undertaking and the completion of this task, and his comments on the presentation of the material were extremely helpful. I would also like to single out the contributions of Dr. Philip E. Merritt of Stanford Research Institute, whose guidance and encouragement were vital in performing most of the work reported in this book and who made many valuable suggestions concerning the final presentation of the material.

I would also like to acknowledge the guidance of D. W. Richardson of the Hughes Aircraft Company, and G. F. Franklin and W. K. Linvill of Stanford University, under whom the concepts of state increment dynamic programming were originally developed. In addition, I would like to thank J. Peschon of Stanford Research Institute, and S. J. Kahne of the University of Minnesota for their comments on the original version of the manuscript. Also, I would like to acknowledge numerous valuable conversations and a large body of outstanding work on topics related to this book by my colleagues at Stanford

Research Institute; I particularly thank P. E. Merritt, J. Peschon, L. Meier, W. G. Keckler, R. M. Dressler, R. S. Ratner, P. J. Wong, and A. J. Korsak. Finally, I would like to thank the programmers and analysts who obtained the numerical results that are such an important part of this book; I would particularly like to mention D. S. Piercy, W. G. Keckler, W. H. Zwisler, and A. S. Chen, all of Stanford Research Institute.

The acknowledgements would be incomplete without a mention of the people responsible for preparing the manuscript. Particular thanks are due to Ellen Campbell and Barbara Bentley for typing the entire final version. Also, a debt of gratitude is owed to Phyllis Craven and Carolyn Langford for preparing the illustrations.

ROBERT E. LARSON

Menlo Park, California
June, 1968

CONTENTS

Chapter One

INTRODUCTION

1.1 SCOPE OF THE BOOK

The purpose of this book is to present a comprehensive treatment of the computational procedure *state increment dynamic programming* and to illustrate its application to a number of practical problems. The development of this procedure is motivated by what Bellman calls "the curse of dimensionality" [Ref. 1]—the excessive computational requirements of the conventional procedures based on dynamic programming. The major accomplishment of state increment dynamic programming is a large reduction in these computational requirements with no sacrifice in the many desirable properties of a dynamic programming solution.

The optimization problems to which dynamic programming applies are variously called dynamic optimization problems, variational control problems, or optimization problems for multistage decision processes. These problems are characterized by the fact that the control decision taken at the present time affects the behavior of the system at future times, and hence the solution is a sequence of decisions over the entire duration of control, not just a decision at the present time instant. Problems of this type occur in many fields, such as aerospace engineering, electrical engineering, mechanical engineering, chemical engineering, operations research, economics, and physics, to mention only a few.

Dynamic programming, which was developed by Richard Bellman [Refs. 1, 2, 3], provides an approach for solving these problems under very general conditions. Basically, this approach converts the simultaneous determination of the entire optimal control sequence into a sequential determination. The conversion is accomplished by using Bellman's principle of optimality [Refs. 1, 2, 3]. The result is an iterative functional equation which can be solved very efficiently by a digital computer. A generalized computational procedure has been developed for solving this equation; in the remainder of the book this procedure is referred to as the *conventional dynamic programming computational procedure*.

The class of problems for which dynamic programming leads to an iterative equation of the desired form is almost limitless. The equations of the dynamic system being controlled can be nonlinear and time-varying. The performance criterion on which the optimization is based can be quite arbitrary. Constraints of a wide variety are permitted. Random effects of a very general nature can be rigorously accounted for using the same analytic formulation. In addition, the solution is always an absolute optimum, never a local minimum or maximum. Furthermore, the solution specifies the optimal control at every state of the system for every instant of time; thus, the dynamic programming solution can be implemented as a feedback (closed-loop) controller, in which the state of the system is constantly measured and the corresponding optimal control is applied.

The main difficulty with applying dynamic programming to practical problems is the excessive computational requirements of the conventional procedure in problems of moderate to high complexity. One limitation is the high-speed memory requirement; this refers to the amount of data that must be stored in the high-speed access memory in order for the computational procedure to work efficiently. A second difficulty is the total amount of computing time required to solve the problem. A third difficulty is the amount of low-speed storage required to store all the results that are generated. In general, the high-speed memory requirement is a more fundamental restriction than either of the other two. In the first place, the high-speed memory requirement usually exceeds the capacity of the computer before the other two requirements become excessive. In the second place, it is difficult and expensive to add high-speed storage capacity to a computer, while it is relatively easy to let the computer run for a longer time or to add low-speed storage capacity in the form of more reels of tape, disc storage, etc. Of the other two requirements, computing time is generally the more restrictive.

State increment dynamic programming [Refs. 4, 5, 6, 7] is a new approach to solving the iterative functional equation obtained from the principle of optimality. This approach leads to a computational procedure that is quite different from the conventional procedure. This new procedure has computational requirements that are significantly less than those for the conventional case. In particular, the high-speed memory requirement is always reduced, generally by orders of magnitude. In one example, the reduction is from 10^6 locations to about 100 locations. The computing time and low-speed memory requirement can also be significantly reduced in many cases. These computational savings are obtained with little sacrifice in the class of problems that can be treated. Also, the desirable properties of the form of the solution are retained.

In Sec. 1.2 of this chapter, the iterative functional equation based on the principle of optimality is presented. Then, the conventional procedure for solving this equation is briefly reviewed in Sec. 1.3. (A detailed description of this procedure is contained in Chapter 2.) In Sec. 1.4, the basic ideas of the state increment dynamic programming approach are presented. Finally, the remainder of the book is outlined in Sec. 1.5.

1.2 THE ITERATIVE FUNCTIONAL EQUATION BASED ON THE PRINCIPLE OF OPTIMALITY

Dynamic programming can be regarded as a point of view for looking at optimization problems which leads to an iterative functional equation that can be solved efficiently on a digital computer. Such an equation can be obtained for a large class of problems, including many that are quite different from the typical deterministic optimal control problem. In most of this book attention will be focused on the deterministic control problem formulated in terms of the calculus of variations. The extension to problems involving uncertainty is discussed in Chapter 11.

In the deterministic control problem three types of variables are present. The first type is the *state variable*. These variables provide a complete description of the dynamic behavior of the system. The second type is the *control variable*. These variables are the decisions that are to be made in an optimum fashion. The third type is the *stage variable*, which determines the order in which controls are applied; it is generally taken to be time.

The equations describing the system being controlled are taken to be a set of nonlinear time-varying differential equations,

$$\dot{\mathbf{x}} = \mathbf{f}(\mathbf{x}, \mathbf{u}, t) \tag{1.1}$$

where $\mathbf{x}$ = n-dimensional state vector
$\mathbf{u}$ = q-dimensional control vector
t = stage variable, assumed to be time

$$\dot{(\,)} = \frac{d}{dt}(\,).$$

In order to represent these equations on a digital computer, some finite integration formula must be used to approximate the differential equation. The simplest approximation is

$$\mathbf{x}(t + \delta t) = \mathbf{x}(t) + \mathbf{f}[\mathbf{x}(t), \mathbf{u}(t), t]\, \delta t, \tag{1.2}$$

where δt = time increment over which control $\mathbf{u}(t)$ is applied.

The performance criterion, which determines the effectiveness of a given control function, is taken to be a cost function that is to be minimized. This cost function has the general variational form, which consists of the integral of a scalar functional of the state variables, control variables, and time plus a scalar functional of the final state and final time.

$$J = \int_{t_0}^{t_f} l[\mathbf{x}(\sigma), \mathbf{u}(\sigma), \sigma]\, d\sigma + \psi[\mathbf{x}(t_f), t_f], \tag{1.3}$$

where t_0 = initial time
t_f = final time
σ = dummy variable for time
l = scalar functional for cost per unit time
ψ = scalar functional for final-value cost.

Again, if this equation is to be represented on a digital computer, the integration must be approximated by a finite formula. The simplest formula is

$$\int_t^{t+\delta t} l[\mathbf{x}(\sigma), \mathbf{u}(\sigma), \sigma]\, d\sigma = l[\mathbf{x}(t), \mathbf{u}(t), t]\, \delta t. \tag{1.4}$$

The constraints in the problem are expressed as

$$\begin{aligned} &\mathbf{x} \in X(t) \\ &\mathbf{u} \in U(\mathbf{x}, t) \end{aligned} \tag{1.5}$$

where X, the set of admissible states, can vary with t, and where U, the set of admissible controls, can vary with $\mathbf{x}$ and t.

The principle of optimality can be applied to this problem to yield an iterative functional equation of the desired form. This theorem is discussed in detail in Chapter 2; briefly, it states that an optimal control sequence has the property that whatever the initial state and control are, the remaining controls in the sequence must be an optimal control sequence with regard to the state resulting from the first control [Ref. 3, p. 15].

In order to write the iterative equation, the minimum cost function $I(\mathbf{x}, t)$, must first be defined. This function determines the minimum cost that is incurred in going to the final time if the present time is t and if the present state is $\mathbf{x}$. It is defined for all admissible states $\mathbf{x} \in X$ and for all time, $t_0 \leq t \leq t_f$. The defining equation is

$$I(\mathbf{x}, t) = \underset{\substack{\mathbf{u}(\sigma) \in U \\ t \leq \sigma \leq t_f}}{\text{Min}} \int_t^{t_f} l[\mathbf{x}(\sigma), \mathbf{u}(\sigma), \sigma]\, d\sigma + \psi[\mathbf{x}(t_f), t_f], \tag{1.6}$$

where $\mathbf{x}(t) = \mathbf{x}$.

The iterative equation can either be derived directly from Eq. (1.6) or else deduced from the principle of optimality. If the approximations of Eqs. (1.2) and (1.4) are used, this equation is obtained as

$$I(\mathbf{x}, t) = \underset{\mathbf{u} \in U}{\text{Min}} \{l[\mathbf{x}, \mathbf{u}, t]\, \delta t + I[\mathbf{x} + \mathbf{f}(\mathbf{x}, \mathbf{u}, t)\, \delta t, t + \delta t]\}. \tag{1.7}$$

The interpretation of this equation is that the minimum cost at a given state $\mathbf{x}$ and the present time t is found by minimizing, through the choice of the present control $\mathbf{u}$, the sum of $l[\mathbf{x}, \mathbf{u}, t]\, \delta t$, the cost over the next time interval δt, plus $I[\mathbf{x} + \mathbf{f}(\mathbf{x}, \mathbf{u}, t)\, \delta t, t + \delta t]$, the minimum cost of going to t_f from the resulting next state, $\mathbf{x} + \mathbf{f}[\mathbf{x}, \mathbf{u}, t]\, \delta t$.

This iterative equation is solved backwards in time because $I(\mathbf{x}, t)$ depends on values of the minimum cost function at future times. Consequently, the iterations begin by specification of the minimum cost function at the final time, t_f. Using Eq. (1.6),

$$I(\mathbf{x}, t_f) = \psi(\mathbf{x}, t_f). \tag{1.8}$$

The minimum cost function for all $\mathbf{x}$ and t can be evaluated by iteratively solving Eq. (1.7) with Eq. (1.8) as a boundary condition. The optimal control at every $\mathbf{x}$ and t, denoted as $\hat{\mathbf{u}}(\mathbf{x}, t)$, is obtained as the value of $\mathbf{u}(t)$ which minimizes Eq. (1.7) for the given $\mathbf{x}$ and t.

1.3 THE CONVENTIONAL COMPUTATIONAL PROCEDURE

The conventional computational procedure is described in detail in Chapter 2. The first step in this procedure is to quantize each state variable, x_i, $i = 1, 2, \ldots, n$; each control variable, u_j, $j = 1, 2, \ldots, q$; and the stage variable, t. Although it is not necessary that this quantization be uniform, the notation is simplified if this is done; the uniform increment in x_i is denoted as Δx_i, the increment in u_j is denoted as Δu_j, and the increment in t is taken to be Δt. As a result of the quantization, the set of admissible states, X, the set of admissible controls, U, and the set of values of t in the interval $t_0 \leq t \leq t_f$ all contain a finite number of elements.

The iterative relation, Eq. (1.7), is applied only at quantized values of $\mathbf{x}$ and t. The quantity δt, the interval over which each control $\mathbf{u}(t)$ is applied, is set equal to Δt, the increment between successive applications of control. If the quantized values of t are indexed as k, $k = 0, 1, 2, \ldots, K$, where $K\,\Delta t = t_f - t_0$, then Eq. (1.7) can be rewritten as

$$I(\mathbf{x}, k) = \min_{\mathbf{u}\in U} \{l[\mathbf{x}, \mathbf{u}, k]\,\Delta t + I[\mathbf{x} + \mathbf{f}(\mathbf{x}, \mathbf{u}, k)\,\Delta t, k + 1]\}. \tag{1.9}$$

The boundary condition is then

$$I(\mathbf{x}, K) = \psi(\mathbf{x}, t_f). \tag{1.10}$$

These equations are used to compute $I(\mathbf{x}, k)$ for all quantized values of $\mathbf{x}$ on the basis of the values of $I(\mathbf{x}, k + 1)$. At given values of $\mathbf{x}$ and k each quantized control, $\mathbf{u} \in U$, is applied. The quantity in brackets in Eq. (1.9) is evaluated for each control. If the state $\mathbf{x} + \mathbf{f}[\mathbf{x}, \mathbf{u}, k]\,\Delta t$ is not one of the quantized states at time $(k + 1)$, then $I[\mathbf{x} + \mathbf{f}(\mathbf{x}, \mathbf{u}, k)\,\Delta t, k + 1]$ is evaluated by interpolating in the n state variables using values of minimum cost at quantized values of $\mathbf{x}$. After the quantity in brackets has been evaluated for each $\mathbf{u}$, $I(\mathbf{x}, k)$ is chosen as the minimum value by comparing these quantities. The optimal control $\hat{\mathbf{u}}(x, k)$ is determined as the value of $\mathbf{u}$ for which the minimum is attained.

It can be seen that this procedure has all the desirable properties referred to in Sec. 1.1. The method does not at all depend on the form of $\mathbf{f}(\mathbf{x}, \mathbf{u}, t)$; this functional can be nonlinear and time-varying, and it can include tabular data. Furthermore, $l(\mathbf{x}, \mathbf{u}, t)$ and $\psi(\mathbf{x}, t_f)$ can have almost any desired form. Also, constraints on $\mathbf{x}$ merely reduce the total number of states at which $I(\mathbf{x}, k)$ must be evaluated, while constraints on $\mathbf{u}$ reduce the total number of admissible controls that must be examined at each $\mathbf{x}$ and k; thus constraints not only present no difficulty, but actually decrease the number of computations.

Because $I(\mathbf{x}, k)$ is determined by exhaustively searching over all possible (quantized) values of $\mathbf{u}$ at each $\mathbf{x}$ and k, it can be shown that an absolute minimum is always found, not just a local minimum or maximum. Furthermore, because optimal control, $\hat{\mathbf{u}}(\mathbf{x}, k)$, is specified at all (quantized) values of $\mathbf{x}$ and k, the solution can be used for feedback (closed-loop) control.

On the other hand, it can be seen that the computational requirements for this method are quite large. Because the next state can lie anywhere in the space of admissible states, the procedure can work efficiently only if minimum cost for each quantized $\mathbf{x} \in X$ at time $(k + 1)$ is stored in the high-speed memory. These data are required so that $I[\mathbf{x} + \mathbf{f}(\mathbf{x}, \mathbf{u}, k)\, \Delta t, k + 1]$ can always be evaluated without referring to a low-speed access memory. Therefore, the high-speed memory requirement is one storage location for each quantized admissible state. If there are N_i quantization levels in x_i, then this requirement is

$$N_h = \prod_{i=1}^{n} N_i. \tag{1.11}$$

If there are 3 state variables ($n = 3$) and 100 quantization levels in each state variable ($N_i = 100, i = 1, 2, 3$), then $N_h = 10^6$. Although this problem is not unreasonably large, 10^6 locations exceeds the high-speed storage capacity of the largest readily available computers.*

If k is the number of time increments, then the total number of values of $\mathbf{x}$ at which $I(\mathbf{x}, k)$ and $\hat{\mathbf{u}}(\mathbf{x}, k)$ are evaluated is N_c, where

$$N_c = N_h \cdot K. \tag{1.12}$$

If Δt_c seconds are required for each computation, then the total computing time becomes T_c, where

$$T_c = N_c \cdot \Delta t_c = N_h \cdot K \cdot \Delta t_c. \tag{1.13}$$

For the problem where $N_h = 10^6$, if $K = 50$ and $\Delta t_c = 50$ μsec, a reasonable set of values, then $T_c = 2500$ seconds, which is about 40 minutes. This amount of time is not trivial in cost, but it is easy to imagine problems with

* As of mid-1967, the maximum high-speed storage capacity of any standard computer is 256,000 locations. However, $2 \cdot 10^6$ or $3 \cdot 10^6$ can be obtained in a few special-purpose installations.

this much complexity where the optimum solution is economically justified. However, we have already seen that the high-speed memory requirement for this problem is excessive. Thus, in this example, as in most complex problems, the high-speed memory requirement is a more restrictive barrier than is computing time.

1.4 STATE INCREMENT DYNAMIC PROGRAMMING

State increment dynamic programming is a new approach to solving iterative functional equations of the form of Eq. (1.7). As shown in Chapters 10 and 11, this approach can be used not only for the deterministic control problem, but also for the stochastic control problem, the optimum estimation problem, and certain discrete-time problems. It is known that there are many other situations where iterative equations of the form of Eq. (1.7) occur and where this approach is applicable.

State increment dynamic programming leads to a computational procedure that has significant advantages in terms of computational requirements over the conventional procedure. First, the high-speed memory requirement is always reduced, generally by orders of magnitude. In certain problems, computing time can also be greatly reduced; in any case, it is never increased by a significant margin. These savings in computational requirements are obtained with little sacrifice in the generality of the problems that can be treated. Furthermore, the desirable properties of the form of the solution are retained.

In most of this book, the approach is applied to the problem formulated in Sec. 1.2. For this case, the iterative equation is as shown in Eq. (1.7). The details of the computational procedure that results from applying the approach are discussed in Chapters 3–6. In this procedure, the quantization of state variables, control variables, and time is done exactly as in the conventional case. Minimum cost and optimal control are computed at all quantized values of $\mathbf{x}$ and t. The minimization in Eq. (1.7) is performed by applying all admissible quantized controls and choosing the minimum value of the quantity in brackets. In this manner, the generality of the problem formulation and the form of the solution are in no way changed.

The reduction in computational requirements is obtained by combining the two basic concepts of the new approach. The first concept is related to the choice of δt in Eq. (1.7). This increment, which is the interval over which control is applied, is *not* set equal to Δt, the time increment between successive computations of optimal control. Instead, it is chosen so that the change in any state variable, x_i, is at most Δx_i, the increment size in that variable. Formally,

$$\delta t = \min_{i=1,2,\ldots,n} \left\{ \left| \frac{\Delta x_i}{f_i[\mathbf{x}, \mathbf{u}, t]} \right| \right\}. \tag{1.14}$$

As a result, the next state, $\mathbf{x} + \mathbf{f}[\mathbf{x}, \mathbf{u}, t]\, \delta t$, is always close to the present state; specifically, it lies on the surface of an n-dimensional hypercube centered at $\mathbf{x}$, with length $2\, \Delta x_i$ along the x_i-axis. The minimum cost at the next state is found using interpolation in $(n - 1)$ of the state variables and time, using values at the quantized states that lie in this hypercube at times $(t + \Delta t)$ and $(t + 2\, \Delta t)$. Consequently, only the minimum costs corresponding to these points need to be stored in the high-speed memory, as opposed to the minimum costs for every admissible quantized state as required by the conventional procedure.

This apparent saving in high-speed memory requirement is not by itself of great value if the computations take place in the same order as in the conventional procedure. If another present state is used, then the quantized states in the hypercube are different. It is thus necessary to retrieve new values of minimum cost from low-speed storage every time a new present state is used. This frequent transfer of data from the low-speed memory to the high-speed memory consumes so much computing time that the procedure is not computationally attractive.

On the other hand, if large amounts of data are transferred infrequently from low-speed to high-speed memory, then (in most modern digital computers), there is little or no loss of computing time, because such transfers can be performed very efficiently, often in parallel with mathematical operations. The second concept in the approach takes advantage of this fact.

This second concept consists of carrying out the computations, not according to the ordering of the time increments, as in the conventional procedure, but in units called blocks. A block covers a small number of quantized states, but many time increments. If the present state $\mathbf{x}$ is in the interior of the block, then the quantized states which are required for interpolation of the minimum cost at the next state are always in this block. It is then possible to compute minimum cost and optimal control throughout the block on the basis of an initial set of minimum costs at the two largest quantized values of time in the block. By modifying the computations at the boundaries of the block as in Chapter 6, it is possible to allow the trajectories any degree of freedom in passing from block to block. Consequently, an efficient computational procedure is obtained in which the transfer of a relatively small amount of initial data enables computations to take place throughout the block. If the block size is so small that frequent referral to low-speed memory is still required, then data for several blocks can be transferred simultaneously.

It should be pointed out that both concepts are necessary to obtain an efficient procedure. If δt is determined as in Eq. (1.14), but the computations take place in the same order as the conventional procedure, frequent referral to the low-speed memory is required. On the other hand, if $\delta t = \Delta t$, but the computations are done in blocks, it is not possible to ensure that the next state is in or even close to the block in which the present state lies. It is then generally necessary to bring in data from the low-speed memory in order to find the minimum cost at the next state; again, this takes excessive computing time.

A major result of state increment dynamic programming is that the high-speed memory requirement is on the order of the number of quantized states contained in a block, rather than being one location for every quantized state in the entire state space. Reductions in this requirement by orders of magnitude are thus possible. For the problem mentioned in Sec. 1.3, where the high-speed memory requirement using the conventional procedure is 10^6 locations, it is shown in Chapter 6 that the requirement for a representative block size is about 100 locations, a reduction by a factor of nearly 10^4.

Similar savings in computing time can be obtained if it is desired to compute optimal control only in a region around a single optimal trajectory, rather than throughout the entire state space. This is all that is required in many real-time control applications. In Sec. 6.6 it is shown how the procedure is then applied only to those blocks that lie in the region of interest. A number of ways are suggested there for determining the region.

1.5 ORGANIZATION OF THE BOOK

The remainder of the book covers four topics:

(1) The Conventional Dynamic Programming Computational Procedure
(2) The State Increment Dynamic Programming Computational Procedure
(3) Practical Applications of State Increment Dynamic Programming
(4) Other Dynamic Programming Computational Procedures

The conventional procedure is described in detail in Chapter 2. This chapter is intended as an introductory treatment so that the reader with no previous background in dynamic programming can obtain a thorough understanding of the method. The reader who has worked with this procedure before may wish to scan this section lightly or skip over it entirely.

The state increment dynamic programming computational procedure is presented in Chapters 3 through 7. The basic concepts are explained in detail in Chapter 3. A general procedure for carrying out the computations within a block is described in Chapter 4. A modified procedure which has additional computational advantages in certain special cases is discussed in Chapter 5; the reader may wish to skip this chapter on first reading. The procedure for processing blocks, including methods for allowing trajectories to pass from block to block, is presented in Chapter 6. Finally, the detailed steps in an illustrative example are worked out in Chapter 7.

Practical applications of state increment dynamic programming are presented in Chapters 8 through 11. A computer program for use in a general class of problems having four or less state variables is described in Chapter 8.

Guidance and control applications are discussed in Chapter 9. Problems in the areas of process control and operations research are discussed in Chapter 10. Problem formulations in which uncertainty is explicitly taken into account are presented in Chapter 11; some applications to such problems are given there.

Other new computational procedures based on dynamic programming that have reduced computational requirements are described in Chapters 8 through 12. Because the emphasis in this book is on state increment dynamic programming, only a few of these procedures are discussed here; a recent survey of many of these procedures can be found in Ref. 8. The most promising of these approaches appears to be Bellman's successive approximations [Refs. 1, 2, 3]; all of Chapter 12 is devoted to this technique. Other procedures that are described in detail include efficient minimization procedures (Chapter 8), forward dynamic programming (Chapter 10), and the transformation of the dimensionality of the high-speed memory requirement from that of the state vector to that of the control vector (Chapter 10). Several other procedures are discussed more briefly in these chapters.

REFERENCES

1. Bellman, R., *Dynamic Programming*, Princeton University Press, Princeton, N.J., 1957.
2. Bellman, R., *Adaptive Control Processes*, Princeton University Press, Princeton, N.J., 1961.
3. Bellman, R., and Dreyfus, S., *Applied Dynamic Programming*, Princeton University Press, Princeton, N.J., 1962.
4. Larson, R. E., *Dynamic Programming with Continuous Independent Variable*, Stanford Electronics Laboratory TR 6302-6, Stanford, California, April, 1964.
5. Larson, R. E., "State Increment Dynamic Programming: Theory and Applications," *Proc. 2nd Allerton Conf. on Ckt. and Systems Theory*, University of Illinois, September, 1964, pp. 643–665.
6. Larson, R. E., "Dynamic Programming with Reduced Computational Requirements," *IEEE Trans. on Automatic Control*, vol. AC-10, no. 2, April, 1965, pp. 135–143.
7. Larson, R. E., "An Approach to Reducing the High-Speed Memory Requirement of Dynamic Programming," *J. Math. Anal. and Appl.*, vol. 11, nos. 1–3, July, 1965, pp. 519–537.
8. Larson, R. E., "Computational Aspects of Dynamic Programming," *1967 IEEE International Convention Record*, Part 3 (March, 1967), pp. 15–26.

Chapter Two

THE CONVENTIONAL DYNAMIC PROGRAMMING COMPUTATIONAL PROCEDURE

2.1 INTRODUCTION

The purpose of this chapter is to present a self-contained, detailed exposition of the conventional dynamic programming computational procedure. Although this chapter is primarily intended for the reader who has had little or no previous experience with this procedure, some of the material, particularly that on utilizing the results of the calculations, may be of interest to other readers.

In the next section the general problem formulation to which the procedure applies is given. Then, a brute-force enumeration procedure for this problem is discussed. In the section after that, the central idea of dynamic programming, Bellman's principle of optimality, is described. Then, the basic iterative functional equation of dynamic programming is derived.

In the next few sections the details of the computational procedure are presented. First, the quantization of the system variables is discussed. Next, the calculations required to start the iterations are indicated. Then, the basic calculations for obtaining the optimal control are described. Finally, an illustrative example is worked out.

The next section discusses the recovery of an optimal trajectory from the results of the dynamic-programming calculations. The following section discusses the interpolation procedures that are used in both the original calculations and the recovery procedure. Then, alternative implementations of a controller based on dynamic programming are described.

The next-to-last section presents in detail the computational requirements of this procedure. The final section summarizes the entire procedure and gives a flow chart for implementing it in a computer program.

2.2 PROBLEM FORMULATION

The optimization problems to which dynamic programming [Refs. 1, 2, 3] applies are variously called dynamic optimization problems, variational control problems, or optimization problems for multistage decision processes.

The essential elements of the problem are the *system equations*, which describe the process being controlled; the *performance criterion*, which evaluates a particular control policy; and the *constraints*, which place restrictions on the system operation.

The *system equations* are a set of relations between three types of variables: the *stage variable*, the *state variables*, and the *control or decision* variables.

The *stage variable* is that variable which determines the order in which events occur in the system. This quantity varies monotonically over the period during which decisions are made; in many cases it is taken to be time. This variable can either be continuous or discrete. If it is continuous, it is denoted as t, defined over the interval $t_0 \leq t \leq t_f$; while if it is discrete, it is defined as the sequence $k = 0, 1, 2, \ldots, K$. In order to implement the procedure of this section, when the stage variable is continuous, it is quantized into increments, denoted as Δt. The quantized values of t that lie in the range over which t is defined, $t_0 \leq t \leq t_f$, can then be indexed by the discrete sequence $k = 0, 1, \ldots, K$, where the value of t corresponding to k is given by

$$t = t_0 + k\,\Delta t \tag{2.1}$$

and where

$$K\,\Delta t = t_f - t_0. \tag{2.2}$$

The upper limit of k, K, can be infinite, but it is generally finite.

The *state variables* are a set of variables that completely describe the system, in the sense that if their values are known for all k, $k = 0, 1, 2, \ldots, K$, then any question about the behavior of the system for this range of k can be answered. The choice of a set of state variables for a particular system is not unique, and the determination of a suitable set is a more or less difficult problem, depending on the extent and nature of the mathematical model of the system. The choice of state variables for the process is a problem that lies outside the scope of this book; the interested reader is referred to other sources [Refs. 4, 5]. Unless otherwise stated, it will be assumed that for any given problem a set of n nonredundant state variables can be specified. These n variables are denoted $x_1, x_2, \ldots, x_n$, and they are often written as an n-dimensional vector x, called the *state vector*.

$$\mathbf{x} = \begin{bmatrix} x_1 \\ x_2 \\ \cdot \\ \cdot \\ \cdot \\ x_n \end{bmatrix}. \tag{2.3}$$

The *control* or *decision variables* are those variables in the process that can be specified directly. These variables influence the process by affecting the state variables in some prescribed fashion. There are, in general, q control variables denoted as $u_1, u_2, \ldots, u_q$. These variables are generally arranged in a q-dimensional vector, $\mathbf{u}$, called the control vector.

$$\mathbf{u} = \begin{bmatrix} u_1 \\ u_2 \\ \cdot \\ \cdot \\ \cdot \\ u_q \end{bmatrix}. \tag{2.4}$$

The *system equations* describe how the state variables of stage $k + 1$ are related to the state variables at stage k and the control variables at stage k. These equations can be expressed as

$$\begin{aligned} x_1(k+1) &= g_1[x_1(k), x_2(k), \ldots, x_n(k), u_1(k), u_2(k), \ldots, u_q(k), k] \\ x_2(k+1) &= g_2[x_1(k), x_2(k), \ldots, x_n(k), u_1(k), u_2(k), \ldots, u_q(k), k] \\ &\;\;\vdots \\ x_n(k+1) &= g_n[x_1(k), x_2(k), \ldots, x_n(k), u_1(k), u_2(k), \ldots, u_q(k), k]. \end{aligned} \tag{2.5}$$

These relations can be written more compactly as

$$\mathbf{x}(k+1) = \mathbf{g}[\mathbf{x}(k), \mathbf{u}(k), k] \tag{2.6}$$

where $\mathbf{g}$ is a n-dimensional vector functional.

The *performance criterion* provides an evaluation of a given control sequence, $\mathbf{u}(0), \mathbf{u}(1), \ldots, \mathbf{u}(K)$. This criterion is either a reward function, in which case it is to be maximized; or else it is a cost function, in which case it is to be minimized. Without loss of generality, it will henceforth be assumed that it is a cost function, which is to be minimized. The performance criterion depends on each value of $\mathbf{u}(k)$ in the control sequence, and also on each value of the state vector, $\mathbf{x}(0), \mathbf{x}(1), \ldots, \mathbf{x}(K)$. If the criterion is denoted as J, it can be written as

$$J = \sum_{k=0}^{K} l[\mathbf{x}(k), \mathbf{u}(k), k]. \tag{2.7}$$

The *constraints* place restrictions on the values that the state variables and control variables can assume. The state vector at stage k is constrained to be in the set $X(k)$. This constraint is expressed mathematically as

$$\mathbf{x} \in X(k). \tag{2.8}$$

The control vector applied at state $\mathbf{x}$, stage k is constrained to be in the set $U(\mathbf{x}, k)$. This constraint is written

$$\mathbf{u} \in U(\mathbf{x}, k). \tag{2.9}$$

The optimization problem can then be stated as follows:

Given:

(i) A system described by Eq. (2.6)
(ii) Constraints that $\mathbf{x} \in X(k)$, $\mathbf{u} \in U(\mathbf{x}, k)$
(iii) An initial state $\mathbf{x}(0)$

Find:

The control sequence $\mathbf{u}(0), \mathbf{u}(1), \ldots, \mathbf{u}(K)$ that minimizes J in Eq. (2.7) while satisfying the constraints.

2.3 OPTIMIZATION VIA ENUMERATION OF ADMISSIBLE CONTROL SEQUENCES

Before discussing the dynamic programming approach to this problem, it is useful to examine a brute-force enumeration procedure. In this method the set of admissible controls, U, is quantized to a finite number of controls. The set U then consists of the elements

$$U = \{\mathbf{u}^{(1)}, \mathbf{u}^{(2)}, \ldots, \mathbf{u}^{(M)}\}, \tag{2.10}$$

where both M, the number of admissible controls, and $\mathbf{u}^{(m)}$, $m = 1, 2, \ldots, M$, the actual values of the admissible controls, can vary with $\mathbf{x}$ and k.

The enumeration procedure then consists of the following steps: At the given state $\mathbf{x}(0)$ every admissible control $\mathbf{u} \in U$ is applied. For each of these controls, the next state is computed from

$$\mathbf{x}(1) = \mathbf{g}[\mathbf{x}(0), \mathbf{u}, 0]. \tag{2.11}$$

If a state $\mathbf{x}(1)$ is an admissible state (if $\mathbf{x}(1) \in X$), then the cost associated with this state is evaluated as

$$L[\mathbf{x}(1), 1] = l[\mathbf{x}(0), \mathbf{u}, 0]. \tag{2.12}$$

If the state $\mathbf{x}(1)$ is *not* admissible, then no further consideration is given to this control sequence, since it violates the constraints.

This process is continued by applying all admissible controls at all of the $\mathbf{x}(1) \in X$, finding the resulting states $\mathbf{x}(2)$ from

$$\mathbf{x}(2) = \mathbf{g}[\mathbf{x}(1), \mathbf{u}, 1] \tag{2.13}$$

and computing the cost of *admissible* states $\mathbf{x}(2)$ as

$$L[\mathbf{x}(2), 2] = l[\mathbf{x}(1), \mathbf{u}, 1] + L[\mathbf{x}(1), 1]. \tag{2.14}$$

In general, when a set of states $\mathbf{x}(k)$ has been defined by the procedure, a new set of states $\mathbf{x}(k+1)$ is defined by applying all $\mathbf{u} \in U$ at all of the $\mathbf{x}(k)$, computing the resulting states from

$$\mathbf{x}(k+1) = \mathbf{g}[\mathbf{x}(k), \mathbf{u}, k] \tag{2.15}$$

and evaluating the cost of admissible states $\mathbf{x}(k + 1)$ from

$$L[\mathbf{x}(k + 1), k + 1] = l[\mathbf{x}(k), \mathbf{u}, k] + L[\mathbf{x}(k), k]. \tag{2.16}$$

The process continues until K, the final value of k, is reached.

This process traces out all trajectories in state space that do not violate the constraints and that begin at $\mathbf{x}(0)$ and end at $k = K$. These trajectories form a tree beginning at $\mathbf{x}(0)$, expanding as k increases. Such a tree is illustrated for a one-dimensional example in Fig. 2.1, where $M = 2$ for all x and k, where X is the interval $0 \leq x \leq 4$ for all k, and where $K = 4$. Note that for $k = 3$

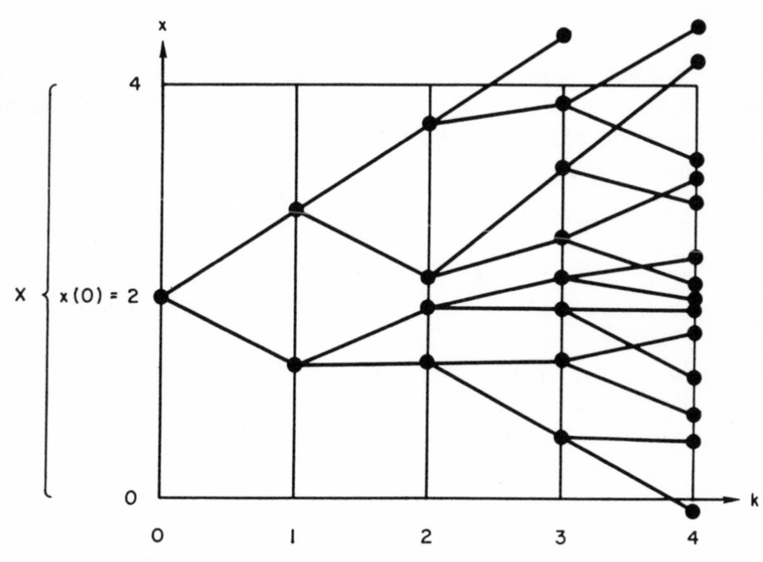

Fig. 2.1. Tree generated by enumeration.

one of the trajectories falls outside of X by taking on too large a value; while at $k = 4$ two trajectories fall outside of X by being too large and one by being too small. Thus, it can be seen that not only is it possible to handle a wide variety of constraints within this framework, but constraints actually serve a useful purpose by reducing the number of trajectories that must be considered.

The minimum cost is evaluated by comparing $L[\mathbf{x}(K), K]$ for the admissible states $\mathbf{x}(K) \in X$ and choosing the minimum value. The optimal control sequence and the optimal trajectory in state space are traced out by following back along the tree the path that led to this value of $L[\mathbf{x}(K), K]$. This direct procedure always determines an absolute minimum rather than a relative minimum or maximum; and if the minimum is not unique, *all* optimal control sequences can be found.

However, straightforward enumeration leads to computational difficulties. Consider a one-dimensional process with $M = 5$ for all $\mathbf{x} \in X$ and k, with $K = 20$. Then, assuming that none of the generated trajectories violate the constraints, the number of trajectories at $k = K = 20$ is given by

$$N_T = M^K = 5^{20} \approx 10^{14}. \tag{2.17}$$

Storing the 10^{14} costs corresponding to these trajectories, comparing them to find the minimum cost, and tracing back along the tree to find the optimal control sequence and the optimal trajectory is an infeasible computational task for any existing computer system.

Although the constraints actually reduce the number of trajectories that must be considered, the number of computations still increases exponentially with K. On the basis of the excessive size of the computational requirements for this simple example, it is clear that the procedure is not practical for problems of any degree of complexity.

2.4 BELLMAN'S PRINCIPLE OF OPTIMALITY

The heart of dynamic programming is Bellman's principle of optimality. This principle makes it possible to construct a computational procedure that retains the desirable properties of the enumeration procedure—such as constraint-handling capability and the computation of an absolute minimum—while requiring much less computational effort.

The principle of optimality can be stated as follows: Given an optimal trajectory from point A to point C, the portion of the trajectory from any intermediate point B to point C must be the optimal trajectory from B to C.

In Fig. 2.2, if the path I–II is the optimal path from A to C, then according to the principle of optimality path II is the optimal path from B to C. The

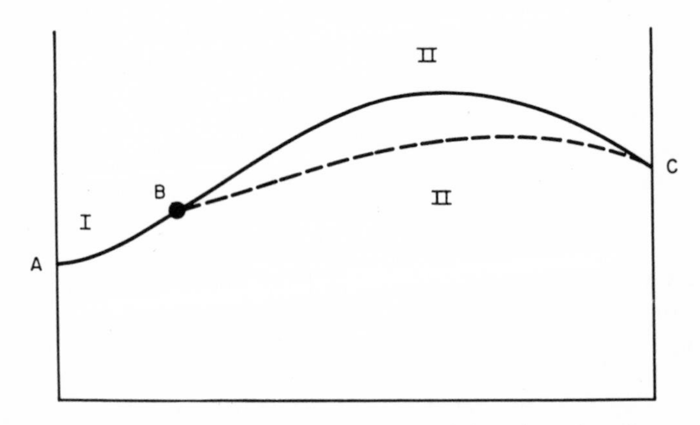

Fig. 2.2. Illustration of the principle of optimality.

proof by contradiction for this case is immediate: Assume that some other path, such as II′, is the optimum path from B to C. Then, path I–II′ has less cost than path I–II. However, this contradicts the fact that I–II is the optimal path from A to C, and hence II must be the optimal path from B to C. A rigorous proof valid for a more general class of problems can be found in Refs. 1 to 3.

2.5 DERIVATION OF THE ITERATIVE FUNCTIONAL EQUATION

When the principle of optimality is applied to a given system, an iterative procedure for determining optimal control is obtained. The basic iterative relation is called the *iterative functional equation*. This equation for the optimization problem described in Sec. 2.2 will now be formally derived.

First, the *minimum cost function* is defined. The minimum cost function expresses the minimum cost that can be obtained by using admissible control for the remainder of the process starting in any admissible state $\mathbf{x}$, $\mathbf{x} \in X$, and at any stage k, $0 \le k \le K$.

$$I(\mathbf{x}, k) = \underset{\substack{\mathbf{u}(j)\in U \\ j=k,\ldots,K}}{\text{Min}} \left\{ \sum_{j=k}^{K} l[\mathbf{x}(j), \mathbf{u}(j), j] \right\} \tag{2.18}$$

where

$$\mathbf{x}(k) = \mathbf{x}.$$

The first step in the derivation is to split the summation inside the braces into two parts.

$$I(\mathbf{x}, k) = \underset{\substack{\mathbf{u}(j)\in U \\ j=k,\ldots,K}}{\text{Min}} \left\{ l[\mathbf{x}, \mathbf{u}(k), k] + \sum_{j=k+1}^{K} l[\mathbf{x}(j), \mathbf{u}(j), j] \right\} \tag{2.19}$$

where $\mathbf{x}$ has been substituted for $\mathbf{x}(k)$ in the first term inside the brackets.

Next, the minimization operation in Eq. (2.18) is also split into two parts, a minimization over $\mathbf{u}(k)$ and a minimization over the controls $\mathbf{u}(k+1)$, $\mathbf{u}(k+2), \ldots, \mathbf{u}(K)$.

$$I(\mathbf{x}, k) = \underset{\mathbf{u}(k)\in U}{\text{Min}} \ \underset{\substack{\mathbf{u}(j)\in U \\ j=k+1,\ldots,K}}{\text{Min}} \left\{ l[\mathbf{x}, \mathbf{u}(k), k] + \sum_{j=k+1}^{K} l[\mathbf{x}(j), \mathbf{u}(j), j] \right\} \tag{2.20}$$

It can be seen that the first term in brackets in Eq. (2.20) depends only on $\mathbf{u}(k)$ and not on any $\mathbf{u}(j)$, $j = k + 1, \ldots, K$. Therefore, the minimization over $\mathbf{u}(j)$, $j > k$, has no affect on this term, and

$$\underset{\mathbf{u}(k)\in U}{\text{Min}} \ \underset{\substack{\mathbf{u}(j)\in U \\ j=k+1,\ldots,K}}{\text{Min}} \{ l[\mathbf{x}, \mathbf{u}(k), k] \} = \underset{\mathbf{u}(k)\in U}{\text{Min}} \{ l[\mathbf{x}, \mathbf{u}(k), k] \} \tag{2.21}$$

The second term in braces in Eq. (2.20) does not depend explicitly on $\mathbf{u}(k)$. However, $\mathbf{u}(k)$ does determine the state $\mathbf{x}(k+1)$ through the state transformation equation

$$\mathbf{x}(k+1) = \mathbf{g}[\mathbf{x}, \mathbf{u}(k), k]. \tag{2.22}$$

Using this relation and recalling the definition of the minimum cost function from Eq. (2.18), then

$$\underset{\mathbf{u}(k)\in U}{\text{Min}} \quad \underset{\substack{\mathbf{u}(j)\in U \\ j=k+1,\ldots,K}}{\text{Min}} \left\{ \sum_{j=k+1}^{K} l[\mathbf{x}(j), \mathbf{u}(j), j] \right\} = \underset{\mathbf{u}(k)\in U}{\text{Min}} \{I[\mathbf{g}(\mathbf{x}, \mathbf{u}(k), k), k+1]\} \tag{2.23}$$

Combining Eqs. (2.21) and (2.23) into Eq. (2.20) and suppressing the index k on $\mathbf{u}(k)$, the functional equation of the principle of optimality for this problem can be written as

$$I(\mathbf{x}, k) = \underset{\mathbf{u}\in U}{\text{Min}} \{l[\mathbf{x}, \mathbf{u}, k] + I\{\mathbf{g}(\mathbf{x}, \mathbf{u}, k), k+1]\}. \tag{2.24}$$

This equation describes an iterative relation for determining $I(\mathbf{x}, k)$ for all $\mathbf{x} \in X$ from knowledge of $I(\mathbf{x}, k+1)$ for all $\mathbf{x} \in X$. The optimal control $\hat{\mathbf{u}}(\mathbf{x}, k)$ is determined as that control for which the quantity in braces in Eq. (2.24) takes on the minimum value.

The iterative functional equation, Eq. (2.24), is related to the principle of optimality in that the minimum cost at state $\mathbf{x}$ and stage k is found by minimizing the sum of the cost at the present stage, k, and the minimum cost in going to the end of the process from the resulting state at the next stage, $k+1$.

2.6 CONSTRAINTS AND QUANTIZATION

One of the most useful properties of the enumeration procedure discussed in Sec. 2.3 is that constraints of a very general nature can be handled. The constraints actually reduce the computational effort rather than increase it. Dynamic programming retains this valuable property.

As in the enumeration procedure, constraints are used to define X, the set of admissible states. For example, inequality constraints of the form

$$\Phi(\mathbf{x}, k) \leq 0 \tag{2.25}$$

can be used to restrict the range of the state variables by relations such as

$$\beta_i^- \leq x_i \leq \beta_i^+ \qquad (i = 1, 2, \ldots, n). \tag{2.26}$$

The quantities β_i^- and β_i^+ can vary with k.

Constraints are also used to define U, the set of admissible controls. Inequality constraints of the form

$$\Phi(\mathbf{x}, \mathbf{u}, t) \leq 0 \tag{2.27}$$

can be used to restrict the range of the control variables

$$\alpha_j^- \leq u_j \leq \alpha_j^+ \qquad (j = 1, 2, \ldots, q). \tag{2.28}$$

The quantities α_j^- and α_j^+ can vary with both $\mathbf{x}$ and k.

Constraints in other forms can generally be manipulated to impose restrictions of the form of Eqs. (2.26) and (2.28). Although the quantities α_j^-, α_j^+, β_i^-, and β_i^+ can vary with $\mathbf{x}$ and k, these variations introduce unnecessary complications into the discussions. Consequently, for the remainder of this chapter these quantities will be assumed to be constant. The extension to the case where these quantities vary presents no formal difficulties.

For a problem to have physical significance, all state and control variables must remain finite. Therefore, if there are no explicit constraints which limit the range of state and control variables, then other considerations, such as the range of values which are of interest, are used to establish constraints of the form of Eqs. (2.26) and (2.28).

In order to apply the dynamic programming computational procedure, there must be a finite number of admissible states and admissible controls. This requirement is usually met by quantizing these variables. Within the range determined by Eq. (2.28), each state variable x_i is quantized with uniform increment Δx_i. These increments could be nonuniform, but again needless notational complications would arise. The extension to the more general case presents no formal difficulties. The quantized values of x_i are thus

$$x_i = \beta_i^- + j\,\Delta x_i \tag{2.29}$$

where

$$j = 0, 1, \ldots, N_i$$
$$N_i\,\Delta x_i = \beta_i^+ - \beta_i^-.$$

The set of all $\mathbf{x}$, where each component x_i is quantized according to Eq. (2.29), will from now on be referred to as X, the set of quantized admissible states.

The control variables can also be quantized according to relations similar to Eq. (2.29). However, in this chapter it is sufficient to assume that there are a finite number of admissible controls. The set of admissible controls, U, can thus be denoted as

$$U = \{\mathbf{u}^{(1)}, \mathbf{u}^{(2)}, \ldots, \mathbf{u}^{(M)}\} \tag{2.30}$$

where M is the total number of admissible controls.

The stage variable has already been quantized and normalized so that k takes on integer values $0, 1, 2, \ldots, K$, where K is the total number of stages.

2.7 INITIALIZATION PROCEDURE

In order to use the iterative functional equation, Eq. (2.24), it is necessary to specify a set of boundary conditions. Because the functional equation expresses the minimum cost function at k in terms of the minimum cost function at $(k + 1)$, the boundary conditions must be specified at the *final* stage, K. Formally, the quantity to be determined is $I(\mathbf{x}, K)$ for all $\mathbf{x} \in X$. In the computational procedure this quantity is defined by specifying a value of the minimum cost function for every quantized state $\mathbf{x} \in X$. If the cost function is given as J from Eq. (2.7),

$$J = \sum_{k=0}^{K} l[\mathbf{x}(k), \mathbf{u}(k), k] \tag{2.31}$$

then $I(\mathbf{x}, K)$ can be determined from

$$I(\mathbf{x}, K) = \operatorname*{Min}_{\mathbf{u} \in U} \{l[\mathbf{x}, \mathbf{u}, K]\}. \tag{2.32}$$

If, as is often the case, no control is applied at $k = K$, and hence the cost function at K depends only on the final state, $\mathbf{x}(K)$, then $I(\mathbf{x}, K)$ can be written directly as

$$I(\mathbf{x}, K) = l(\mathbf{x}, K). \tag{2.33}$$

In many problems this determination is not completely straightforward. One difficulty that occurs frequently in practice is that the final state is constrained to fall in C, a subset of X. The set C is determined as the solution of a set of equations involving the final state and final stage.

$$\boldsymbol{\theta}[\mathbf{x}(K), K] = 0. \tag{2.34}$$

These equations are referred to as the final conditions or stopping conditions. The set C may contain one or many points in X. The final stage K may or may not be fixed. All these cases can be handled within the framework of dynamic programming, provided that only those trajectories are considered where the final state $\mathbf{x}(K)$ satisfies the final conditions, Eq. (2.34).

2.8 CALCULATION OF OPTIMAL CONTROL

Once $I(\mathbf{x}, K)$ has been determined for all $\mathbf{x} \in X$, it is possible to compute optimal control by iterative application of the functional equation, Eq. (2.24).

Consider a quantized state $\mathbf{x} \in X$ at stage $(K - 1)$. At this state each of the admissible controls $\mathbf{u}^{(m)} \in U$ is applied. For each of these controls the cost of control over the next stage can be determined as

$$L_1^{(m)} = l[\mathbf{x}, \mathbf{u}^{(m)}, K - 1] \qquad (m = 1, 2, \ldots, M). \tag{2.35}$$

Next, for each of these controls the next state at stage K is determined from the system equations, Eq. (2.6).

$$\mathbf{x}^{(m)}(K) = \mathbf{g}[\mathbf{x}, \mathbf{u}^{(m)}, K - 1] \qquad (m = 1, 2, \ldots, M). \tag{2.36}$$

The next step is to compute the minimum cost at stage K for each of the stages $\mathbf{x}^{(m)}$. However, in general a particular state $\mathbf{x}^{(m)}$ will not lie on one of the quantized states $\mathbf{x} \in X$ at which the optimal cost $I(\mathbf{x}, K)$ is defined. In fact, it may lie outside of the range of admissible states determined by Eq. (2.26). In the latter case the control is rejected as a candidate for the optimal control for this state and stage.

If a next stage $\mathbf{x}^{(m)}$ does fall within the range of allowable states, but not on a quantized value, then it is necessary to use some type of interpolation procedure to compute the minimum cost function at these points. In general, the interpolation procedure consists of using a low-order polynomial in the n state variables to approximate the minimum cost function in small regions of state space. The coefficients of the polynomial are determined in terms of the known values of the minimum cost function at quantized states, $\mathbf{x} \in X$. The determination of the coefficients is made according to some criterion, such as least-squares fit. Computational procedures for calculating the coefficients are relatively simple and well-known [Refs. 3, 6, 7]. Interpolation formulas are discussed in more detail in Sec. 2.11.

Assume, then, that the values of the minimum cost at the states $\mathbf{x}^{(m)}$ can be expressed as a function of the values of the optimal cost at quantized states $\mathbf{x} \in X$.

$$I[\mathbf{x}^{(m)}, K] = P[\mathbf{x}^{(m)}, K, I(\mathbf{x}, K)] \qquad (\text{all } \mathbf{x} \in X.) \tag{2.37}$$

The total cost of applying control $\mathbf{u}^{(m)}$ at state $\mathbf{x}$, stage $(k - 1)$, can then be written as

$$J_1^{(m)} = l[\mathbf{x}, \mathbf{u}^{(m)}, K - 1] + I[\mathbf{x}^{(m)}, K]. \tag{2.38}$$

This quantity is exactly the function which is to be minimized by choice of $\mathbf{u}^{(m)}$ in the functional equation, Eq. (2.24). The minimization can be achieved by simply comparing the M quantities. According to the functional equation, the minimum value will be the minimum cost at state $\mathbf{x}$, stage $(K - 1)$.

$$I[\mathbf{x}, K - 1] = \operatorname*{Min}_{\mathbf{u}^{(m)} \in U} \{l[\mathbf{x}, \mathbf{u}^{(m)}, K - 1] + I[\mathbf{x}^{(m)}, K]\}. \tag{2.39}$$

The optimal control at this state and stage, $\hat{\mathbf{u}}[\mathbf{x}, K - 1]$, is the control $\mathbf{u}^{(m)}$ for which the minimum in Eq. (2.39) is actually taken on.

This procedure is repeated at each quantized $\mathbf{x} \in X$ at stage $(K - 1)$. When this has been done, $I(\mathbf{x}, K - 1)$ and $\hat{\mathbf{u}}(\mathbf{x}, K - 1)$ are known for all $\mathbf{x} \in X$. It is now possible to compute $I(\mathbf{x}, K - 2)$ and $\hat{\mathbf{u}}(\mathbf{x}, K - 2)$ for all $\mathbf{x} \in X$ based on knowledge of $I(\mathbf{x}, K - 1)$.

The general iterative procedure continues this process. Suppose that $I(\mathbf{x}, k + 1)$ is known for all $\mathbf{x} \in X$. Then $I(\mathbf{x}, k)$ and $\hat{\mathbf{u}}(\mathbf{x}, k)$ are computed for all $\mathbf{x} \in X$ from

$$I(\mathbf{x}, k) = \operatorname*{Min}_{\mathbf{u}^{(m)} \in U} \{l[\mathbf{x}, \mathbf{u}^{(m)}, k] + I[\mathbf{x}^{(m)}, k + 1]\} \tag{2.40}$$

where $\mathbf{x}^{(m)}$ is determined from

$$\mathbf{x}^{(m)} = \mathbf{g}[\mathbf{x}, \mathbf{u}^{(m)}, k] \tag{2.41}$$

and where $I[\mathbf{x}^{(m)}, k + 1]$ is computed by interpolation on the known values $I[\mathbf{x}, k + 1]$ for all $\mathbf{x} \in X$

$$I[\mathbf{x}^{(m)}, k + 1] = P[\mathbf{x}^{(m)}, k + 1, I(\mathbf{x}, k + 1)] \qquad (\text{all } \mathbf{x} \in X). \tag{2.42}$$

The optimal control $\hat{\mathbf{u}}(\mathbf{x}, k)$ is the control for which Eq. (2.40) takes on the minimum. The iterative procedure begins by computing $\hat{\mathbf{u}}(\mathbf{x}, K - 1)$ and $I(\mathbf{x}, K - 1)$ from the given boundary conditions, $I(\mathbf{x}, K)$, and it continues until $\hat{\mathbf{u}}(\mathbf{x}, 0)$ and $I(\mathbf{x}, 0)$ have been computed.

A flow chart illustrating this procedure appears in Sec. 2.14.

2.9 AN ILLUSTRATIVE EXAMPLE

In this section a simple optimization problem will be solved by dynamic programming in order to illustrate the basic computational procedure. The system equation is taken to be

$$x(k + 1) = x(k) + u(k) \tag{2.43}$$

where $x(k)$ = scalar state variable at stage k
$u(k)$ = scalar control variable at stage k.

The performance criterion, which is to be minimized, is

$$J = \sum_{k=0}^{9} [x^2(k) + u^2(k)] + 2.5[x(10) - 2]^2. \tag{2.44}$$

The state variable is constrained to lie in the interval

$$0 \le x \le 8 \tag{2.45}$$

while the control variable is bounded by

$$-2 \le u \le 2. \tag{2.46}$$

The final state of the system, at $k = 10$, is constrained to fall in the interval

$$0 \le x(10) \le 2. \tag{2.47}$$

The state variable is quantized in uniform increments of one, i.e., $\Delta x = 1$.

The control variable is also quantized in uniform increments of one, so that the set of admissible controls is

$$U = \{-2, -1, 0, 1, 2\}. \tag{2.48}$$

With these quantizations there is no need to perform any interpolations; this makes the calculations easier for the reader to follow.

The grid of quantized values of x and t at which computations are to be made is shown in Fig. 2.3. When the computational procedure is about to begin, all final states with $x(10) > 2$ are forbidden by the constraint of Eq. (2.47). This is denoted by placing $\times$'s at all these states. A small circle is placed at $x(10) = 0$, $x(10) = 1$, and $x(10) = 2$ to indicate that a minimum

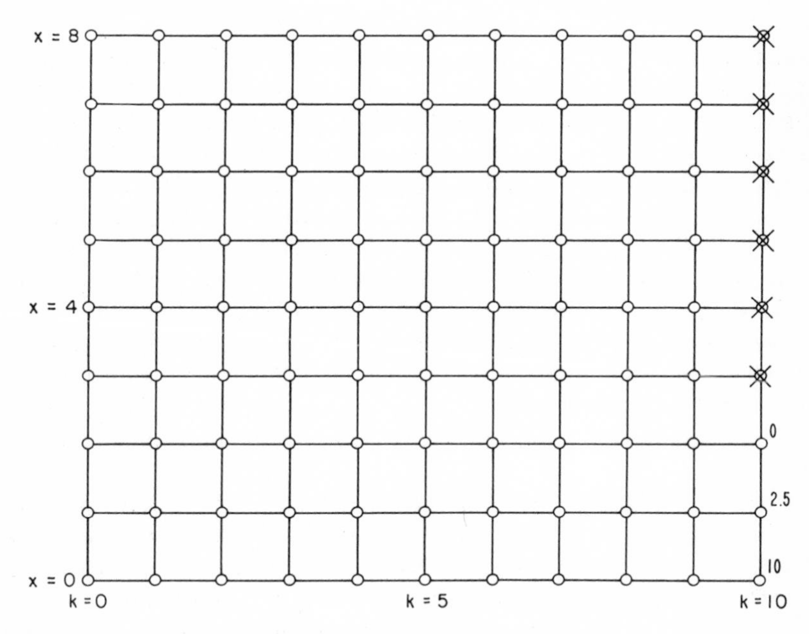

Fig. 2.3. Initial conditions for dynamic programming computational procedure.

cost can be computed for those points. From Eq. (2.44), it can be seen that these minimum costs are

$$\begin{aligned} I(0, 10) &= 10 \\ I(1, 10) &= 2.5 \\ I(2, 10) &= 0. \end{aligned} \tag{2.49}$$

These values are placed in Fig. 2.3 just above and to the right of the grid points to which they correspond.

With these initial values of minimum cost established, it is possible to apply iteratively the functional equation based on the principle of optimality, Eq. (2.24). Making the identifications

$$l[x(k), u(k), k] = [x^2(k) + u^2(k)] \qquad (k = 0, 1, 2, \ldots, 9) \quad (2.50)$$

and

$$g[x(k), u(k), k] = x(k) + u(k) \qquad (k = 0, 1, 2, \ldots, 9) \tag{2.51}$$

this iterative relation becomes

$$I(x, k) = \operatorname*{Min}_{u \in U} \{x^2 + u^2 + I(x + u, k + 1)\} \qquad (k = 0, 1, 2, \ldots, 9). \tag{2.52}$$

The set of admissible controls, U, is given in Eq. (2.48). The initial values in Eq. (2.47) are used as initial conditions, $I(x, 10)$. If an admissible control results in a next state which is not admissible (the next state either fails to satisfy Eq. (2.45) or else it has an × placed on it), then that control is not considered further. If all admissible controls are rejected on this basis, then an × is placed at such a point, and it is considered an inadmissible state for this stage. A state which is an element of the set X, but which becomes inadmissible for this reason, is said to be a *non-terminable* state [Refs. 8, 9].

Equation (2.52) is first applied at $k = 9$, then at $k = 8$, then at $k = 7$, etc., until $k = 0$ is reached. Each time an admissible optimal control is found, a small circle is drawn at that grid point, the optimal control is written to the right and below the circle, and the minimum cost is written to the right and above it. These numbers constitute the dynamic programming solution to the problem.

The computations at a given state x for $k = 9$ take place as follows: each control $u \in U$ is applied. The next state is computed for each control. If the next state is admissible, the minimum cost of that state is found from Eq. (2.49). The cost over the next stage, $[x^2(9) + u^2(9)]$, is computed. The sum of the two costs is stored. If the next state is *not* admissible, nothing is stored. From among the stored costs, the minimum is picked. This cost becomes the minimum cost at this point, while the optimal control is the control corresponding to the minimum cost.

For $x(9) = 0$, the computations can be summarized in tabular form as follows:

Table 2.1. Computations at $x = 0$, $k = 9$

u	$g(x, u, k)$	$l(x, u, k)$	$I(g, k + 1)$	*Total cost*
−2	−2	×	×	×
−1	−1	×	×	×
0	0	$0^2 + 0^2 = 0$	10	10
1	1	$0^2 + 1^2 = 1$	2.5	3.5
2	2	$0^2 + 2^2 = 4$	0	4

The ×'s in the table indicate that the next state is not admissible. Comparing the total costs for admissible next states, it is seen that the minimum is 3.5, which corresponds to $u = 1$. Note that this minimum utilizes *neither* the minimum "immediate cost," $l(x, u, k)$, which in this case corresponds to $u = 0$, *nor* the minimum cost at the next state, $I(g, k + 1)$, which corresponds to $u = 2$. It is the *sum* of the two which must be minimized.

For $x(9) = 1$, the computations can be abbreviated in the following form

Table 2.2. Computations at $x = 1$, $k = 9$

u	$g(x, u, k)$	$l(x, u, k)$	$I(g, k+1)$	*Total cost*
−2	−1	×	×	×
−1	0	$1^2 + (-1)^2 = 2$	10	12
0	1	$1^2 + 0^2 = 1$	2.5	3.5
1	2	$1^2 + 1^2 = 2$	0	2
2	3	×	×	×

The minimum cost is 2, again corresponding to $u = 1$.

For $x(9) = 2$,

Table 2.3. Computations at $x = 2$, $k = 9$

u	$g(x, u, k)$	$l(x, u, k)$	$I(g, k+1)$	*Total cost*
−2	0	$2^2 + (-2)^2 = 8$	10	18
−1	1	$2^2 + (-1)^2 = 5$	2.5	7.5
0	2	$2^2 + 0^2 = 4$	0	4
1	3	×	×	×
2	4	×	×	×

The minimum cost is 4, corresponding to $u = 0$.

For x (9) = 3,

Table 2.4. Computations at $x = 3$, $k = 9$

u	$g(x, u, k)$	$l(x, u, k)$	$I(g, k+1)$	*Total cost*
−2	1	13	2.5	15.5
−1	2	10	0	10
0	3	×	×	×
1	4	×	×	×
2	5	×	×	×

The minimum cost is 10, corresponding to $u = -1$

For $x(9) = 4$,

Table 2.5. Computations at $x = 4$, $k = 9$

u	$g(x, u, k)$	$l(x, u, k)$	$I(g, k+1)$	*Total cost*
−2	2	20	0	20
−1	3	×	×	×
0	4	×	×	×
1	5	×	×	×
2	6	×	×	×

The minimum cost is 20, corresponding to $u = -2$, the only control which results in an admissible next state.

For $x(9) \geq 4$, there is no control which results in an admissible next state. Consequently, these states must be regarded as not admissible. This completes the computations at $k = 9$. The results are summarized in Fig. 2.4.

The computations at $k = 8$ are then performed using Eq. (2.42). The values of minimum cost at $k = 9$ are used as $I(g, k + 1)$ in this relation. The

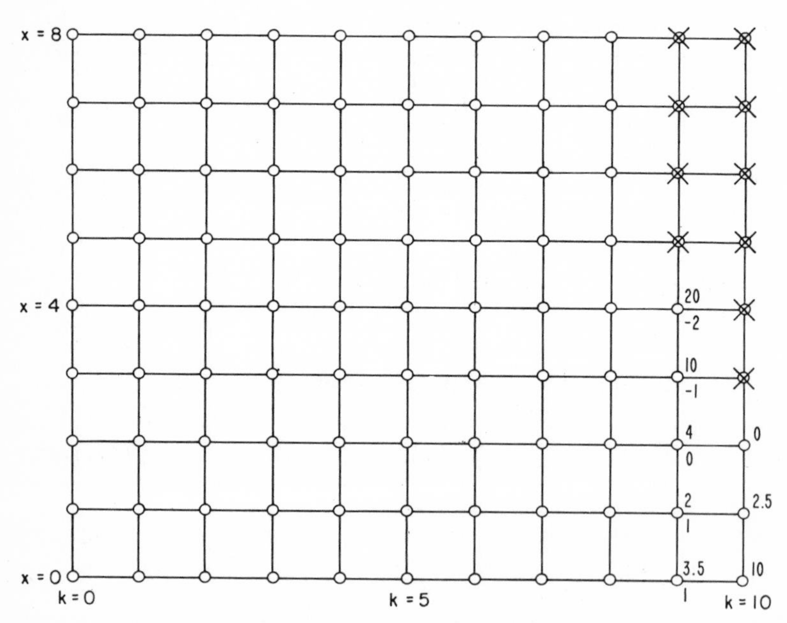

Fig. 2.4 Results after computations at $k = 9$.

computations take place exactly as at $k = 9$. For example, at $x(8) = 3$,

Table 2.6. Computations at $x = 3$, $k = 8$

u	$g(x, u, k)$	$l(x, u, k)$	$I(g, k + 1)$	*Total cost*
−2	1	13	2	15
−1	2	10	4	14
0	3	9	10	19
1	4	10	20	30
2	5	×	×	×

The minimum cost is 14, corresponding to $u = -1$.

After the computations at $k = 8$ have been completed, the resulting minimum costs are used in the computations at $k = 7$. This procedure, in which optimal control and minimum cost at stage k are computed using minimum costs at $(k + 1)$, continues until $k = 0$ is reached. The results are pictured in Fig. 2.5. The reader is encouraged to try to duplicate these results as a test of his understanding of the method.

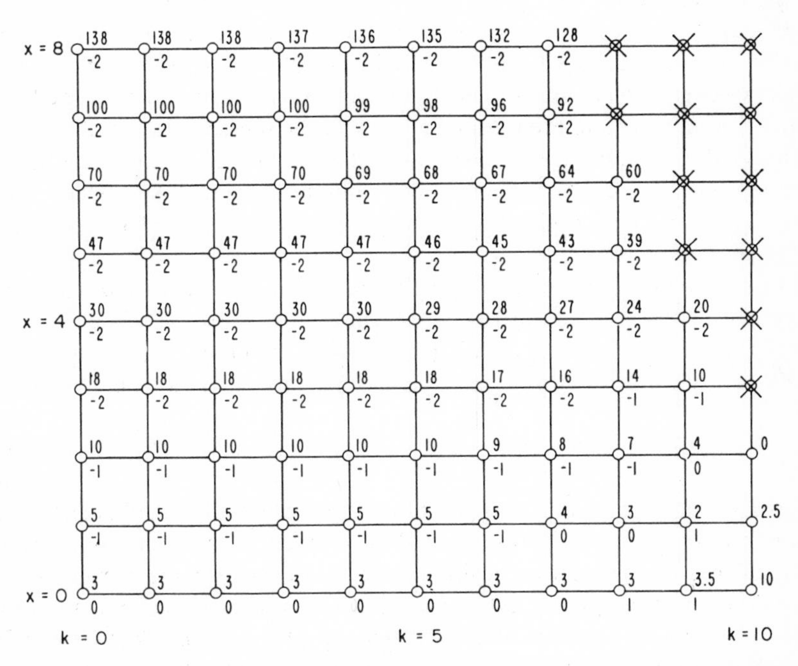

Fig. 2.5 Complete results of dynamic programming computational procedure.

2.10 RECOVERY OF AN OPTIMAL TRAJECTORY

The dynamic programming solution is a specification of $\hat{\mathbf{u}}(\mathbf{x}, k)$ and $I(\mathbf{x}, k)$ for all quantized $\mathbf{x} \in X$ and for $k = 0, 1, 2, \ldots, K$. However, the original problem was to find the optimum sequence of controls starting from the given $\mathbf{x}(0)$. This sequence can easily be determined from the dynamic programming solution. The first control in the sequence is evaluated as

$$\hat{\mathbf{u}}(0) = \hat{\mathbf{u}}[\mathbf{x}(0), 0]. \tag{2.53}$$

The next state along the sequence is then found as

$$\hat{\mathbf{x}}(1) = \mathbf{g}[\mathbf{x}(0), \hat{\mathbf{u}}(0), 0]. \tag{2.54}$$

This state may or may not be a quantized state. If it is, then the next control in the optimum sequence is evaluated directly as

$$\hat{\mathbf{u}}(1) = \hat{\mathbf{u}}[\hat{\mathbf{x}}(1), 1]. \tag{2.55}$$

If it is not a quantized state, then values of $\hat{\mathbf{u}}(\mathbf{x}, 1)$ for a number of quantized states, $\mathbf{x} \in X$, must be found and a suitable interpolation formula used. These interpolations follow a procedure similar to that indicated for the minimum cost function in Eq. (2.37) and discussed further in the next section.

The recovery procedure, in which the next state is computed on the basis of the present state and the optimal control at the present state, and then the corresponding optimal control is evaluated from this next state utilizing the function $\hat{\mathbf{u}}(\mathbf{x}, k)$, continues until the complete control sequence, $\hat{\mathbf{u}}(0)$, $\hat{\mathbf{u}}(1), \ldots, \hat{\mathbf{u}}(K)$, and optimal trajectory, $\mathbf{x}(0), \hat{\mathbf{x}}(1), \ldots, \hat{\mathbf{x}}(K)$, have been obtained.

For the solution shown in Fig. 2.5, the optimum control sequence and optimal trajectory from $\mathbf{x}(0) = 8$ are shown in Table 2.7. The controls $\hat{u}(\hat{x}, k)$, $k = 0, 1, 2, \ldots, 10$, are read directly from the figure (no interpolations are required), while the states along the trajectory are obtained from

$$\hat{x}(k+1) = \hat{x}(k) + \hat{u}(k) \qquad (k = 0, 1, 2, \ldots, 9). \tag{2.56}$$

An important feature of this solution is that optimal control is specified, not just along this optimal trajectory, but for every admissible state at every stage. Thus, in solving this problem, the solutions to a great many other problems have been found. Bellman refers to the process of solving an entire set of problems in order to solve the original problem as *invariant imbedding*. A most important consequence of this approach is that a *feedback control* or *closed-loop control* solution is obtained. This simply means that the optimal

Table 2.7. Optimum Control Sequence and Optimal Trajectory From $x(0) = 8$

k	$\hat{x}$	$\hat{u}$	$l(\hat{x}, \hat{u}, k)$
0	8	−2	68
1	6	−2	40
2	4	−2	20
3	2	−1	5
4	1	−1	2
5	0	0	0
6	0	0	0
7	0	0	0
8	0	1	1
9	1	1	2
10	2	—	0

$$\text{Total Cost} = \sum_{k=0}^{10} l(\hat{x}, \hat{u}, k) = 138.$$

control is specified as a function of both state and stage, not just as a function of the stage, as is the case in *open-loop control.* The importance of this type of solution is that in the case of deviations from the original optimal trajectory, as, for example, might occur if an uncontrolled input were applied to the system or if an incorrect value of control was inadvertently used, a true optimal control can be found for the remaining stages; if the solution were a function of stage only, the remainder of the control sequence would be incorrect unless further computations were performed.

As an illustration of this, suppose that the initial state is $x(6) = 1$. The optimum control sequence and optimal trajectory are as in Table 2.8. Assume now that the control $\hat{u}(6) = 0$ is applied instead. The optimal control

Table 2.8. Optimum Control Sequence and Optimal Trajectory from $x(6) = 1$

k	$\hat{x}$	$\hat{u}$	$l(\hat{x}, \hat{u}, k)$
6	1	−1	2
7	0	0	0
8	0	1	1
9	1	1	2
10	2	—	0

$$\text{Total Cost} = \sum_{k=0}^{10} l(\hat{x}, \hat{u}, k) = 5.$$

Table 2.9. Optimum Control Sequence and Optimal Trajectory from $x(7) = 1$

k	$\hat{x}$	$\hat{u}$	$l(\hat{x}, \hat{u}, k)$
7	1	0	1
8	1	0	1
9	1	1	2
10	2	—	0

$$\text{Total Cost} = \sum_{k=0}^{10} l(\hat{x}, \hat{u}, k) = 4.$$

sequence from the resulting state, which is $\hat{x}(7) = 1$, is as shown in Table 2.9. If the control sequence for $\hat{u}(7)$, $\hat{u}(8)$, and $\hat{u}(9)$ from Table 2.8 is applied instead of the true optimum sequence shown in Table 2.9, then the final state, $\hat{x}(10)$, is inadmissible—thus, in this case the open-loop control sequence would actually be inadmissible. If a control $\hat{u}(7)$ were applied such that $\hat{x}(8)$ were back on the open-loop optimal trajectory—a task that would require additional computation—the cost of the resulting trajectory from $\hat{x}(7) = 1$ is 5, rather than 4 as found in Table 2.9.

2.11 INTERPOLATION PROCEDURES

In Sec. 2.8 it was indicated that an interpolation of $I(\mathbf{x}, k)$ in the n state variables is required in the basic dynamic programming computational procedure. In Sec. 2.10 the need for a similar interpolation of $\hat{\mathbf{u}}(\mathbf{x}, k)$ in the utilization of results is mentioned. The purpose of this section is to describe briefly how these interpolations are performed.

The standard numerical analysis technique for interpolation is multi-dimensional polynomial approximation. A considerable number of methods and formulas can be found in the literature [Refs. 6, 7]. For practical reasons, however, only the simplest of these formulas have been used extensively in dynamic programming applications.

The most sophisticated methods involve using a high-order polynomial over a large number of data points. However, such methods generally require excessive computing time. Not only is a considerable amount of time necessary for calculating the coefficients of the polynomial, but also much time is used in evaluating the high-order polynomial every time an interpolation is required.

A second approach is to cover many data points, but to use a low-order polynomial. In this case the polynomial must be fitted to the data according to some criterion, such as least-squares error. Unfortunately, such an approach usually results in a considerable loss of accuracy. Since the polynomial does not pass through all the data points, interpolated values can differ considerably from the known values at near-by data points. Furthermore, much computing time is required in computing these coefficients on the basis of many data points.

A third approach is to use a high-order polynomial, but to cover only a small number of data points. However, experience has indicated that the optimal cost function $I(\mathbf{x}, k)$ tends to vary fairly smoothly in a small region, and the use of a high-order polynomial tends to introduce unrealistically large variations between data points. Limited success has been obtained in approximating $\hat{\mathbf{u}}(\mathbf{x}, k)$ by a high-order polynomial over even relatively small regions.

The only remaining alternative, use of a low-order polynomial over a small region, is the most widely practiced procedure in dynamic programming. The simplest approach is to approximate the function by the value at the nearest quantized state. For a one-dimensional example with x quantized in uniform increments Δx, from $x = 0$ to $x = \mathrm{A}\,\Delta x$, the equations for interpolating $I(x, k)$ are

$$I(x, k) = I(a\,\Delta x, k), \tag{2.57}$$

for

$$(a - \tfrac{1}{2})\,\Delta x < x \leq (a + \tfrac{1}{2})\,\Delta x.$$

A similar equation can be written for $\hat{u}$.

The next simplest approach is to use a multi-dimensional linear approximation between data points, where the linear approximation is exact at the data points. The equations for the one-dimensional example specified previously are

$$I(x, k) = I(a\,\Delta x, k) + \frac{I[(a+1)\,\Delta x, k] - I(a\,\Delta x, k)}{\Delta x}(x - a\,\Delta x) \tag{2.58}$$

for

$$a\,\Delta x \leq x \leq (a+1)\,\Delta x.$$

Again, a similar formula exists for $\hat{u}(x, k)$.

The highest-order formula that receives extensive use in practice is the multi-dimensional quadratic approximation. For the one-dimensional case above, where exact fit is made at the data points, the formula is

$$\begin{aligned} I(x, k) = {} & I(a\Delta x, k) \\ & + \frac{-\tfrac{1}{2}I[(a+2)\,\Delta x, k] + 2I[(a+1)\,\Delta x, k] - \tfrac{3}{2}I(a\,\Delta x, k)}{\Delta x}(x - a\,\Delta x) \\ & + \frac{\tfrac{1}{2}I[(a+2)\,\Delta x, k) - I[(a+1)\,\Delta x, k] + \tfrac{1}{2}I(a\,\Delta x, k)}{(\Delta x)^2}(x - a\,\Delta x)^2 \end{aligned} \tag{2.59}$$

for

$$(a + \tfrac{1}{2})\,\Delta x \leq x \leq (a + \tfrac{3}{2})\,\Delta x.$$

In other cases, particularly multi-dimensional problems, more data points are used than can be fit exactly by the interpolation formula. It is then necessary to determine the coefficients in the formula by least-squares fit techniques; the coefficients are selected by minimizing the sum of the squares of the deviations of the interpolation formula from the actual values at data points. Detailed discussions of this procedure can be found in Refs. 6 and 7.

Because these interpolation formulas are not exact, errors are introduced both into the calculation of $I(\mathbf{x}, k)$ and $\hat{\mathbf{u}}(\mathbf{x}, k)$ from the iterative functional equation and into the recovery of the optimal trajectory from values of $\hat{\mathbf{u}}(\mathbf{x}, k)$. In determining which formulas to use in a specific application, it is necessary to consider not only the instantaneous effects of these errors, but also the propagation of these errors into subsequent calculations. Bellman [Ref. 1] has studied the propagation of these errors in the computation of the iterative functional equation; he has shown that, under relatively modest assumptions, the effects of the errors made at a given stage diminish as the number of stages computed increases. This property of the iterative functional equation is extremely desirable, and it adds to the desirability of dynamic programming as a computational tool. By using results on the stability of difference equations [Refs. 6, 7], it is possible to obtain similar results about the errors in trajectory recovery.

2.12 IMPLEMENTATION OF A CONTROLLER BASED ON DYNAMIC PROGRAMMING

As discussed in Sec. 2.10, the dynamic programming solution, $\hat{\mathbf{u}}(\mathbf{x}, k)$, leads to a feedback control configuration. One method of implementing a controller based on this solution is to simply store all the values of $\hat{\mathbf{u}}(\mathbf{x}, k)$ in a memory, monitor the state and stage of the system, and look up the appropriate value of $\hat{\mathbf{u}}(\mathbf{x}, k)$ as required. This type of implementation is attractive because the dynamic programming calculations can be done off-line, and the only operation that needs to be done during the control interval is retrieval of the appropriate optimal control. The system configuration is as shown in Fig. 2.6.

An alternative scheme is to carry out the dynamic programming computations in real time. In this case the state and stage are again monitored, but the entire optimum control sequence is re-computed by an on-line computer; this re-computation may be required as often as once every stage, but if deviations from the computed trajectory are small the number of re-computations can be much less. Generally, a nominal trajectory based on either the last-computed trajectory, the results of pre-computations, or operating experience, is used to decrease the size of the sets X and U over

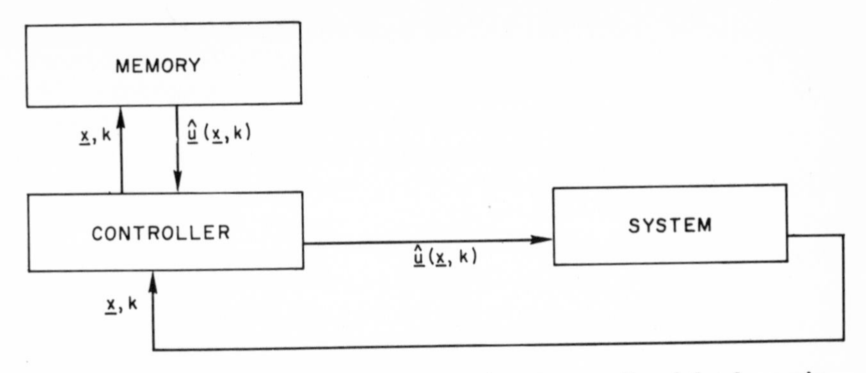

Fig. 2.6 A controller based on retrieving the results of the dynamic programming computation from a memory.

which computations are made; this type of approach is considered in more detail in Sec. 6.6. This implementation is pictured in Fig. 2.7; note that a feedback control solution is still obtained.

Still another method of utilizing the dynamic programming results is as a guide in developing a controller realization. One approach is to search for simple functions that closely approximate $\hat{\mathbf{u}}(\mathbf{x}, k)$. In any case, the minimum cost function for the original problem, $I(\mathbf{x}, k)$, serves to evaluate the degradation of any controller realization from the optimal controller.

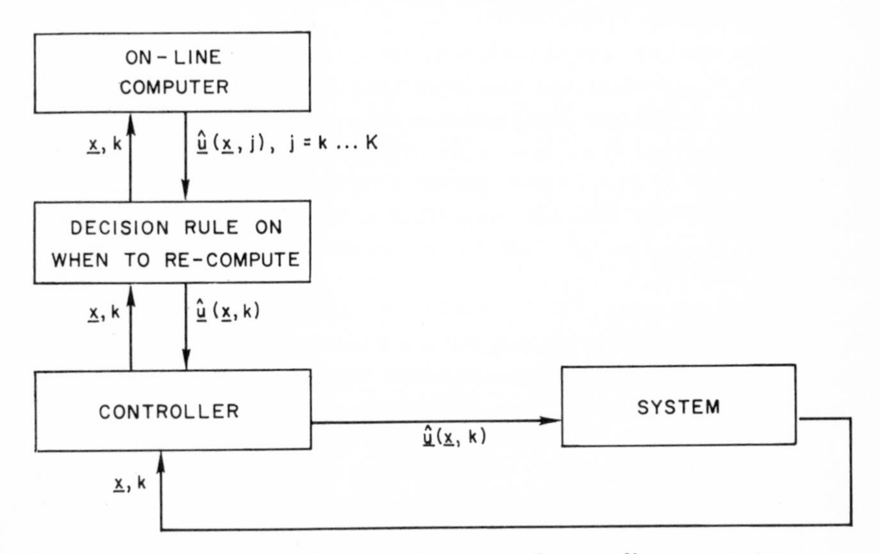

Fig. 2.7 A controller based on use of an on-line computer.

2.13 COMPUTATIONAL REQUIREMENTS

Before using dynamic programming to solve an actual problem on a digital computer, it is necessary to determine how much computational effort is required to solve the problem. The requirements are given in terms of the amount of memory needed to store both program and data and the amount of time required to perform the computations. If, for a given computer, either there is not sufficient memory available or if the cost of computing time exceeds the economic value of the results, then the problem should not be run on that computer. If this is true for all available computers, then the problem, while solvable in principle, must be regarded as unsolvable in a practical sense.

The most commonly encountered barrier to the use of dynamic programming is the high-speed memory requirement. This requirement refers to the number of locations in the high-speed access memory (core memory) which must be available during the computations. In addition to the locations needed for the program, the compiler, and other special functions—a number of locations which may be a large fraction of the total number available—the dynamic programming procedure requires that sufficient data be stored to specify $I(\mathbf{x}, k)$ for all $\mathbf{x} \in X$ at a single value of k. In general, this is done by storing one value of $I(\mathbf{x}, k)$ for every quantized $\mathbf{x} \in X$ and by using a simple interpolation formula as discussed in Sec. 2.11. The number of locations required is then

$$N_h = \prod_{i=1}^{n} N_i \tag{2.60}$$

where N_i = number of quantized values of ith variable
n = total number of state variables.

If values of minimum cost are always to be retained in the high-speed memory, the number of locations required is increased to $2N_h$. If computations are taking place at stage k, the N_h values $I(\mathbf{x}, k + 1)$ must be stored for interpolation purposes. However, as the values $I(\mathbf{x}, k)$ are generated, they also must be stored so that computations using these results can be performed at stage $(k - 1)$. The total number of locations required to store both $I(\mathbf{x}, k - 1)$ and $I(\mathbf{x}, k)$ is thus $2N_h$.

As pointed out in Sec. 1.3, the number N_h can be extremely large for even moderately sized problems. For example, if there are 3 state variables ($n = 3$), and if there are 100 quantization levels in each variable ($N_i = 100, i = 1, 2, 3$), then

$$N_h = (100)^3 = 10^6. \tag{2.61}$$

The number 10^6 greatly exceeds the total high-speed storage capacity of the largest existing computers, which is about 256,000 locations.* Of course, as

* As many as 2 or 3 × 10^6 storage locations are available with some computers as of this writing; however, at the present time this is extremely costly.

much as half of this number is reserved for program, compiler, and other special functions. Thus, this reasonably sized problem saturates the high-speed memory of any existing computer by an order of magnitude.

A second storage problem arises in retaining the results of the computation. If there are N_h quantized states at each stage and if there are K stages, then the number of values of $\hat{\mathbf{u}}(\mathbf{x}, k)$ and of $I(\mathbf{x}, k)$ which are computed is N_c where

$$N_c = N_h \cdot K. \tag{2.62}$$

Although this number can be extremely large, currently available low-speed memory devices, such as magnetic tape and bulk disk storage, are capable of storing this much information reliably and at a reasonable cost. In the example where $N = 10^6$, which saturates the high-speed memory of any available computer, a value of $K = 50$ implies that $N_c = 50 \cdot 10^6$, a large but feasible number.

The computing time requirement is also related to the vast number of results that are obtained. If N_c values of $\hat{\mathbf{u}}(\mathbf{x}, k)$ and $I(\mathbf{x}, k)$ are obtained and if it takes Δt_c seconds for each computation, then the total amount of computing time is T_c,

$$T_c = N_c \cdot \Delta t_c = N_h \cdot K \cdot \Delta t_c. \tag{2.63}$$

A reasonable value of Δt_c for a fast computer on the problems described above might be 50 μsec. Then, if $N = 10^6$, $K = 50$,

$$T_c = 50 \cdot 10^6 \cdot 50 \cdot 10^{-6} = 2500 \text{ secs.} \tag{2.64}$$

This time, about forty minutes, is justifiable for many problems of this complexity.

It is useful to compare this computing time with that required for enumeration. If there are 5 possible values of $\hat{\mathbf{u}}(\mathbf{x}, k)$ at each $\mathbf{x}$ and k and if $K = 50$, then the number of computations is

$$N_T \approx 10^{35}. \tag{2.65}$$

If the amount of time for each computation is taken to be $t_T = 1$ μsec, a rather charitable estimate compared to the value of $\Delta t_c = 50$ μsec, then the total computing time for the enumeration is

$$T_T = N_T \cdot t_T = 10^{35} \cdot 10^{-6} = 10^{29} \text{ seconds.} \tag{2.66}$$

This value of T_T, which is more than 10^{25} times greater than the value of T_c for this problem, is approximately $3 \cdot 10^{21}$ years. Similar calculations verify that in all but the very simplest examples dynamic programming requires less computing time than enumeration by many orders of magnitude.

Despite the great advantage over enumeration, the preceding results indicate that even in relatively simple problems, the computational requirements exceed the capacity of even the largest and fastest computers available.

However, it is worth noting that in most cases the high-speed memory requirement becomes excessive before the other requirements do. Furthermore, it is considerably more difficult to add high-speed memory capacity than it is to overcome the other computational difficulties—in particular, it is relatively easy and inexpensive to use more reels of tape or to let the computer run longer, but it is very difficult and expensive to add more banks of high-speed core storage. Consequently, it is customary to associate excessive computational requirements primarily with the high-speed memory requirement.

2.14 SUMMARY AND FLOW CHART OF PROCEDURE

In this chapter the class of optimization problems which dynamic programming solves was described. The generality of this class of problems, which includes nonlinear system equations, arbitrary performance criteria, and multiple constraints, was emphasized. Next, Bellman's principle of optimality was discussed. Then, it was shown that these optimization problems could be solved by an iterative functional equation based on the principal of optimality.

The next part of the chapter described in detail the conventional computational procedure based on the functional equations. This procedure is summarized in the flow chart in Fig. 2.8. First, state, control, and stage variables are quantized. Constraints are used to restrict the admissible values of state and control variables. The stage variable is indexed as k, $k = 0, 1, 2, \ldots, K$. The set of quantized admissible states, X, is indexed as $\mathbf{x}^{(j)}$, $j = 1, 2, \ldots, N$. The set of admissible controls, U, is indexed as $\mathbf{u}^{(m)}$, $m = 1, 2, \ldots, M$. The initial conditions for the functional equation are determined as $I[\mathbf{x}^{(j)}, K]$, $j = 1, 2, \ldots, N$. The functional equation is used to compute the minimum cost function, $I(\mathbf{x}, k)$, as

$$I[\mathbf{x}^{(j)}, k] = \min_{m=1,2,\ldots,M} \{l[\mathbf{x}^{(j)}, \mathbf{u}^{(m)}, k] + I[\mathbf{x}^{(m)}, k + 1]\} \qquad (2.67)$$

where

$$\mathbf{x}^{(m)} = \mathbf{g}(\mathbf{x}^{(j)}, \mathbf{u}^{(m)}, k] \qquad (2.68)$$

and where $I[\mathbf{x}^{(m)}, k + 1]$ is determined by using the interpolation formula and the known values $I[\mathbf{x}^{(j)}, k + 1]$. The optimal control is determined as

$$\hat{\mathbf{u}}[\mathbf{x}^{(j)}, k] = \mathbf{u}^{(\hat{m})} \qquad (2.69)$$

where $\hat{m}$ is the index of m for which the quantity in brackets in Eq. (2.67) is minimized. As illustrated in the flow chart, the procedure consists of determining $I[\mathbf{x}^{(j)}, k]$ end $\hat{\mathbf{u}}[\mathbf{x}^{(j)}, k)$ in terms of $I[\mathbf{x}^{(j)}, k + 1]$, for every quantized state $\mathbf{x}^{(j)}$, $j = 1, 2, \ldots, N$, beginning at stage $(K - 1)$ and continuing until stage 0 is reached. The nature of the results and how they can be utilized effectively was discussed in a separate section.

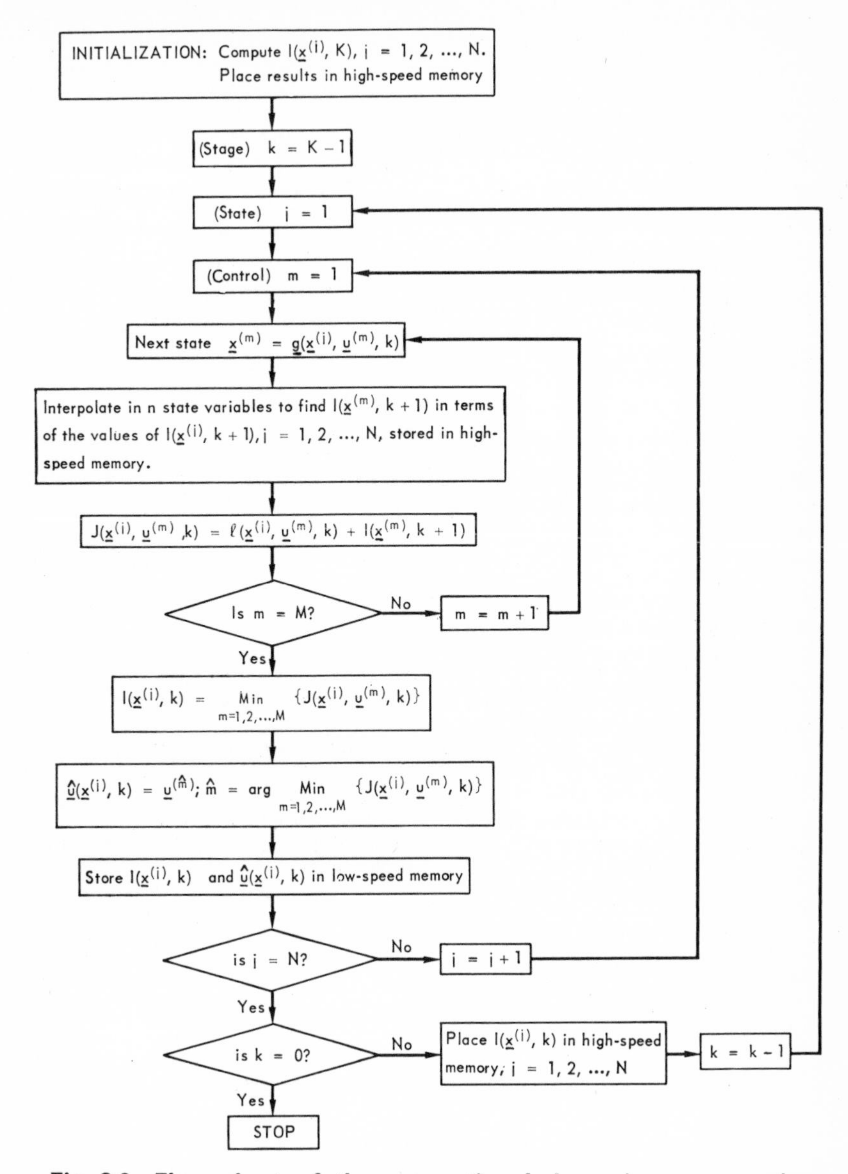

Fig. 2.8 Flow chart of the conventional dynamic programming computational procedure.

The chapter concluded with an examination of the computational aspects of dynamic programming. A discussion of interpolation schemes suitable for dynamic programming computations was presented. Next, the high-speed memory requirement, the low-speed memory requirement, and computing time were discussed; expressions for these quantities were derived. It was shown that even in problems of moderate size, these computational requirements can be quite large; it was also pointed out that in most problems the high-speed memory requirement is a more severe restriction than the others. The fact that computational requirements increase very rapidly with the dimension of the problem is called by Bellman "the curse of dimensionality," and these difficulties have so far constituted the main barrier to the widespread application of dynamic programming.

REFERENCES

1. Bellman, R., *Dynamic Programming*, Princeton University Press, Princeton, N.J., 1957.
2. Bellman, R., *Adaptive Control Processes*, Princeton University Press, Princeton, New Jersey, 1961.
3. Bellman, R., and Dreyfus, S., *Applied Dynamic Programming*, Princeton University Press, Princeton, N.J., 1962.
4. Zadeh, L. A., and DeSoer, C. A., *Linear System Theory: The State Space Approach*, McGraw-Hill Book Company, Inc., New York, 1963.
5. Peschon, J., *Disciplines and Techniques of Systems Control*, Blaisdell Publishing Company, New York, 1965.
6. Hamming, R. W., *Numerical Methods for Scientists and Engineers*, McGraw-Hill Book Company, Inc., New York, 1962.
7. Hildebrand, F. B., *Introduction to Numerical Analysis*, McGraw-Hill Book Company, Inc., New York, 1956.
8. Kahne, S. J., "Feasible Control Computations Using Dynamic Programming," *Proc. Third IFAC Congress*, London, June 1966, Paper No. 18G, Session 18.
9. Larson, R. E., "Comments on 'Feasible Control Computations Using Dynamic Programming'," *Proc. Third IFAC Congress*, London, June 1966.

Chapter Three

BASIC CONCEPTS OF STATE INCREMENT DYNAMIC PROGRAMMING

3.1 INTRODUCTION

State increment dynamic programming [Refs. 1, 2, 3, 4] is an optimization procedure that has all the desirable properties of the conventional dynamic programming computational procedure and yet has much smaller computational requirements. In particular, the high-speed memory requirement is always reduced significantly, and often by orders of magnitude. Similar savings in computing time and low-speed memory can also be achieved in certain cases.

The complete computational procedure of state increment dynamic programming is presented in the next four chapters. The purpose of this chapter is to state the general problem formulation and to indicate the basic concepts used in the procedure.

3.2 PROBLEM FORMULATION

The optimization problem to which the state increment dynamic programming computational procedure is applied is most conveniently formulated in the continuous-time case. The stage variable, t, thus varies continuously over the interval $t_0 \leq t \leq t_f$. The system equations become a set of nonlinear time-varying differential equations.

$$\dot{\mathbf{x}} = \mathbf{f}(\mathbf{x}, \mathbf{u}, t) \tag{3.1}$$

where $\mathbf{x}$ = n-dimensional state vector

$\mathbf{u}$ = q-dimensional control vector

t = stage variable, usually taken to be time

$\mathbf{f}$ = n-dimensional vector functional

$$\dot{(\ )} = \frac{d}{dt}(\ ).$$

The performance criterion J is a cost function, which is to be minimized. This cost function is the sum of an integral over time of a scalar function of state variables, control variables, and time plus a scalar function depending on the final state and final time

$$J = \int_{t_0}^{t_f} l[\mathbf{x}(\sigma), \mathbf{u}(\sigma), \sigma]\, d\sigma + \psi[\mathbf{x}(t_f), t_f] \tag{3.2}$$

where t_0 = initial time
t_f = final time
σ = dummy variable for time
J = cost function
l = loss function; cost function per unit time
ψ = final value term in cost function.

Constraints are specified on both state and control variables in the form

$$\begin{aligned} \mathbf{x} &\in X(t) \\ \mathbf{u} &\in U(\mathbf{x}, t) \end{aligned} \tag{3.3}$$

where $X(t)$ = set of allowable states at time t
$U(\mathbf{x}, t)$ = set of admissible controls at state $\mathbf{x}$, time t.

The optimization problem can then be stated as follows:

Given:

(i) A system equation as in Eq. (3.1)
(ii) Constraints on state and control variables as in Eq. (3.3)
(iii) An initial state, $\mathbf{x}(t_0)$

Find:

A control function, $\mathbf{u}(t)$, $t_0 \leq t \leq t_f$, such that the performance criterion in Eq. (3.2) is minimized and all the constraints are satisfied.

The solution to this problem is found by assuming that $\mathbf{u}(t)$, $t_0 \leq t \leq t_f$, is piecewise constant over intervals of length δt. The system differential equation, Eq. (3.1), is then approximated by

$$\mathbf{x}(t + \delta t) = \mathbf{x}(t) + \mathbf{f}[\mathbf{x}(t), \mathbf{u}(t), t]\, \delta t. \tag{3.4}$$

The change in the performance criterion over the interval from t to $(t + \delta t)$ is approximated by

$$\int_{t}^{t+\delta t} l[\mathbf{x}(\sigma), \mathbf{u}(\sigma), \sigma]\, d\sigma = l[\mathbf{x}(t), \mathbf{u}(t), t]\, \delta t. \tag{3.5}$$

More complex formulas could be used, but in the remainder of the book attention will be confined to these expressions.

If δt were always a fixed value, Δt, the stage variable t would be effectively quantized into uniform increments. Then, the system differential equation, Eq. (3.1), could be converted into a difference equation by use of Eq. (3.4), and the integral performance criterion, Eq. (3.2), could be converted into a summation by use of Eq. (3.5). It would then be possible to apply the computational procedure of the preceding chapter. However, δt is *not* fixed, and, as will be shown shortly, the technique for choosing it is one of the basic elements of the state increment dynamic programming computational procedure.

3.3 THE ITERATIVE FUNCTIONAL EQUATION BASED ON THE PRINCIPLE OF OPTIMALITY

An iterative functional equation similar to the form of Eq. (2.24) can be developed for the problem formulated in the previous section. This derivation follows exactly the steps outlined in Sec. 2.5.

The first step in the derivation is to define the minimum cost function. This quantity, written as $I(\mathbf{x}, t)$, is the minimum cost that can be incurred in going from state $\mathbf{x}$, time t to the final time, t_f. Formally,

$$I(\mathbf{x}, t) = \underset{\substack{\mathbf{u}(\sigma)\in U \\ t\leqslant\sigma\leqslant t_f}}{\text{Min}} \left\{ \int_t^{t_f} l[\mathbf{x}(\sigma), \mathbf{u}(\sigma), \sigma]\, d\sigma + \psi[\mathbf{x}(t_f), t_f] \right\} \tag{3.6}$$

where

$$\mathbf{x}(t) = \mathbf{x}.$$

The desired equation can be derived from Eq. (3.6) by splitting the integral into two parts, one from t to $(t + \delta t)$ and the other from $(t + \delta t)$ to t_f

$$I(\mathbf{x}, t) = \underset{\substack{\mathbf{u}(\sigma)\in U \\ t\leqslant\sigma\leqslant t_f}}{\text{Min}} \left\{ \int_t^{t+\delta t} l[\mathbf{x}(\sigma), \mathbf{u}(\sigma), \sigma]\, d\sigma + \int_{t+\delta t}^{t_f} l[\mathbf{x}(\sigma), \mathbf{u}(\sigma), \sigma]\, d\sigma + \psi[x(t_f), t_f] \right\}. \tag{3.7}$$

The minimization operation is also split into two parts, one part from t to $(t + \delta t)$ and the other from $(t + \delta t)$ to t_f.

$$I(\mathbf{x}, t) = \underset{\substack{\mathbf{u}(\sigma)\in U \\ t\leq\sigma\leq t+\delta t}}{\text{Min}} \; \underset{\substack{\mathbf{u}(\sigma)\in U \\ t+\delta t\leq\sigma\leq t_f}}{\text{Min}} \left\{ \int_t^{t+\delta t} l[\mathbf{x}(\sigma), \mathbf{u}(\sigma), \sigma]\, d\sigma + \int_{t+\delta t}^{t_f} l[\mathbf{x}(\sigma), \mathbf{u}(\sigma), \sigma]\, d\sigma + \psi[\mathbf{x}(t_f), t_f] \right\}. \tag{3.8}$$

Since the first integral in Eq. (3.8) depends only on values of σ where $t \leq \sigma \leq t + \delta t$, the minimization over σ, $\sigma > t + \delta t$, has no effect.

$$\underset{\substack{\mathbf{u}(\sigma)\in U \\ t\leq\sigma\leq t+\delta t}}{\text{Min}} \; \underset{\substack{\mathbf{u}(\sigma)\in U \\ t+\delta t\leq\sigma\leq t_f}}{\text{Min}} \left\{ \int_t^{t+\delta t} l[\mathbf{x}(\sigma), \mathbf{u}(\sigma), \sigma]\, d\sigma \right\} = \underset{\substack{\mathbf{u}(\sigma)\in U \\ t\leq\sigma\leq t+\delta t}}{\text{Min}} \left\{ \int_t^{t+\delta t} l[\mathbf{x}(\sigma), \mathbf{u}(\sigma), \sigma]\, d\sigma \right\}. \tag{3.9}$$

If the given state $\mathbf{x}$ is substituted for $\mathbf{x}(t)$ and if Eq. (3.5) is applied, then

$$\underset{\substack{\mathbf{u}(\sigma)\in U \\ t\leq\sigma\leq t+\delta t}}{\text{Min}} \; \underset{\substack{\mathbf{u}(\sigma)\in U \\ t+\delta t\leq\sigma\leq t_f}}{\text{Min}} \left\{ \int_t^{t+\delta t} l[\mathbf{x}(\sigma), \mathbf{u}(\sigma), \sigma]\, d\sigma \right\} = \underset{\mathbf{u}(t)\in U}{\text{Min}} \{ l[\mathbf{x}, \mathbf{u}(t), t]\, \delta t \}. \tag{3.10}$$

In the remaining terms, which vary over values of σ from $(t + \delta t)$ to t_f, the only effect on the control applied from t to $(t + \delta t)$ is to determine the state at time $(t + \delta t)$. If the expression for $\mathbf{x}(t + \delta t)$ in Eq. (3.4) is utilized and if the definition of the minimum cost function from Eqs. (3.6) is applied, the minimization of the remaining terms can be written as

$$\underset{\substack{\mathbf{u}(\sigma)\in U \\ t\le\sigma\le t+\delta t}}{\text{Min}} \quad \underset{\substack{\mathbf{u}(\sigma)\in U \\ t\le\delta t\le\sigma\le t_f}}{\text{Min}} \left\{\int_{t+\delta t}^{t_f} l[\mathbf{x}(\sigma), \mathbf{u}(\sigma), \sigma]\, d\sigma + \psi[\mathbf{x}(t_f), t_f]\right\}$$
$$= \underset{\mathbf{u}(t)\in U}{\text{Min}} \{I[\mathbf{x} + \mathbf{f}[\mathbf{x}, u(t), t]\, \delta t, t + \delta t]\}. \tag{3.11}$$

Finally, combining Eqs. (3.10) and (3.11) and dropping the argument of $\mathbf{u}(t)$, the iterative functional equation becomes

$$I(\mathbf{x}, t) = \underset{\mathbf{u}\in U}{\text{Min}} \{l[\mathbf{x}, \mathbf{u}, t]\, \delta t + I[\mathbf{x} + \mathbf{f}(\mathbf{x}, \mathbf{u}, t)\, \delta t, t + \delta t]\}. \tag{3.12}$$

The optimal control at state $\mathbf{x}$, time t, denoted as $\hat{\mathbf{u}}(\mathbf{x}, t)$, can be evaluated directly as that control for which the minimum value is taken on in Eq. (3.12). As a boundary condition for Eq. (3.12),

$$I(\mathbf{x}, t_f) = \psi(\mathbf{x}, t_f). \tag{3.13}$$

The above results are completely equivalent to the corresponding equations in Sec. 2.5, except that δt in the above equations is not fixed at Δt, but is instead allowed to vary. As will be seen in subsequent chapters, the solution to the problem is obtained by determining $\hat{\mathbf{u}}(\mathbf{x}, t)$ and $I(\mathbf{x}, t)$ at all quantized $\mathbf{x} \in X$ for a selected set of values of t; thus, a feedback control solution is again obtained.

3.4 CONSTRAINTS AND QUANTIZATION

In the state increment dynamic programming computational procedure constraints are handled exactly as in Sec. 2.6; they are used to restrict the set of admissible states, X, and the set of admissible controls, U. For example, inequality constraints of the form

$$\Phi(\mathbf{x}, t) \le 0 \tag{3.14}$$

can be used to bound the state variables

$$\beta_i^- \le x_i \le \beta_i^+ \qquad (i = 1, 2, \ldots, n). \tag{3.15}$$

Inequality constraints of the form

$$\Phi(\mathbf{x}, \mathbf{u}, t) \le 0 \tag{3.16}$$

can be used to restrict the control variables

$$\alpha_j^- \le u_j \le \alpha_j^+ \qquad (j = 1, 2, \ldots, q). \tag{3.17}$$

The quantities β_i^- and β_i^+ can vary with t, while the quantities α_j^- and α_j^+ can vary with $\mathbf{x}$ and t. Most other constraints can be used to restrict the range of state variables and control variables through relations such as Eqs. (3.15) and (3.17).

Within the allowable range determined by Eq. (3.15), each state variable x_i is quantized to a finite number of values, N_i. Although it is not necessary to do so, it is convenient to assume that the quantization is in uniform increments, Δx_i. The quantized values of x_i are then given by

$$x_i = \beta_i^- + j_i \Delta x_i \tag{3.18}$$

where $j_i = 1, 2, \ldots, N_i$

$N_i \Delta x_i = \beta_i^+ - \beta_i^-$

$i = 1, 2, \ldots, n$

The set of state vectors for which each component has the form of Eq. (3.18) is called the set of quantized admissible states, X.

The control variables also can be quantized in a similar manner. However, it is necessary only that there be a finite number of admissible controls. The set of admissible controls, U, is indexed as

$$U = \{\mathbf{u}^{(1)}, \mathbf{u}^{(2)}, \ldots, \mathbf{u}^{(M)}\}. \tag{3.19}$$

The choice of the $\mathbf{u}^{(m)} \in U$ is made in accordance with the individual problem under consideration.

3.5 DETERMINATION OF THE TIME INCREMENT δt

A fundamental difference between state increment dynamic programming and conventional dynamic programming is in the method for determining δt, the time interval *over* which a given control is applied. In conventional dynamic programming the total time interval over which optimization is performed. $t_0 \leq t \leq t_f$, is quantized into uniform increments, Δt, and optimal control is computed only at these quantized values of t. In the computation of optimal control according to the iterative functional equation, the next state is always taken to occur Δt seconds after the present state. Consequently, every admissible control is considered to be applied for Δt seconds. Therefore, in conventional dynamic programming,

$$\delta t = \Delta t \tag{3.20}$$

i.e., the time over which a given control is applied is fixed at Δt, the time interval between successive computations of optimal control.

In state increment dynamic programming, on the other hand, the determination of these two time intervals is made independently. The time interval between successive computations of optimal control may or may not be fixed, according to the nature of the problem.† However, δt varies with each control which is applied. *The interval δt is determined as the minimum time interval required for any one of the n state variables to change by one increment.* This is the source of the name "state increment dynamic programming"; instead of control being applied until time changes by a specified increment, as in conventional dynamic programming, control is applied until one of the *states* changes by a specified increment.

If Δx_i is the increment in the ith state variable and if control $\mathbf{u}$ is being applied at state $\mathbf{x}$ and time t, then δt can be expressed as

$$\delta t = \underset{i=1,2,\ldots,n}{\text{Min}} \left\{ \frac{\Delta x_i}{|f_i(\mathbf{x}, \mathbf{u}, t)|} \right\} \tag{3.21}$$

where $f_i(\mathbf{x}, \mathbf{u}, t)$ is the ith component of $\mathbf{f}(\mathbf{x}, \mathbf{u}, t)$, the system differential equation vector. The interpretation of Eq. (3.21), the basic equation of state increment dynamic programming, can be seen by noting that from Eq. (3.1) $f_i(\mathbf{x}, \mathbf{u}, t)$ is the rate at which x_i, the ith state variable, is changing. The amount of time taken for x_i to change by one increment, Δx_i, is $\Delta x_i/|f_i(\mathbf{x}, \mathbf{u}, t)|$, where the absolute value sign is necessary to allow change in either the positive or negative direction. The minimum time necessary for any one of the n state variables to change by one increment is clearly the minimum of these n quantities.

In a problem with one state variable ($n = 1$), there is no need to perform a minimization in determining δt. For such an example,

$$\delta t = \frac{\Delta x_1}{|f(x_1, u, t)|}\,. \tag{3.22}$$

The determination of δt at a specific x_1 and t, denoted as x_1^* and t^*, is illustrated in Fig. 3.1 for two cases, $f(x_1^*, u, t^*) > 0$ and $f(x_1^*, u, t) < 0$. The case $f(x_1^*, u, t^*) = 0$ is handled according to special procedures described in Chapters 4 and 5.

In an example with two state variables ($n = 2$), the determination of δt requires a minimization over two possible values. If the two state variables are x_1 and x_2 and if control $\mathbf{u}^{(m)}$ is applied at state $\mathbf{x}$ and time t, then δt is given by either

$$\delta t = \frac{\Delta x_1}{|f_1(\mathbf{x}, \mathbf{u}^{(m)}, t)|} \tag{3.23}$$

or

$$\delta t = \frac{\Delta x_2}{|f_2(\mathbf{x}, \mathbf{u}^{(m)}, t)|} \tag{3.24}$$

† In Chapter 4 this time interval is fixed at Δt, exactly as in conventional dynamic programming; but in Chapter 5, which treats a special class of problems, the time interval varies.

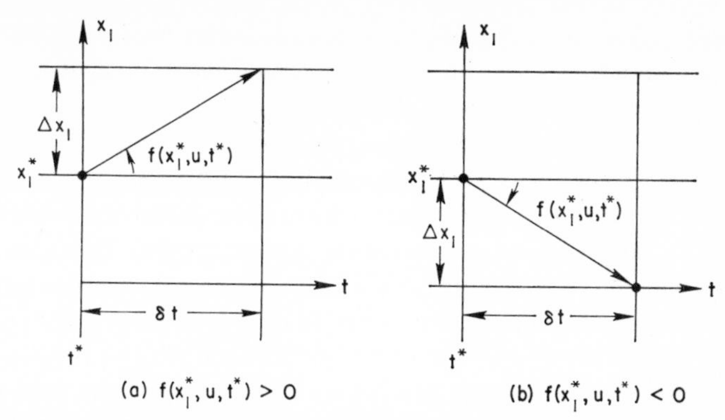

Fig. 3.1. Determination of δt in one-dimensional example.

depending on which is smaller. In Fig. 3.2 the next state that results from applying each of five controls, $\mathbf{u}^{(m)}$, $m = 1, 2, 3, 4, 5$, at state $\mathbf{x}^*$, time t^* is shown. In each case the next state is found by applying control until the trajectory intersects a rectangular boundary, centered at the present state and with dimensions $2\,\Delta x_1$ by $2\,\Delta x_2$. For $m = 2, 4$, state x_1 changes by Δx_1 in less time than x_2 changes by Δx_2; therefore, δt is given by Eq. (3.23). For $m = 1$, 3, 5, state x_2 changes by Δx_2 in less time than x_1 changes by Δx_1, and hence δt is determined by Eq. (3.24). Again, if both $f_i(\mathbf{x}^*, \mathbf{u}, t^*) = 0$, $i = 1$ and 2, then δt is determined by one of the special procedures in Chapters 4 and 5.

In general, δt is determined by applying control until the trajectory reaches an n-dimensional hypercube centered at the present state and with length $2\,\Delta x_i$ along the ith coordinate. As a result, the next state is always close to the

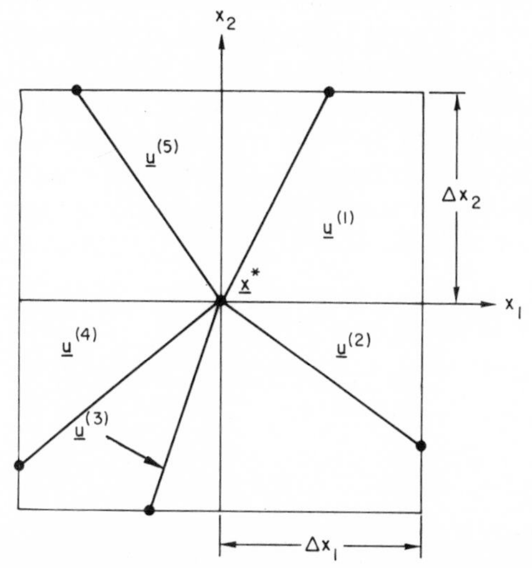

Fig. 3.2. Determination of δt in two dimensions.

present state in the sense that the two states can differ by at most Δx_i along the x_i coordinate. In the general case of conventional dynamic programming, on the other hand, because control is always applied for the fixed increment, Δt, the next state can occur anywhere in the space of admissible states.

It is this difference that enables state increment dynamic programming to reduce the computational requirements from those of conventional dynamic programming. If the next state is close to the present state, as in state increment dynamic programming, then in order to perform the interpolation of the minimum cost function, as required in the iterative functional equation, Eq. (3.12), it is necessary to store values of the minimum cost function at only those quantized states near the present state. However, if the next state can occur anywhere in the space of admissible states, as in conventional dynamic programming, then, as shown in Chapter 2, it is necessary to store values of the minimum cost function for every quantized state $\mathbf{x} \in X$. The mechanism through which this basic concept is exploited to produce great reductions in computational requirements is discussed in the next section.

3.6 CONCEPT OF THE BLOCK

The previous section discussed how in state increment dynamic programming it is possible to reduce the amount of data which must be stored in order to apply the iterative functional equation, Eq. (3.12), at a specific $\mathbf{x} \in X$ and t, $t_0 \leq t \leq t_f$. In Chapter 2, this amount of data is referred to as the high-speed memory requirement. However, a significant *overall* saving in high-speed memory requirement can be achieved only by processing the data so as to obtain maximum utilization of the reduction for a single calculation. In state increment dynamic programming this is achieved by doing computations in units called blocks.

Blocks are defined by partitioning the $(n + 1)$-dimensional space containing the n state variables and time into rectangular sub-units. Each block covers w_i increments along the x_i-axis and ΔT seconds along the t-axis. A particular block is denoted by the largest values of the coordinates that are contained within the block. The block $B(j_0, j_1, j_2, \ldots, j_n)$ contains values of $\mathbf{x}$ and t such that

$$(j_i - 1)w_i\,\Delta x_i \leq x_i - \beta_i^- \leq j_i w_i\,\Delta x_i \qquad (i = 1, 2, \ldots, n) \qquad (3.25)$$
$$(j_0 - 1)\,\Delta T \leq t - t_0 \leq j_0\,\Delta T$$

where $j_i = 1, 2, \ldots, J_i$

$J_i w_i\,\Delta x_i = \beta_i^+ - \beta_i^-$

$i = 1, 2, \ldots, n$

$j_0 = 1, 2, \ldots, J_0$

$J_0\,\Delta T = t_f - t_0.$

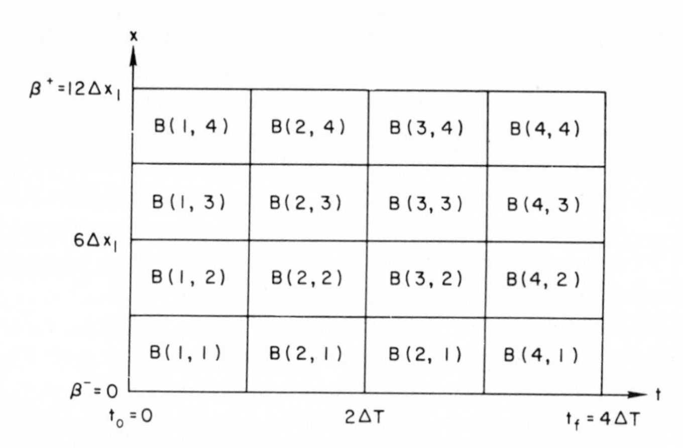

Fig. 3.3. Blocks for one-dimensional example.

As indicated by the above equations, the boundaries between blocks are considered to be in *both* blocks.

The numbers w_i are taken to be small integers, usually between 2 and 5. The value of ΔT, on the other hand, is taken to be considerably larger than the average value of δt as determined by Eq. (3.21).

For a one-dimensional example the blocks are two-dimensional rectangles. A typical set of blocks is illustrated in Fig. 3.3, where $w_1 = 3$, $\beta_1^+ = 12\,\Delta x_1$, $\beta_1^- = 0$, $t_f = 4\,\Delta T$, and $t_0 = 0$. Note that each block contains $w_1 + 1 = 4$ quantized values of x_1.

For a two-dimensional ($n = 2$) example each block is a three-dimensional rectangular solid. A few blocks for a hypothetical problem are illustrated in Fig. 3.4.

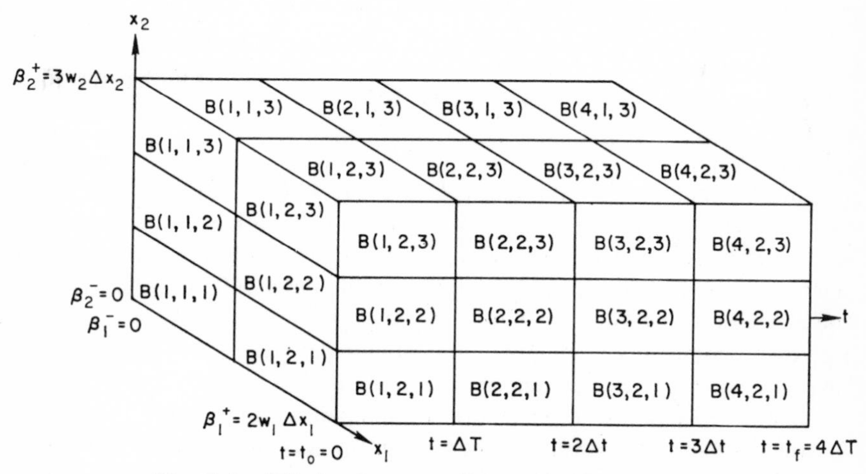

Fig. 3.4. Blocks for two-dimensional example.

It should be pointed out that although the blocks are defined over the entire rectangular region, $\beta_i^- \leq x_i \leq \beta_i^+, i = 1, 2, \ldots, n$, it is possible in any time interval to consider X, the set of allowable states, to be confined to any subset of blocks. The subset of blocks in which allowable states occur can vary from time interval to time interval. In this manner time-varying constraints on the states can be implemented. By restricting the blocks in the interval $t_f - \Delta T \leq t - t_0 \leq t_f$, it is also possible to implement final-value constraints in the initialization procedure. In addition, constraints can be placed on the allowable states within a block, if desired.

The reason for defining blocks can now be stated. In state increment dynamic programming computations are done for one block at a time on the assumption that optimal trajectories never leave the block. Since each block contains only $(w_i + 1)$ increments along state x_i, there are a relatively small number of quantized states contained within a single block. Furthermore, the method of choosing δt implies that, unless the present state is on the boundary of the block, the quantized states needed for interpolation of the minimum cost function at the next state are within the block. Therefore, the number of high-speed storage locations required for the computation of optimal control in one block is quite small. Also, since ΔT is relatively large, a large number of computations of optimal control take place within the block. The result is that many computations can take place using a small number of high-speed storage locations. The exact computational procedure to be followed within a block is described in Chapter 4 for the general optimization problem of Sec. 3.2 and in Chapter 5 for a special case.

It should be pointed out that it is unrealistic to assume that optimal trajectories always remain inside a block over a time interval ΔT. There are, however, many satisfactory techniques available for implementing interblock transitions; these methods are discussed in detail in Chapter 6.

3.7 SUMMARY

In this chapter it has been shown that the optimization problem which state increment dynamic programming solves is the same as the one solved by conventional dynamic programming. In both procedures an iterative functional equation, based on Bellman's principle of optimality, is used to determine optimal control. Constraints are implemented in the same way, and quantization of state and control variables is the same for both cases. For a given problem the results of the two procedures are equivalent, and both have the same degree of generality and applicability.

The difference between the two is in the detailed computation of optimal control. In state increment dynamic programming δt, the time increment over which a given control is applied, is not fixed, but instead is determined as the interval required for any one of the state variables to change by a fixed increment. As a result, the next state is always close to the present state. Consequently, the interpolation of the minimum cost function can be done using only values for quantized states near the present state, rather than values for every admissible quantized state, $\mathbf{x} \in X$.

This basic reduction in high-speed memory requirement is extended to the entire procedure by computing optimal control in units called blocks. A block covers relatively few quantized states, but a relatively long time interval. Consequently, optimal control can be computed at a considerable number of points using only a few high-speed storage locations.

The detailed computational procedure within a block for the general optimization problem is described in Chapter 4. The resulting reduction in computational requirements is also discussed. Chapter 5 presents a modified procedure, which has additional computational advantages, but which is applicable only to a certain class of problems. The complete procedure, including methods for implementing transitions of optimal trajectories between blocks, is discussed in Chapter 6. An illustrative example is worked in Chapter 7.

REFERENCES

1. Larson, R. E., *Dynamic Programming with Continuous Independent Variable*, Stanford Electronics Laboratory TR 6302-6, Stanford, California, April 1964.
2. Larson, R. E., "State Increment Dynamic Programming: Theory and Applications," *Proc. 2nd Allerton Conf. on Ckt. and Systems Theory*, U. of Illinois, Sept. 1964, pp. 643–665.
3. Larson, R. E., "Dynamic Programming with Reduced Computational Requirements," *IEEE Trans. on Automatic Control*, Vol. AC-10, No. 2, April 1965, pp. 135–143.
4. Larson, R. E., "An Approach to Reducing the High-Speed Memory Requirement of Dynamic Programming," *J. Math. Anal. and Appl.*, vol. 11, nos 1–3, July, 1965, pp. 519–537.

Chapter Four

COMPUTATIONAL PROCEDURE WITHIN A BLOCK IN THE GENERAL CASE

4.1 INTRODUCTION

In the previous chapter the basic equations and concepts of state increment dynamic programming have been presented. The purpose of this chapter is to describe in detail a procedure for performing the computations within a given block. The procedure is applicable to the general dynamic programming problem of Sec. 3.2, whereas the procedure described in the next chapter can be used only in special cases.

It is assumed throughout this chapter that the next state always lies in the same block as the present state, i.e., the n-dimensional hypercube on the surface of which the next state lies is always contained in the same block as the present state. This condition is assured if the present state is in the interior of the block and not on the boundary. The extremely important question of how to perform calculations when the present state lies on the boundary of the block is discussed in detail in Chapter 6.

4.2 QUANTIZATION OF STATE AND STAGE VARIABLES

The region covered by block $B(j_0, j_1, j_2, \ldots, j_n)$ was defined in Chapter 3 as those values of $\mathbf{x}$ and t for which

$$\begin{aligned} (j_i - 1)w_i\,\Delta x_i \le x_i - \beta_i^- \le j_i w_i\,\Delta x_i \qquad (i = 1, 2, \ldots, n) \\ (j_0 - 1)\,\Delta T \le t - t_0 \le j_0 \end{aligned} \tag{4.1}$$

where $j_i = 1, 2, \ldots, J_i$

$J_i w_i\,\Delta x_i = \beta_i^+ - \beta_i^-$

$i = 1, 2, \ldots, n$

$j_0 = 1, 2, \ldots, J_0$

$J_0 = t_f - t_0.$

As noted in Chapter 3, the value of ΔT, the interval in t covered by a block, is considerably larger than the average value of δt, the increment in t over which a given control is applied. In the procedure described in this chapter, within each block t is quantized into uniform increments, Δt. The size of Δt is about the same as the average value of δt. The increments Δt and ΔT are related by

$$\Delta T = S\,\Delta t \tag{4.2}$$

where S is an integer. In most applications S is between 5 and 15.

Within a block computations of optimal control and minimum cost are made only at quantized values of t. These values can be written as

$$t = t_0 + (j_0 - 1)\,\Delta T + s\,\Delta t \tag{4.3}$$

where

$$s = 0, 1, 2, \ldots, S$$
$$S\,\Delta t = \Delta T.$$

In order to simplify the notation, the variable t_0' is defined within a given block as

$$t_0' = t_0 + (j_0 - 1)\,\Delta T. \tag{4.4}$$

The quantized points are then

$$t = t_0' + s\,\Delta t \tag{4.5}$$

where s is defined as in Eq. (4.3). In the remainder of this chapter the quantized

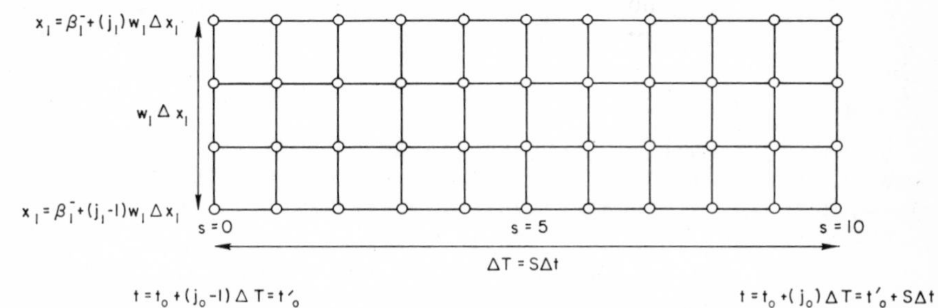

Fig. 4.1. Quantization within a block $B(j_0, j_1)$ for a one-dimensional example.

values of t will often be indicated by the index s; the corresponding value of t can always be determined by either Eq. (4.3) or Eqs. (4.4) and (4.5).

As a result of this quantization, the increment in t between successive computations of optimal control is fixed at Δt, exactly as in conventional dynamic programming. Note, however, as has been previously stressed, δt is determined independently.

The n state variables are already quantized into increments Δx_i, $i = 1, 2, \ldots, n$. Optimal control is thus computed both at quantized states, $\mathbf{x} \in X$, and at quantized values of t, exactly as in conventional dynamic programming. This complete quantization provides a convenient means for indexing the various points at which optimal control is computed.

The quantized states and quantized times in a single block are illustrated in Fig. 4.1 for a one-dimensional example. The block size is $w_1\,\Delta x_1 = 3\,\Delta x_1$

by $S\,\Delta t = 10\,\Delta t$. Consequently, there are $(w_1 + 1) = 4$ quantized states and $(S + 1) = 11$ quantized values of t in the block.

4.3 ORDER OF COMPUTATIONS WITHIN A BLOCK

Because of the interpolation procedure used in this chapter, the computation of optimal control and minimum cost at a given **x** and t generally requires the minimum costs for all quantized states in the block at the next *two* quantized values of t. Thus, when a specific block is about to be processed,

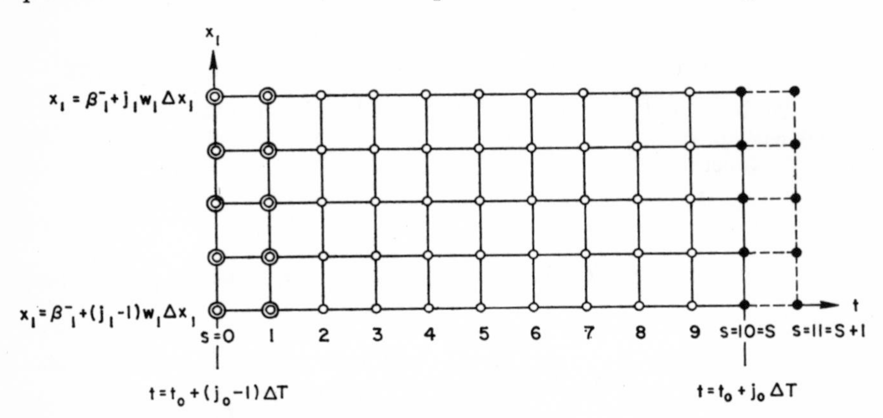

Fig. 4.2. Computations in block $B(j_0, j_1)$.

values of the minimum cost function for every quantized state in the block are stored in high-speed memory at both $s = S$, i.e., $t = t_0' + S\,\Delta t$, the largest value of t in the block, and at $s = S + 1$, i.e., $t = t_0' + (S + 1)\,\Delta t$, the value of t that lies one increment outside this block. If $t_0' + S\,\Delta t = t_f$, i.e., if this block covers the last time interval, then these values are generated by a special procedure, as described in Chapter 6; if not, these values are retrieved from the results of computations in $B(j_0 + 1, j_1, j_2, \ldots, j_n)$, the block covering the same region of state space, but the next higher time interval. The points at which these values are specified for the block of Fig. 4.1 are indicated by solid circles in Fig. 4.2.

Within the block, computations are first performed at $s = S - 1$, i.e., $t = t_0' + (S - 1)\,\Delta t$. For this value of t, optimal control and minimum cost are computed for all quantized states on the basis of the minimum costs at $s = S$ and $s = S + 1$. When these calculations have been completed, computations are then made for all quantized states at $s = S - 2$ on the basis of minimum costs at $s = S - 1$ and $s = S$. This process continues, in order of decreasing s, with optimal control and minimum costs being computed at s on the basis of minimum costs at $(s + 1)$ and $(s + 2)$, until $s = 0$ is reached. The points at which these computations are made in the block of Fig. 4.1 are indicated by open circles in Fig. 4.2. The minimum costs at $s = 0$ and $s = 1$ are stored for later use as initial values for block $B(j_0 - 1, j_1, j_2, \ldots, j_n)$, the block covering the same region of state space, but the next lower time interval. These points for the block of Fig. 4.1 are indicated by two concentric open circles in Fig. 4.2.

4.4 COMPUTATIONS AT A GIVEN STATE AND STAGE

The computations at given values of $\mathbf{x}$ and $t = t_0' + s\,\Delta t$ are based on the iterative functional equation

$$I(\mathbf{x}, t_0' + s\,\Delta t) = \min_{\mathbf{u}\in U} \{l(\mathbf{x}, \mathbf{u}, t_0' + s\,\Delta t)\,\delta t + I[\mathbf{x} + \mathbf{f}(\mathbf{x}, \mathbf{u}, t_0' + s\,\Delta t)\,\delta t, t_0' + s\,\Delta t + \delta t]\} \quad (4.6)$$

As discussed in Chapter 3, the minimization is performed by applying a discrete set of controls and choosing the minimum value by direct comparison. This set of controls is denoted as

$$U = \{\mathbf{u}^{(1)}, \mathbf{u}^{(2)}, \ldots, \mathbf{u}^{(M)}\} \quad (4.7)$$

The value of δt for a specific $\mathbf{u}^{(m)} \in U$ is determined as

$$\delta t^{(m)} = \min_{i=1,2,\ldots,n} \left\{ \left| \frac{\Delta x_i}{f_i(\mathbf{x}, \mathbf{u}^{(m)}, t_0' + s\,\Delta t)} \right|, \Delta t \right\} \quad (4.8)$$

where $\delta t^{(m)}$ is now also constrained to be less than Δt.

For this value of $\delta t^{(m)}$, the next state when control $\mathbf{u}^{(m)}$ is applied, $\mathbf{x}^{(m)}$, can be written as

$$\mathbf{x}^{(m)}(t_0' + s\,\Delta t + \delta t^{(m)}) = \mathbf{x} + \mathbf{f}(\mathbf{x}, \mathbf{u}^{(m)}, t_0' + s\,\Delta t)\,\delta t^{(m)} \quad (4.9)$$

As emphasized repeatedly, this next state can differ from the present state by at most one increment, Δx_i, in the ith state variable. This can be clearly seen by rewriting Eq. (4.9) as

$$\mathbf{x}^{(m)} - \mathbf{x} = \mathbf{f}(\mathbf{x}, \mathbf{u}^{(m)}, t_0' + s\,\Delta t)\,\delta t^{(m)} \quad (4.10)$$

and then observing from Eq. (4.9) that

$$|f_i(\mathbf{x}, \mathbf{u}^{(m)}, t_0' + s\,\Delta t)|\,\delta t^{(m)} \leq \Delta x_i \qquad (i = 1, 2, \ldots, n) \quad (4.11)$$

As a consequence of this result, the interpolation of the minimum cost function at the next state can be accomplished by storing only values at quantized states that differ from the present state by at most one increment in each state variable. Since equality in Eq. (4.11) occurs for the value or values of i which actually minimize Eq. (4.8), then at most $(n - 1)$ of the state variables can take on nonquantized values. However, the value of t for the next state, $t_0' + s\,\Delta t + \delta t^{(m)}$, will in general be a nonquantized value. Thus, the interpolation is in n variables, where $(n - 1)$ of the variables are state variables and the other variable is t.

In conventional dynamic programming, the interpolation of the minimum cost function was also n-dimensional, namely in the n state variables. For the present procedure, these same formulas can be used in the interpolations along the $(n - 1)$ state variables. As discussed in Sec. 2.11 this interpolation is generally linear or quadratic.

It has been noted that Eq. (4.8) differs from Eq. (3.21) in that $\delta t^{(m)}$ is constrained to be less than or equal to Δt. This is done so that changes in t can be held to a reasonable value. As noted in the previous section, values of minimum cost are available at $t_0' + (s + 1)\,\Delta t$ and $t_0' + (s + 2)\,\Delta t$. Detailed interpolation formulas for the case where $\delta t^{(m)} < \Delta t$ are given in the next section. If $\delta t^{(m)}$ is determined as Δt, then Eq. (4.11) can be a strict inequality for all i, $i = 1, 2, \ldots, n$. In this case the interpolation procedure still requires values of minimum cost only at quantized states that lie within one increment of the present state along any state variable axis, but all n components of the next state vector can be at nonquantized values.

Using these interpolation formulas to obtain

$$I[\mathbf{x} + \mathbf{f}(\mathbf{x}, \mathbf{u}^{(m)}, t_0' + s\,\Delta t)\,\delta t^{(m)}, t_0' + s\,\Delta t + \delta t^{(m)}]$$

for a given control $\mathbf{u}^{(m)}$, the minimization in Eq. (4.6) can be performed by evaluating the quantity in brackets for each quantized control $\mathbf{u}^{(m)} \in U$ and choosing the minimum value by a direct comparison. The optimal control is determined as

$$\hat{\mathbf{u}}(\mathbf{x}, t_0' + s\,\Delta t) = \mathbf{u}^{(\hat{m})} \tag{4.12}$$

where $\hat{m}$ is the index of the control for which the minimum in Eq. (4.6) is attained.

4.5 INTERPOLATION PROCEDURES

In this section interpolation formulas for evaluating $I[\mathbf{x} + \mathbf{f}(\mathbf{x}, \mathbf{u}^{(m)}, t_0' + s\,\Delta t)\,\delta t^{(m)}, t_0' + s\,\Delta t + \delta t^{(m)}]$ are discussed. Many of these formulas make use of the fact that, for many of the quantized states which differ from $\mathbf{x}$ by one increment or less in all state variables, minimum cost has already been computed at the present time, $t_0' + s\,\Delta t$. The mechanism through which this occurs is discussed in Chapter 6.

The formulas used in most applications are linear in t and either linear or quadratic in the state variables. If $\delta t^{(m)} = \Delta t$, an n-dimensional interpolation in the n state variables is made; formulas of the type used in the conventional procedure and described in Sec. 2.11 can be used. If $\delta t^{(m)} < \Delta t$, then the interpolation in the $(n - 1)$ state variables is again based on formulas of the type discussed in Sec. 2.11. Any of a number of alternative methods for performing the linear interpolation in t can be used.

In one method for the linear interpolation in t, if the next state, $\mathbf{x} + \mathbf{f}(\mathbf{x}, \mathbf{u}^{(m)}, t_0' + s\,\Delta t)\,\delta t^{(m)}$, is near quantized states where minimum cost has already been computed at the present time, $t_0' + s\,\Delta t$, then a linear interpolation in t is made using values of minimum cost and quantized states for $t = t_0' + s\,\Delta t$ and $t = t_0' + (s + 1)\,\Delta t$. If the next state does not lie in such a region then linear extrapolation using values of minimum cost at quantized states for $t = t_0' + (s + 1)\,\Delta t$ and $t = t_0' + (s + 2)\,\Delta t$ is performed. This procedure is illustrated for a one-dimensional example in Fig. 4.3.

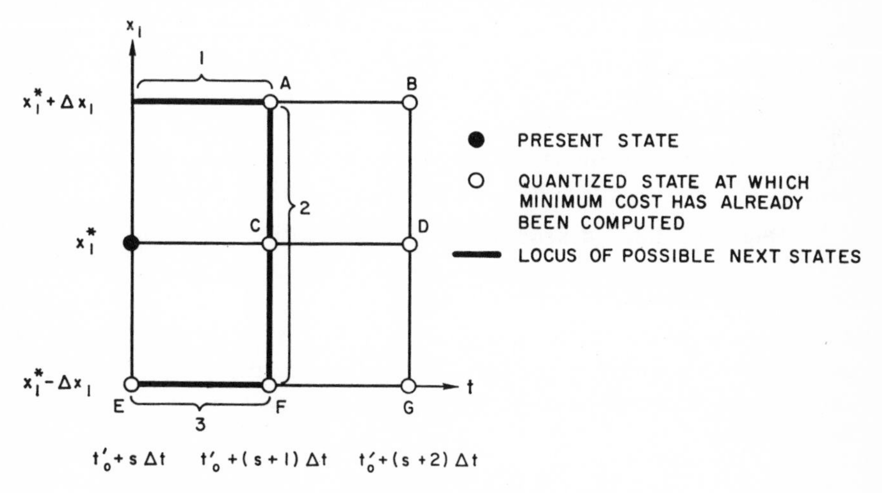

Fig. 4.3. First procedure for linear interpolation in t.

Interpolation formulas in regions 1 and 3:

1: $$I(x_1^* + \Delta x_1, t_0' + s\,\Delta t + \delta t^{(m)}) = A + \frac{A - B}{\Delta t}(\Delta t - \delta t^{(m)})$$

$$0 \le \delta t^{(m)} \le \Delta t$$

3: $$I(x_1^* - \Delta x_1, t_0' + s\,\Delta t + \delta t^{(m)}) = E + \frac{(F - E)}{\Delta t}\,\delta t^{(m)}$$

$$0 \le \delta t^{(m)} \le \Delta t$$

In a second method, the linear interpolation in t utilizes the formula for interpolation in the n state variables at the next quantized time, $t = t_0' + (s + 1)\,\Delta t$. A quantized state that is very close to $\mathbf{x}$, the present state, is selected. The minimum cost function is then assumed to be the product of a linear function of t and the interpolation formula in the n state variables at

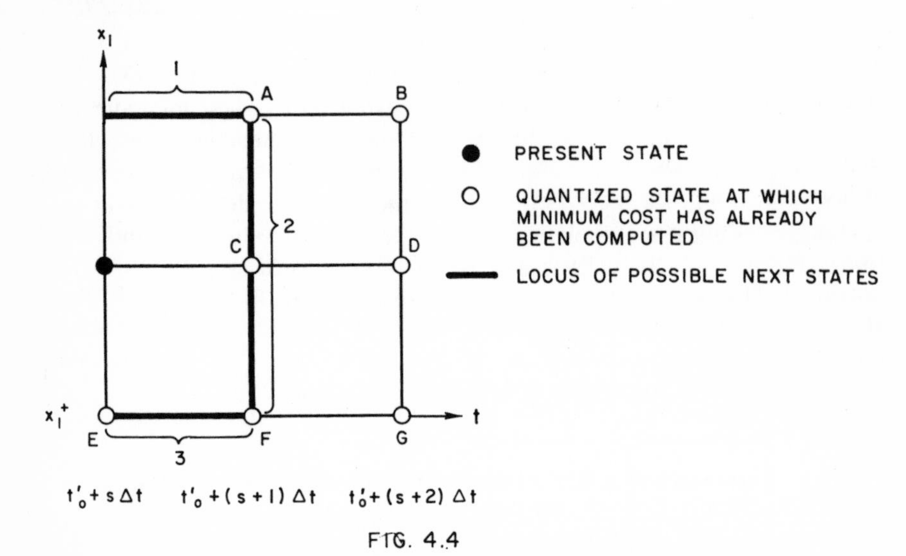

Fig. 4.4. Second procedure for linear interpolation in t.

Interpolation formulas in regions 1 and 3:

$$1: \qquad I(x_1^* + \Delta x_1, t_0' + s\,\Delta t + \delta t^{(m)}) = A\left(\frac{E}{F} + \frac{(F - E)}{F\,\Delta t}\,\delta t^{(m)}\right)$$

$$3: \qquad I(x_1^* - \Delta x_1, t_0' + s\,\Delta t + \delta t^{(m)}) = F\left(\frac{E}{F} + \frac{(F - E)}{F\,\Delta t}\,\delta t^{(m)}\right)$$

$$= E + \frac{(F - E)}{\Delta t}\,\delta t^{(m)}$$

$t = t_0' + (s + 1)\,\Delta t$. The linear function of t is selected so that the interpolation formula is exact for the selected state. If $\mathbf{x}^+$ is the selected state, the formula is

$$I[\mathbf{x} + \mathbf{f}(\mathbf{x}, \mathbf{u}^{(m)}, t_0' + s\,\Delta t)\,\delta t^{(m)}, t_0' + s\,\Delta t + \delta t^{(m)}]$$
$$= (c_1 + c_2\,\delta t^{(m)})\{I[\mathbf{x} + \mathbf{f}(\mathbf{x}, \mathbf{u}^{(m)}, t_0' + s\,\Delta t)\,\delta t^{(m)}, t_0' + (s + 1)\,\Delta t]\} \quad (4.13)$$

where $I[\mathbf{x} + \mathbf{f}(\mathbf{x}, \mathbf{u}^{(m)}, t_0' + s\,\Delta t)\,\delta t^{(m)}, t_0' + (s + 1)\,\Delta t]$ is obtained by using the interpolation formula in n state variables at time $t_0' + (s + 1)\,\Delta t$, and where c_1 and c_2 are given by

$$c_1 = \frac{I(\mathbf{x}^+, t_0' + s\,\Delta t)}{I[\mathbf{x}^+, t_0' + (s+1)\,\Delta t]}$$

$$c_2 = \frac{1 - c_1}{\Delta t}. \tag{4.14}$$

This formula is utilized in the computer program for problems with four state variables described in Chapter 8. The formula in the one-dimensional case is illustrated in Fig. 4.4. One very useful property of this procedure is that values of minimum cost need be stored only at $t = t_0' + (s + 1)\,\Delta t$, not at both this value and $t = t_0' + (s + 2)\,\Delta t$.

Other procedures of this type have been suggested [Refs. 1–4], but these two have been the most-used in previous applications.

Still other procedures involving a higher order fit in t could be used. However, such procedures require the storage of the minimum cost function for quantized states at more than two quantized values of t. The increased accuracy is seldom worth the increased computational effort.

4.6 SUMMARY AND FLOW CHART OF PROCEDURES

The computations for a given block, $B(j_0, j_1, j_2, \ldots, j_n)$, are illustrated in the flow chart, Fig. 4.5. The quantized states in the block are ordered as $\mathbf{x}^{(r)}$, $r = 1, 2, \ldots, R$. This ordering, which is essentially arbitrary (see Chapter 6), determines the sequence of states at which optimal control is computed for a given value of t.

The least value of t within the block, $t = t_0 + (j_0 - 1)\,\Delta t$, is denoted as t_0'. The quantized values of t for the block are thus written as

$$t = t_0' + s\,\Delta t \tag{4.15}$$

where $s = 0, 1, 2, \ldots, S$.

The initial data for the block consists of a specification of the minimum cost at all quantized states $\mathbf{x}^{(r)}$ for $t = t_0' + S\,\Delta t$ and $t = t_0' + (S + 1)\,\Delta t$. These data are stored in two arrays, one at each value of t. The arrays are denoted as $I[\mathbf{x}^{(r)}, t_0' + S\,\Delta t]$ and $I[\mathbf{x}^{(r)}, t_0' + (S + 1)\,\Delta t]$ respectively, where $r = 1, 2, \ldots, R$ in both cases.

The processing of data begins at $s = S - 1$. At every value of s, a computation of optimal control and minimum cost is made for each $\mathbf{x}^{(r)}$, $r = 1, 2, \ldots, R$. The arrays $I[\mathbf{x}^{(r)}, t_0' + (s + 1)\,\Delta t]$ and $I[\mathbf{x}^{(r)}, t_0' + (s + 2)\,\Delta t]$ are used for interpolation purposes.

For a fixed s and fixed $\mathbf{x}^{(r)}$, each admissible control $\mathbf{u}^{(m)} \in U$ is applied. For a given s, $\mathbf{x}^{(r)}$, and $\mathbf{u}^{(m)}$, each component of $\mathbf{f}[\mathbf{x}^{(r)}, \mathbf{u}^{(m)}, t_0' + s\,\Delta t]$, the

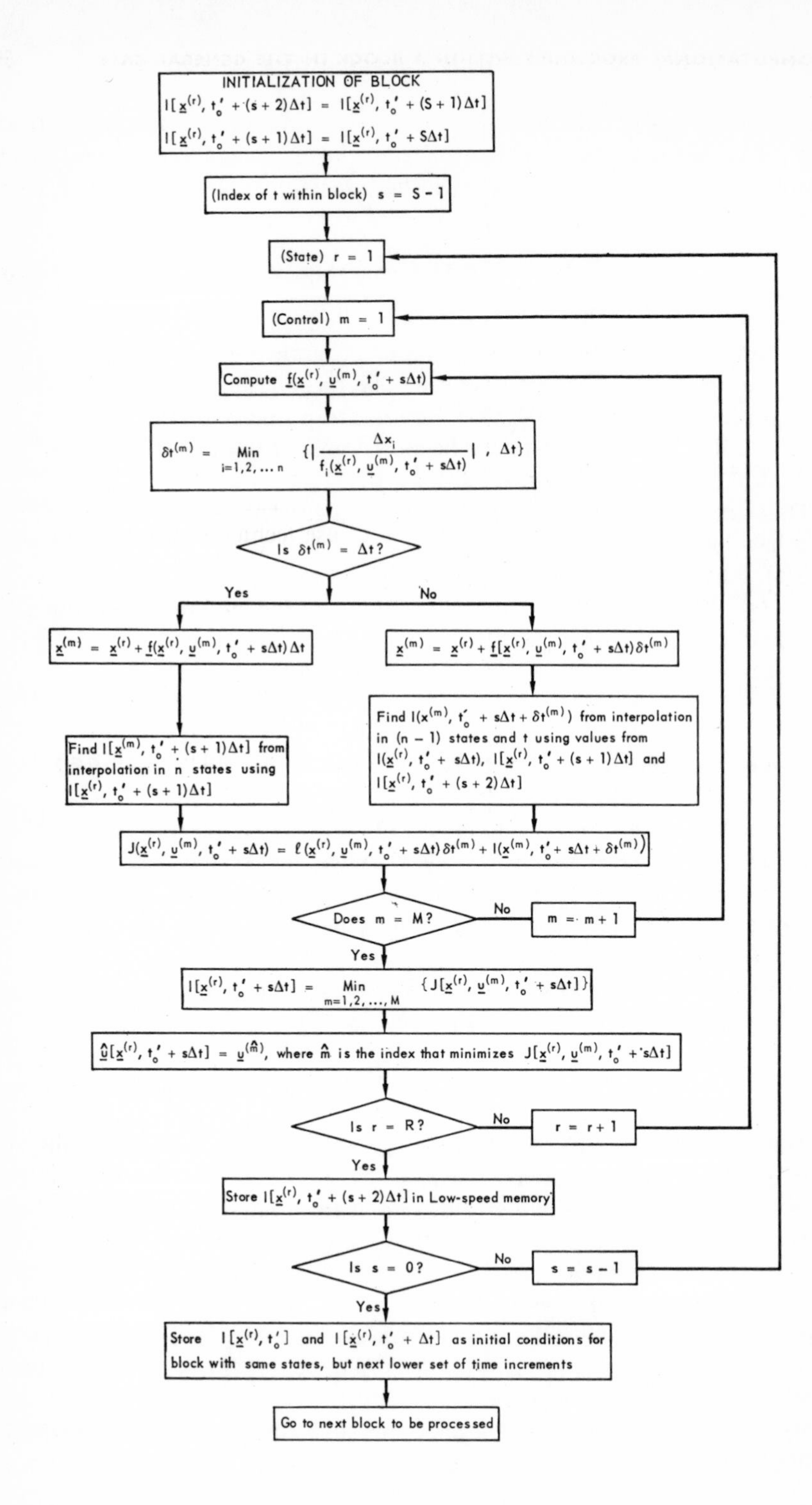

Fig. 4.5 Flow chart for computations within a block in the general case.

state differential equation vector, is examined, and $\delta t^{(m)}$ is determined from Eq. (4.8). Actually, because multiplication takes place in a computer much faster than division, it is more effective to precompute $(1/\Delta x_i)$, $i = 1, 2, \ldots, n$ and evaluate δt from

$$(\delta t^{(m)})^{-1} = \operatorname*{Max}_{i=1,2,\ldots,n} \left\{ \left(\frac{1}{\Delta x_i}\right) |f_i(\mathbf{x}, \mathbf{u}, t)|, \frac{1}{\Delta t} \right\} \tag{4.16}$$

and

$$\delta t^{(m)} = \frac{1}{(\delta t^{(m)})^{-1}} \tag{4.17}$$

In this manner only one division is required rather than n. However, in order to be consistent with the previous material, $\delta t^{(m)}$ will be evaluated in the flow chart using Eq. (4.8). Nevertheless, the reader who is interested in writing an actual program is encouraged to use techniques such as this one for obtaining the most efficient program.

If $\delta t = \Delta t$, the next state is

$$\mathbf{x}^{(m)} = \mathbf{x}^{(r)} + \mathbf{f}[\mathbf{x}^{(r)}, \mathbf{u}^{(m)}, t_0' + s\,\Delta t]\,\Delta t. \tag{4.18}$$

The minimum cost at this state is found by interpolation in the n state variables based on the values $I[\mathbf{x}^{(r)}, t_0' + (s+1)\,\Delta t]$, $r = 1, 2, \ldots, R$.

If $\delta t < \Delta t$, then Eq. (4.9) is used directly to determine the next state as

$$\mathbf{x}^{(m)} = \mathbf{x}^{(r)} + \mathbf{f}[\mathbf{x}^{(r)}, \mathbf{u}^{(m)}, t_0' + s\,\Delta t]\,\delta t^{(m)}. \tag{4.19}$$

The minimum cost at this state is determined by interpolation in $(n-1)$ state variables and t, using those values in the array $I(\mathbf{x}^{(r)}, t_0' + s\,\Delta t)$ that have already been computed and values from the arrays $I[\mathbf{x}^{(r)}, t_0' + (s+1)\,\Delta t]$ and $I[\mathbf{x}^{(r)}, t_0' + (s+2)\,\Delta t]$, $r = 1, 2, \ldots, R$.

The quantity in brackets in Eq. (4.6) is evaluated for a given $\mathbf{u}^{(m)}$ as

$$J(\mathbf{x}^{(r)}, \mathbf{u}^{(m)}, t_0' + s\,\Delta t) = l(\mathbf{x}^{(r)}, \mathbf{u}^{(m)}, t_0' + s\,\Delta t)\,\delta t^{(m)} + I(\mathbf{x}^{(m)}, t_0' + s\,\Delta t + \delta t^{(m)}). \tag{4.20}$$

The minimum cost, $I(\mathbf{x}^{(r)}, t_0' + s\,\Delta t)$, is evaluated by comparing the M values of $J(\mathbf{x}^{(r)}, \mathbf{u}^{(m)}, t_0' + s\,\Delta t)$ and choosing the minimum value. The control which leads to this minimum cost is denoted as the optimal control, $\hat{\mathbf{u}}(\mathbf{x}^{(r)}, t_0' + s\Delta t)$.

After the computations have been performed for every $\mathbf{x}^{(r)}$ at a given value of s, s is decreased by 1 and the procedure is repeated. Decreasing s by 1 has the effect of replacing the array $I[\mathbf{x}^{(r)}, t_0' + (s+2)\,\Delta t]$ by $I[\mathbf{x}^{(r)}, t_0' + (s+1)\,\Delta t]$ and replacing $I[\mathbf{x}^{(r)}, t_0' + (s+1)\,\Delta t]$ by the newly computed array $I[\mathbf{x}^{(r)}, t_0' + s\,\Delta t]$. The old array $I[\mathbf{x}^{(r)}, t_0' + (s+2)\,\Delta t]$, which is no longer needed for interpolation purposes, is transferred to the low-speed memory.

These iterations begin with $s = S - 1$ and continue until computations have been performed at $s = 0$. At this time the computations for the entire block are completed. The arrays $I(\mathbf{x}^{(r)}, t_0')$ and $I(\mathbf{x}^{(r)}, t_0' + \Delta t)$ are stored as the initial values for the block $B(j_0 - 1, j_1, j_2, \ldots, j_n)$, which contains the same states, but the next lower set of time increments. The next block to be processed is then called.

REFERENCES

1. Larson, R. E., *Dynamic Programming with Continuous Independent Variable*, Stanford Electronics Laboratory TR 6302-6, Stanford, California, April 1964.
2. Larson, R. E., "State Increment Dynamic Programming: Theory and Applications," *Proc. 2nd Allerton Conf. on Ckt. and Systems Theory*, U. of Illinois, September 1964, pp. 643–665.
3. Larson, R. E., "Dynamic Programming with Reduced Computational Requirements," *IEEE Trans. on Automatic Control*, Vol. AC-10, No. 2, April,1965, pp. 136–143.
4. Larson, R. E., "An Approach to Reducing the High-Speed Memory Requirement of Dynamic Programming," *J. Math. Anal. and Appl.*, vol. 8, nos. 1–3, July, 1965, pp. 519–537.

Chapter Five

COMPUTATIONS WITHIN A BLOCK FOR THE CASE OF COMPLETE STATE-TRANSITION CONTROL

5.1 INTRODUCTION

In this chapter a second procedure for performing the computations within a block is described. This procedure requires less computational effort than that described in the previous chapter, but it can only be applied to a special class of problems. Since the computational savings are quite significant and since the class of problems is of considerable practical interest, the complete details of the procedure are presented here. However, the reader being exposed to state increment dynamic programming for the first time may wish to skip directly to the next chapter, in which case there is no loss of continuity in the presentation.

5·2 DEFINITION OF COMPLETE STATE-TRANSITION CONTROL

In order to use this procedure, the system equations must have a property called *complete state-transition control.* A consequence of this property is that for all admissible states $\mathbf{x} \in X$ and for all t, $t_0 \leq t \leq t_f$, each control which is considered as a candidate for optimal control changes at most one state variable. Formally, if $U^{(c)}$ is the set of admissible controls from which optimal control is chosen, then $\mathbf{u}^{(c)} \in U^{(c)}$ implies that $\mathbf{f}(\mathbf{x}, \mathbf{u}^{(c)}, t)$ has at most one non-zero component.

The conditions under which a suitable subset $U^{(c)}$ of U can be defined are discussed below. However, it should be clear that a number of significant simplifications are introduced when such a set exists. For example, in the determination of δt from

$$\delta t = \underset{i=1,2,\ldots,n}{\mathrm{Min}} \left\{ \frac{\Delta x_i}{|f_i(\mathbf{x}, \mathbf{u}, t)|} \right\} \tag{5.1}$$

no minimization is necessary, since the quantity $\Delta x_i / |f_i(\mathbf{x}, \mathbf{u}, t)|$ remains finite only for the single component of $\mathbf{f}(\mathbf{x}, \mathbf{u}, t)$ which is non-zero. Also, since only one state variable changes, interpolation of the minimum cost function need be performed only in one dimension. These and other simplifications will be discussed in detail in subsequent sections.

The use of Eq. (5.1) constrains the change in the state variable i for which $f_i(\mathbf{x}, \mathbf{u}, t)$ is non-zero to be exactly Δx_i units in magnitude. Thus the trajectories which are produced are piecewise constant in $(n - 1)$ of the state variables while the other one changes by Δx_i. An example of the kind of approximation which is obtained is shown in Fig. 5.1. In this example there are two state

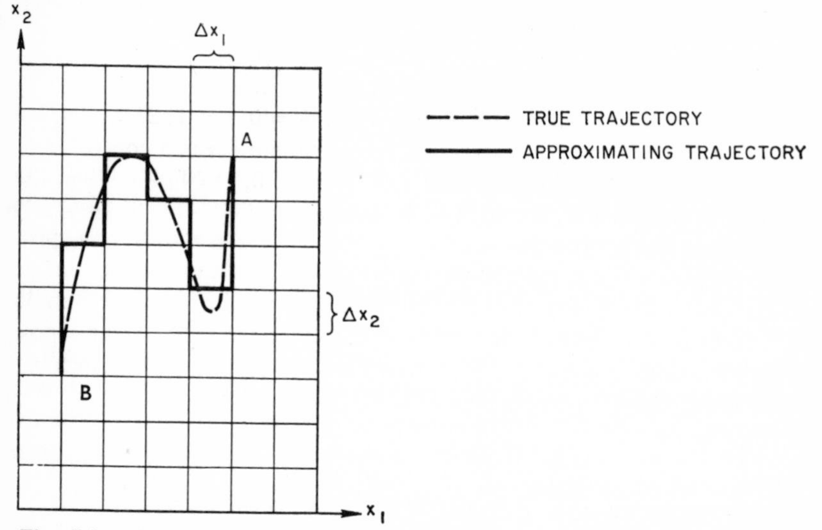

Fig. 5.1. Approximation of a true trajectory using controls from the set $U^{(c)}$.

variables, x_1 and x_2, with corresponding increment sizes Δx_1 and Δx_2. The true trajectory from A to B is shown as a dotted line, while the approximating trajectory, which follows along the grid lines, is shown as a solid line.

The error between the true trajectory and the approximating trajectory can be limited to at most Δx_1 in x_1 and Δx_2 in x_2 respectively. Clearly, by choosing sufficiently small values of Δx_1 and Δx_2, this error can be made arbitrarily small. In the case of n state variables, an approximating trajectory of this form can be constructed which at any value of t differs from the true trajectory by at most Δx_i in the ith state variable. This difference can be made arbitrarily small provided that the Δx_i, $i = 1, 2, \ldots, n$, are suitably small.

In order to obtain this arbitrary degree of accuracy, it is *not* sufficient to use an arbitrary set $U^{(c)}$, where each $\mathbf{u}^{(c)} \in U^{(c)}$ changes at most one state variable. If an admissible control $\mathbf{u} \in U$ changes a given state variable, x_i, in a given direction (positive or negative), then regardless of the remaining components of $\mathbf{f}(\mathbf{x}, \mathbf{u}, t)$, there must exist at that value of $\mathbf{x}$ and t a control $\mathbf{u}^{(c)} \in U^{(c)}$ which changes the state variable, $x_{\bar{i}}$, in that direction, *while holding the remaining states constant.* In general terms, the controls $\mathbf{u}^{(c)} \in U^{(c)}$ must be capable of changing each state variable in any direction that the admissible controls $\mathbf{u} \in U$ can.

These ideas can be made precise by defining a property called *complete state-transition control.*

Definition. A set of system equations, $\dot{\mathbf{x}} = \mathbf{f}(\mathbf{x}, \mathbf{u}, t)$, is said to have the complete state-transition control property if there exists a subset U' of the set of admissible controls, U, such that for every state $\mathbf{x} \in \mathbf{x}$ and every t, $t_0 \leq t \leq t_f$,

(i) if for some $\mathbf{u} \in U$, $f_{\bar{i}}(\mathbf{x}, \mathbf{u}, t) > 0$, then there exists a control $\mathbf{u}' \in U'$ for which $f_{\bar{i}}(\mathbf{x}, \mathbf{u}', t) > 0$ and $f_i(\mathbf{x}, \mathbf{u}', t) = 0, i = 1, 2, \ldots, n, i \neq \bar{i}$.
(ii) if for some $\mathbf{u} \in U$, $f_{\bar{i}}(\mathbf{x}, \mathbf{u}, t) < 0$, then there exists a control $\mathbf{u}' \in U'$ for which $f_{\bar{i}}(\mathbf{x}, \mathbf{u}', t) < 0$ and $f_i(\mathbf{x}, \mathbf{u}', t) = 0, i = 1, 2, \ldots, n, i \neq \bar{i}$.
(iii) if for some $\mathbf{u} \in U$, $\mathbf{f}(\mathbf{x}, \mathbf{u}, t) = 0$, then there exists a control $\mathbf{u}' \in U'$ for which $\mathbf{f}(\mathbf{x}, \mathbf{u}', t) = 0$.

If in a given problem the system equations have the above property, then the procedure for complete state-transition control can be applied.

The most important case where this property occurs is that in which the set of admissible controls, U, permits the vector $\mathbf{f}(\mathbf{x}, \mathbf{u}, t)$ to have any direction whatsoever; formally, for any arbitrary vector $\mathbf{a}$, there exists a control $\mathbf{u} \in U$ for which $\mathbf{f}(\mathbf{x}, \mathbf{u}, t) = \alpha\mathbf{a}$, where α is some scalar. In this case the controls $\mathbf{u}' \in U'$ are found by solving for all i, $i = 1, 2, \ldots, n$.

$$\mathbf{f}(\mathbf{x}, \mathbf{u}', t) = \alpha\boldsymbol{\epsilon}_i \tag{5.2}$$

where $\boldsymbol{\epsilon}_i$ is the unit vector along the ith state variable and where α can take on any positive or negative scalar value.

A necessary condition which is particularly useful in determining whether or not Eq. (5.2) can be satisfied is that q, the number of control variables, is greater than or equal to n, the number of state variables. The satisfaction of Eq. (5.2) is a sufficient condition for the complete state-transition control property. In Chapter 9 an aircraft control problem that meets both conditions is discussed.

The number of elements in the set U' can be finite or infinite. In order to actually apply the computational procedure a finite subset of controls $U^{(c)}$, must be selected. In general, this subset contains exactly one control that changes a particular state variable in a particular direction. If all n state variables can change in either direction, then $U^{(c)}$ consists of $(2n + 1)$ elements.

$$U^{(c)} = \{\mathbf{u}^{(-n)}, \mathbf{u}^{(-n+1)}, \mathbf{u}^{(-n+2)}, \ldots, \mathbf{u}^{(-1)}, \mathbf{u}^{(0)}, \mathbf{u}^{(1)}, \ldots, u^{(n-1)}\, u^{(n)}\} \tag{5.3}$$

where control $\mathbf{u}^{(m)}$ changes state variable m in the positive direction, $m = 1, 2, \ldots, n$; control $\mathbf{u}^{(-m)}$ changes state variable m in the negative direction, $m = 1, 2, \ldots, n$; and control $\mathbf{u}^{(0)}$ changes no state variable. For the remainder of this chapter it will be assumed that $U^{(c)}$ has the above form.

5.3 DEFINITION OF AN OPTIMAL POINT

In the procedure with complete state-transition control, unlike the procedure for the general case, the variable t is not quantized within a block. Optimal control is computed only at a finite number of values of t, but these values can occur at any value of t in the block. Consequently, it is necessary to have some method for keeping track of the values of optimal control and minimum cost as they are generated. This need is met by the concept of an *optimal point.*

Definition. An optimal point consists of the data which result from a single computation of optimal control. These data specify the state $\mathbf{x}$ and the value of t at which the computation took place, and the resulting optimal control and minimum cost.

The data is generally stored as a $(2n + 2)$-dimensional vector, $\mathbf{p}$

$$\mathbf{p} = [\mathbf{x}^T, t, \hat{\mathbf{u}}^T(\mathbf{x}, t), I(\mathbf{x}, t)] \tag{5.4}$$

where $\mathbf{x}$ = the state vector at which computation took place
$\mathbf{x}^T$ = transpose of $\mathbf{x}$
t = the value of t at which the computation took place
$\hat{\mathbf{u}}(\mathbf{x}, t)$ = optimal control at this state $\mathbf{x}$ and this value of t
$I(\mathbf{x}, t)$ = minimum cost at this state $\mathbf{x}$ and this value of t.

Each optimal point is given an index, e.g., $\mathbf{p}(a)$, and the corresponding data takes on this index as a subscript, e.g., I_a. Generally, the index is an integer based on the order in which optimal points are generated.

5.4 COMPUTATIONS IN THE n-DIMENSIONAL CASE

If the set of admissible controls is taken to be $U^{(c)}$ in Eq. (5.3), then the value of $\mathbf{f}(\mathbf{x}, \mathbf{u}, t)$ for each of the controls $\mathbf{u} \in U^{(c)}$ at a given $\mathbf{x}$ and t is as follows:

$$\begin{aligned}
f_i(\mathbf{x}, \mathbf{u}^{(m)}, t) &= 0, && (i \neq m) \\
f_m(\mathbf{x}, \mathbf{u}^{(m)}, t) &> 0 && (m = 1, 2, \ldots, n) \\
f_i(\mathbf{x}, \mathbf{u}^{(-m)}, t) &= 0, && (i \neq m) \\
f_m(\mathbf{x}, \mathbf{u}^{(-m)}, t) &< 0, && (m = 1, 2, \ldots, n) \\
\mathbf{f}(\mathbf{x}, \mathbf{u}^{(0)}, t) &= 0.
\end{aligned} \tag{5.5}$$

When a given block is about to be processed, one optimal point is provided for each quantized state in the block. These optimal points are stored as the initial set Q. In general, the set Q, which consists of one optimal point for each quantized state in the block, contains the optimal point at a particular quantized state which has the *least* value of t.

From the set Q the point $\mathbf{p}^*$ is chosen as that point in Q which has the *largest* value of t. A new optimal point is then computed at the state $\mathbf{x}^*$, but at a value of t less than t^*.

The controls $\mathbf{u}^{(m)}$ and $\mathbf{u}^{(-m)}$, $m = 1, 2, \ldots, n$, are applied at $t = \bar{t}^{(m)}$ and $t = \bar{t}^{(-m)}$ respectively, where, as will be shown presently, $\bar{t}^{(m)}$ and $\bar{t}^{(-m)}$ are determined from values of t at existing optimal points. The interval in t over which each of these controls is applied can be written from Eqs. (5.1) and (5.5) as

$$\delta t^{(m)} = \frac{\Delta x_m}{f_m(\mathbf{x}^*, \mathbf{u}^{(m)}, \bar{t}^{(m)})} \qquad m = 1, 2, \ldots, n$$

$$\delta t^{(-m)} = \frac{\Delta x_m}{-f_m(\mathbf{x}^*, \mathbf{u}^{(-m)}, \bar{t}^{(-m)})} \qquad m = 1, 2, \ldots, n. \tag{5.6}$$

The next state for each of these controls is

$$\begin{aligned} \mathbf{x}^{(m)} &= \mathbf{x}^* + \mathbf{f}(\mathbf{x}^*, \mathbf{u}^{(m)}, \bar{t}^{(m)})\, \delta t^{(m)} \\ &= \mathbf{x}^* + \boldsymbol{\epsilon}_m\, \Delta x_m \qquad (m = 1, 2, \ldots, n) \\ \mathbf{x}^{(-m)} &= \mathbf{x}^* + \mathbf{f}(\mathbf{x}^*, \mathbf{u}^{(-m)}, \bar{t}^{(-m)})\, \delta t^{(-m)} \\ &= \mathbf{x}^* - \boldsymbol{\epsilon}_m\, \Delta x_m \qquad (m = 1, 2, \ldots, n) \end{aligned} \tag{5.7}$$

where $\boldsymbol{\epsilon}_m$ is the unit vector along the mth state coordinate.

Each next state defined in this manner is a quantized state. If the state $\mathbf{x}^*$ is in the interior of the block, then all of the next states are also in this block. In this case the next state is taken to occur at an optimal point in Q. If the state $\mathbf{x}^*$ lies on the boundary of the block, some of the next states lie outside the block, and one of the interpolation procedures described in Chapter 6 must be used.

If $\mathbf{x}^*$ is in the interior of the block, the optimal point in Q at which the next state occurs for control $\mathbf{u}^{(m)}$, $m = 1, 2, \ldots, n$ is denoted as $\mathbf{p}^{(m)}$, and the optimal point for $\mathbf{u}^{(-m)}$, $m = 1, 2, \ldots, n$, is denoted as $\mathbf{p}^{(-m)}$

$$\mathbf{p}^{(m)} = (\mathbf{x}^{(m)T}, t^{(m)}, \hat{\mathbf{u}}^{(m)T}, I^{(m)}, \ldots) \qquad (m = 1, 2, \ldots, n)$$

$$\mathbf{p}^{(-m)} = (\mathbf{x}^{(-m)T}, t^{(-m)}, \hat{\mathbf{u}}^{(-m)T}, I^{(-m)}, \ldots) \qquad (m = 1, 2, \ldots, n). \tag{5.8}$$

The quantities $\bar{t}^{(m)}$ and $\bar{t}^{(-m)}$ are written in terms of $t^{(m)}$ and $t^{(-m)}$ as

$$\begin{aligned} \bar{t}^{(m)} &= t^{(m)} - \delta t^{(m)} \qquad (m = 1, 2, \ldots, n) \\ \bar{t}^{(-m)} &= t^{(-m)} - \delta t^{(-m)} \qquad (m = 1, 2, \ldots, n). \end{aligned} \tag{5.9}$$

In the actual computational procedure, Eq. (5.7) is not rederived every time a new optimal point is calculated. Instead, these results are used directly to find the points in Q which correspond to $\mathbf{p}^{(m)}$ and $\mathbf{p}^{(-m)}$, $m = 1, 2, \ldots, n$. Then, $\delta t^{(m)}$ and $\delta t^{(-m)}$ are computed from Eq. (5.6). If $\mathbf{f}(\mathbf{x}, \mathbf{u}, t)$ does not depend on t, then knowledge of $\bar{t}^{(m)}$ and $\bar{t}^{(-m)}$ is not necessary at this time. However, if it does, then some method of obtaining these values is required. If $\mathbf{f}(\mathbf{x}, \mathbf{u}, t)$ changes slowly as t varies over an interval δt, as is generally the case, then a good approximation is

$$\mathbf{f}(\mathbf{x}, \mathbf{u}^{(m)}, \bar{t}^{(m)}) = \mathbf{f}(\mathbf{x}, \mathbf{u}^{(m)}, t^{(m)}) \qquad (m = 1, 2, \ldots, n)$$

$$\mathbf{f}(\mathbf{x}, \mathbf{u}^{(-m)}, \bar{t}^{(-m)}) = \mathbf{f}(\mathbf{x}, \mathbf{u}^{(-m)}, t^{(-m)}) \qquad (m = 1, 2, \ldots, n) \quad (5.10)$$

If this approximation is not sufficiently accurate, then some iterative procedure must be used.

When $\delta t^{(m)}$ and $\delta t^{(-m)}$ have been found, $\bar{t}^{(m)}$ and $\bar{t}^{(-m)}$ are computed from Eq. (5.9). The time at which control $\mathbf{u}^{(0)}$ is applied, $\bar{t}^{(0)}$, is taken to be the minimum of these times.

$$\bar{t}^{(0)} = \operatorname*{Min}_{m=1,2,\ldots,n} \{\bar{t}^{(m)}, \bar{t}^{(-m)}\}. \quad (5.11)$$

The next state for $\mathbf{u}^{(0)}$ is taken to be at $\mathbf{p}^*$. Therefore, the interval $\delta t^{(0)}$ is

$$\delta t^{(0)} = t^* - \bar{t}^{(0)}. \quad (5.12)$$

Now, for each control $\mathbf{u} \in U^{(c)}$ values of $\bar{t}$ and δt have been selected so that the next state lies exactly at an existing optimal point in Q. Consequently, there is no interpolation necessary to obtain the minimum costs at these points. However, because the values of $\bar{t}$ can differ, it is necessary to do an interpolation in t.

The cost of applying each control at $\mathbf{x} = \mathbf{x}^*$, $t = \bar{t}^{(0)}$ is computed as

$$J(\mathbf{x}^*, \mathbf{u}^{(0)}, \bar{t}^{(0)}) = l(\mathbf{x}^*, \mathbf{u}^{(0)}, \bar{t}^{(0)})\, \delta t^{(0)} + I^*$$

$$J(\mathbf{x}^*, \mathbf{u}^{(m)}, \bar{t}^{(0)}) = l(\mathbf{x}^*, \mathbf{u}^{(m)}, \bar{t}^{(m)})\, \delta t^{(m)} + \Delta J^{(m)} + I^{(m)} \qquad (m = 1, 2, \ldots, n)$$

$$J(\mathbf{x}^*, \mathbf{u}^{(-m)}, \bar{t}^{(0)}) = l(\mathbf{x}^*, \mathbf{u}^{(-m)}, \bar{t}^{(m)})\, \delta t^{(-m)} + \Delta J^{(-m)} + I^{(-m)} \qquad (m = 1, 2, \ldots, n). \quad (5.13)$$

The terms $\Delta J^{(m)}$ and $\Delta J^{(-m)}$ in Eqs. (5.37) and (5.38) account for the necessary interpolation in t. This interpolation is one-dimensional, as contrasted to the n-dimensional interpolation required by the procedure described in Chapter 4. This interpolation can be performed at the present state, $\mathbf{x}^*$; the next state, $\mathbf{x}^{(m)}$ or $\mathbf{x}^{(-m)}$; or both. Alternately, $\Delta J^{(m)}$ and

$\Delta J^{(-m)}$ can be regarded as the cost of holding the state fixed at t^* from $\bar{t}^{(0)}$ to $\bar{t}^{(m)}$ or $\bar{t}^{(-m)}$ respectively. In this case

$$\Delta J^{(m)} = l(\mathbf{x}^*, \mathbf{u}^{(0)}, \bar{t}^{(0)})(\bar{t}^{(m)} - \bar{t}^{(0)}) \qquad (m = 1, 2, \ldots, n)$$

$$\Delta J^{(-m)} = l(\mathbf{x}^*, \mathbf{u}^{(0)}, \bar{t}^{(0)})(\bar{t}^{(-m)} - \bar{t}^{(0)}) \qquad (m = 1, 2, \ldots, n). \quad (5.14)$$

The final step is to find the minimum cost from among the quantities in Eq. (5.13).

$$I(\mathbf{x}^*, \bar{t}^{(0)}) = \operatorname*{Min}_{m=-n,-n+1,\ldots,n} \{J(\mathbf{x}^*, \mathbf{u}^{(m)}, \bar{t}^{(0)})\} \quad (5.15)$$

The quantity $\hat{m}$ is determined as the value of m which minimizes Eq. (5.15). The interpolation term is then subtracted out, and a new optimal point is established at $t = \bar{t}^{(\hat{m})}$

$$\begin{aligned} \mathbf{p}(s) &= (\mathbf{x}^{*T}, \bar{t}^{(\hat{m})}, \mathbf{u}^{(\hat{m})T}, I^{(\hat{m})} - \Delta J^{(\hat{m})}, \mathbf{p}^{(\hat{m})}) && \text{if} \quad \hat{m} \neq 0 \\ &= (\mathbf{x}^{*T}, \bar{t}^{(0)}, \mathbf{u}^{(0)T}, I^*, \mathbf{p}^*) && \text{if} \quad \hat{m} = 0. \quad (5.16) \end{aligned}$$

The index for the next state is added to the data for the optimal point, since the optimal trajectories go from optimal point to optimal point.

When the point $\mathbf{p}(s)$ is established, it replaces $\mathbf{p}^*$ in the set Q, and the above procedure is repeated for the newly modified Q. This process continues until, for the $\mathbf{p}^*$ which has just been chosen, t^* is less than $t_0 + (j_0 - 1)\,\Delta T$. At this time at least one optimal point has been computed for each quantized state in the block $B(j_0 - 1, j_1, j_2, \ldots, j_n)$, the block containing the same states and the next lower interval in t. These points are stored as the initial points for this block, and computations are done for the next block in sequence.

5.5 AN ILLUSTRATIVE EXAMPLE FOR THIS PROCEDURE

The procedure of this chapter is more clearly understood in terms of a one-dimensional example. Because the illustrative example of Chapter 7 uses the procedure of the preceding chapter, the application of the present procedure to a one-dimensional problem will be indicated here.

The example is a simplified version of the problem of determining optimal trajectories for the supersonic transport (SST). A more realistic version of the problem is solved by this procedure in Chapter 9.

For this simple example the single state variable is taken to be altitude, h. The single control variable is γ, the flight path angle with respect to horizontal. The state vector and control vector can thus be identified as

$$\mathbf{x} = h$$

$$\mathbf{u} = \gamma. \quad (5.17)$$

The system differential equation can be deduced from Fig. 5.2. It is assumed that thrust is adjusted so that the velocity is maintained at the fixed value v. If the aircraft is flying at an angle γ with respect to the horizontal, the time rate of change of h is equal to the component of v along the h-axis. Therefore, the system equation is

$$\frac{dh}{dt} = v \sin \gamma \tag{5.18}$$

The performance criterion, which is to be minimized, is taken to be the

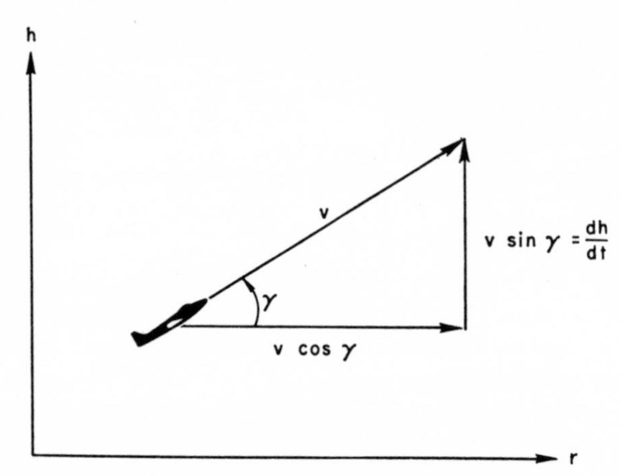

Fig. 5.2. Simplified SST trajectory optimization problem.

total fuel consumed during the flight. The loss function, l, can thus be written as

$$l(h, \gamma, t) = W_f(h, \gamma, t) \tag{5.19}$$

where W_f = fuel flow per unit time.

The fuel flow function, W_f, is assumed to be specified in the above form, either as a given function or as tabular data. A more general form of this function is discussed in Chapter 10.

The cost function, which is to be minimized, is the integral of fuel flow

$$J = \int_{t_0}^{t_f} l[h, \gamma, \sigma]\, d\sigma = \int_{t_0}^{t_f} W_f[h, \gamma, \sigma]\, d\sigma \tag{5.20}$$

where σ is a dummy variable for time, t.

The constraints in the problem are on the maximum and minimum values of altitude

$$0 = \beta^- \le h \le \beta^+ \tag{5.21}$$

and on the maximum and minimum values of flight path angle

$$\alpha^- \le \gamma \le \alpha^+.$$

The set of admissible controls is taken to be

$$U = \{\gamma^{(-1)}, \gamma^{(0)}, \gamma^{(1)}\} \tag{5.22}$$

where

$$\begin{aligned} \gamma^{(-1)} &< 0 \\ \gamma^{(0)} &= 0 \\ \gamma^{(1)} &> 0 \end{aligned} \tag{5.23}$$

Because the number of control variables, $q = 1$, is equal to the number of state variables, $n = 1$, the necessary condition for Eq. (5.2) to be satisfied is met. It can be verified from the system equation, Eq. (5.18), that the set of

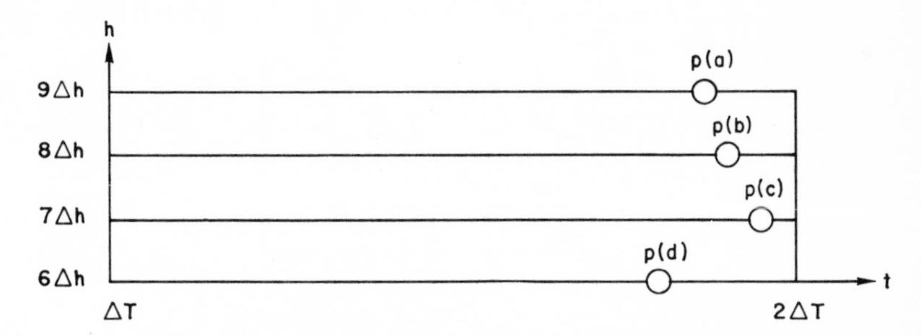

Fig. 5.3. Initial conditions for block $B(2, 3)$**.**

admissible controls in Eq. (5.22) has the necessary properties of the set U' in the definition of complete state-transition control. Therefore, this set will be used as the set $U^{(c)}$ in the subsequent calculations.

The computations for a given block are illustrated in Figs. 5.3 to 5.5. The block size covers $w\,\Delta h = 3\,\Delta h$ along h and ΔT along t. It is assumed that $\beta^- = 0$, $t_0 = 0$, so that the particular block pictured, block $B(2,3)$, covers the intervals $6\,\Delta h \leq h \leq 9\,\Delta h$ and $\Delta T \leq t \leq 2\,\Delta T$.

As an initial condition for the block, one optimal point is provided for each quantized state in the block. These points are denoted as $\mathbf{p}(a)$, $\mathbf{p}(b)$, $\mathbf{p}(c)$, and $\mathbf{p}(d)$ in Fig. 5.3. The initial set Q is taken as these four initial optimal points.

From the set Q the point $\mathbf{p}^*$ is chosen as that point in Q which has the *largest* value of t. In Fig. 5.3, $\mathbf{p}^* = \mathbf{p}(c)$. A new optimal point is now computed at h^*, the state corresponding to $\mathbf{p}^*$, but at a value of t *less than* t^*, the value of t for $\mathbf{p}^*$.

Each control $\gamma \in U$, $\gamma \neq \gamma^{(0)}$, is applied at a value of t such that the next state occurs exactly at one of the optimal points on Q. In Fig. 5.3, control $\gamma^{(1)}$ is applied at $t = \bar{t}^{(1)}$ such that the next state, $(h^* + \Delta h)$, occurs at $t = t_b$, while control $\gamma^{(-1)}$ is applied at $t = \bar{t}^{(-1)}$ such that the next state, $(h^* - \Delta h)$, occurs at $t = t_d$. The values $\bar{t}^{(1)}$ and $\bar{t}^{(-1)}$ are determined in terms of t_b and t_d. From Eq. (5.1) the interval δt for each of these controls is

$$\delta t^{(1)} = \frac{\Delta h}{v \sin \gamma^{(1)}}$$

$$\delta t^{(-1)} = \frac{\Delta h}{-v \sin \gamma^{(-1)}} \,. \tag{5.24}$$

The values $\bar{t}^{(1)}$ and $\bar{t}^{(-1)}$ can then be written as

$$\bar{t}^{(1)} = t_b - \delta t^{(1)}$$

$$\bar{t}^{(-1)} = t_d - \delta t^{(-1)}. \tag{5.25}$$

These values are shown in Fig. 5.4. There is no general rule determining the relation between $\bar{t}^{(1)}$ and $\bar{t}^{(-1)}$.

For $\gamma^{(0)} = 0$, Eq. (5.1) cannot be used to determine δt. In this case, the next state is taken to be at $\mathbf{p}^*$, a member of Q. The value of t at which control $\gamma^{(0)}$ is applied is chosen as $\bar{t}^{(0)}$, where

$$\bar{t}^{(0)} = \text{Min}\,\{\bar{t}^{(1)}, \bar{t}^{(-1)}\}. \tag{5.26}$$

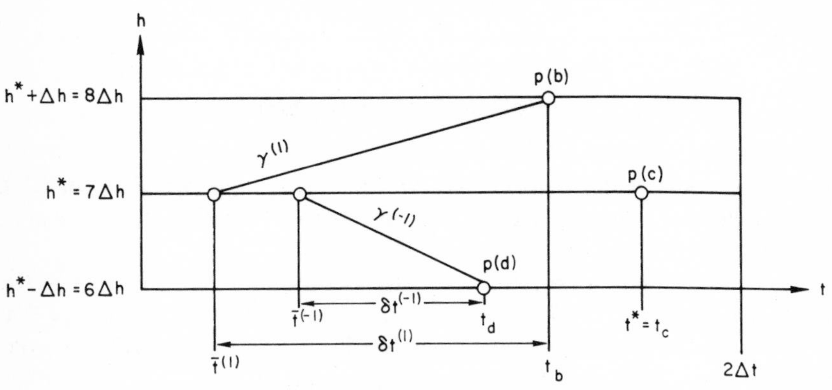

Fig. 5.4. Computations when $p^* = p(c)$.

By choosing $t^{(0)}$ to be one of $t^{(1)}$ and $t^{(-1)}$, one interpolation of the minimum cost function is eliminated. It then follows that

$$\delta t^{(0)} = t^* - \bar{t}^{(0)}. \tag{5.27}$$

The next step is to utilize the functional equation based on the principle of optimality. The minimum cost is computed at $h = h^*$, $t = \bar{t}^{(0)}$ as

$$\begin{aligned} I(h^*, \bar{t}^{(0)}) &= \underset{\gamma^{(m)} \in U}{\text{Min}}\, \{J(h^*, \gamma^{(m)}, \bar{t}^{(0)})\} \\ &= \underset{\gamma^{(m)} \in U}{\text{Min}}\, \{W_f(h^*, \gamma^{(m)}, \bar{t}^{(0)})\, \delta t^{(m)} + \Delta J^{(m)} + I(h^{(m)}, \bar{t}^{(0)} + \delta t^{(m)})\} \end{aligned} \tag{5.28}$$

where $J(h^*, \gamma^{(m)}, \bar{t}^{(0)})$ is the quantity in brackets in the fundamental iterative equation [Eq. (5.15)] when control $\gamma^{(m)}$ is applied, and where $h^{(m)}$ is the next state for this control.

As already noted, the next states $h^{(m)}$ are determined as

$$\begin{aligned} h^{(1)} &= h^* + \Delta h \\ h^{(0)} &= h^* \\ h^{(-1)} &= h^* - \Delta h. \end{aligned} \tag{5.29}$$

For the two controls which are applied at $\bar{t}^{(0)}$, the next state is at an optimal point, for which the minimum cost is known exactly. Consequently, there is no interpolation required, i.e., $\Delta J^{(1)} = \Delta J^{(0)} = 0$. Therefore,

$$J(h^*, \gamma^{(1)}, \bar{t}^{(0)}) = W_f(h^*, \gamma^{(1)}, \bar{t}^{(0)})\, \delta t^{(1)} + I_b \tag{5.30}$$

$$J(h^*, \gamma^{(0)}, \bar{t}^{(0)}) = W_f(h^*, \gamma^{(0)}, \bar{t}^{(0)})\, \delta t^{(0)} + I_c. \tag{5.31}$$

For the control $\gamma^{(-1)}$, the next state is at an optimal point if $\gamma^{(-1)}$ is applied at $\bar{t}^{(-1)}$, but not if it is applied at $\bar{t}^{(0)}$. Consequently, some interpolation in t must be made in order to evaluate the difference between a cost at $\bar{t}^{(-1)}$ and one at $\bar{t}^{(0)}$. Thus,

$$J(h^*, \gamma^{(-1)}, \bar{t}^{0)}) = W_f(h^*, \gamma^{(-1)}, \bar{t}^{(-1)})\, \delta t^{(-1)} + \Delta J^{(-1)} + I_d. \tag{5.32}$$

The quantity $\Delta J^{(-1)}$ can be evaluated by interpolating in t using values of the minimum cost function at h^*, $h^{(-1)}$ or both. Another alternative is to compute $\Delta J^{(-1)}$ as the cost of applying that control which keeps the state constant at h^* over the interval from $\bar{t}^{(0)}$ to $\bar{t}^{(-1)}$. In this case

$$\Delta J^{(-1)} = W_f(h^*, \gamma^{(0)}, t^{(0)})(\bar{t}^{(-1)} - t^{(0)}). \tag{5.33}$$

The quantity $W_f(h^*, \gamma^{(0)}, t^{(0)})$ has already been computed in Eq. (5.31). In the context of the aircraft control problem, this interpolation can be regarded as the cost of an "incremental cruise" at altitude h^* from $t = \bar{t}^{(0)}$ to $t = \bar{t}^{(-1)}$.

The final step in the computations is to compare the three quantities in Eqs. (5.30), (5.31), and (5.32), and to choose the minimum value. If the minimum value is *not* the expression involving the interpolation—in this case, if the minimum value is $J(h^*, \gamma^{(1)}, \bar{t}^{(0)})$ or $J(h^*, \gamma^{(0)}, \bar{t}^{(0)})$—then a new optimal point is established at $h = h^*$, $t = \bar{t}^{(0)}$. However, if the expression containing the interpolation is the minimum value—in this case, $J(h^*, \gamma^{(-1)}, \bar{t}^{(0)})$—then the interpolation is subtracted out, and the optimal point is established at $\bar{t}^{(-1)}$ rather than at $\bar{t}^{(0)}$. Formally, a new optimal point, $\mathbf{p}(e)$, is obtained from the following equations:

$$I(h^*, \bar{t}^{(0)}) = \min_{m=-1,0,1} \{J(h^*, \gamma^{(m)}, \bar{t}^{(0)})\}. \tag{5.34}$$

If $\hat{m}$ is the value of m for which the quantity in parentheses in Eq. (5.34) is minimized, then

$$\begin{aligned}\mathbf{p}(e) &= [h^*, \bar{t}^{(0)}, \gamma^{(1)}, I(h^*, \bar{t}^{(0)}), b], \qquad (\hat{m} = 1)\\ &= [h^*, \bar{t}^{(0)}, \gamma^{(0)}, I(h^*, \bar{t}^{(0)}), c], \qquad (\hat{m} = 0)\\ &= [h^*, \bar{t}^{(-1)}, \gamma^{(-1)}, I(h^*, \bar{t}^{(0)}) - \Delta J^{(-1)}, d], \quad (\hat{m} = -1). \quad (5.35)\end{aligned}$$

In the above expressions for $\mathbf{p}(e)$, an extra term has been added which specifies the index of the next state. This next state is always an existing optimal point. In addition to simplifying the interpolation procedure during

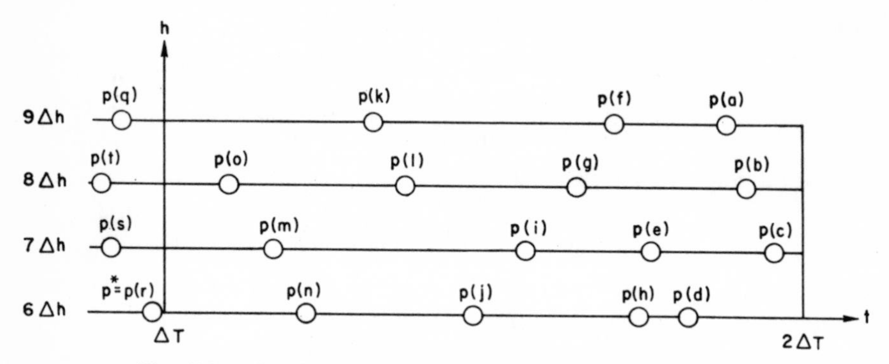

Fig. 5.5. Optimal points computed in block $B(2, 3)$**.**

the computation of optimal control, this property makes it possible to completely specify an optimal trajectory by simply listing in order the indices of the optimal points through which it passes.

The procedure continues by replacing $\mathbf{p}^*$ in Q by the optimal point that has just been computed and repeating the above steps for the new set Q. If the point $\mathbf{p}^*$ is on the boundary of block, i.e., if $h^* = 6\,\Delta h$ or $h^* = 9\,\Delta h$, one of the methods discussed in Chapter 6 is used to provide data for the interpolation procedure. Optimal points are computed in this manner, with the newly computed point replacing the old $\mathbf{p}^*$ in Q each time, until optimal points cover the entire block. The computation of the block is stopped when t^*, the time coordinate of the $\mathbf{p}^*$ which has just been chosen, is less than ΔT, the lower bound of the interval in t which the block $B(2, 3)$ covers. When this occurs, at least one optimal point has been computed for each state in block $B(1, 3)$, the block covering the same states, but the next lower interval in t. The optimal points which have been computed at this time are shown in Fig. 5.5, where computations for this block are stopped when $\mathbf{p}^* = \mathbf{p}(r)$. After this time, the next block in sequence is processed, and the points which have been computed in block $B(1, 3)$ are used as initial conditions when that block is processed.

5.6 COMPUTATIONAL ADVANTAGES OF THE PROCEDURE

In Sec. 5.1 it was mentioned that in the cases where this procedure can be applied, namely problems with complete state-transition control, significant computational reductions result from its use. In this section these computational savings are summarized.

One reduction that occurs is the elimination of the minimization operation in the determination of δt. Since only one state is changing for a given control, δt can be written directly as

$$\delta t = \frac{\Delta x_m}{|f_m(\mathbf{x}, \mathbf{u}, t)|} \tag{5.36}$$

where x_m is the state variable which is changing.

An important reduction occurs in the interpolation of the minimum cost function. This interpolation is in only one dimension, t, rather than in n dimensions as in both conventional dynamic programming and the general case of state increment dynamic programming.

In using the results of the computations, the procedure with complete state-variable control requires no interpolation at all. This is because the optimal trajectories go directly from optimal point to optimal point, and hence a complete optimal trajectory can be specified by merely listing in order the indices of the optimal points through which it passes. In conventional dynamic programming an n-dimensional interpolation is required for this purpose.

Still another reduction occurs because the interval in t between successive computations of optimal control is the variable quantity, δt, rather than a fixed value Δt. Thus, when δt as determined from Eq. (5.36) is large, the interval between computations of optimal control is large. Consequently, when the state differential equation vector, $\mathbf{f}(\mathbf{x}, \mathbf{u}, t)$, is small in magnitude, fewer computations are performed. The number of computations is thus cut down in the regions where the optimal control tends to vary slowly.

5.7 SUMMARY AND FLOW CHART OF PROCEDURE

When a given block is about to be processed, an initial set of optimal points, one for every quantized state in the block, is provided. This set becomes the initial set Q, and it is placed in the high-speed memory. From the set Q the point $\mathbf{p}^*$ is selected as the optimal point in Q having the largest value of t. A new optimal point is computed at $\mathbf{x}^*$, the state corresponding to $\mathbf{p}^*$, but at a value of t less than t^*, the t associated with $\mathbf{p}^*$.

The optimal points $\mathbf{p}^{(m)}$ and $\mathbf{p}^{(-m)}$, $m = 1, 2, \ldots, n$, are selected as the points in Q at states $\mathbf{x}^* + \boldsymbol{\epsilon}_m \Delta x_m$ respectively. These optimal points are used as the next states for controls $\mathbf{u}^{(m)}$ and $\mathbf{u}^{(-m)}$, $m = 1, 2, \ldots, n$. The values of t for these optimal points are used to determine $\delta t^{(m)}$ and $\delta t^{(-m)}$

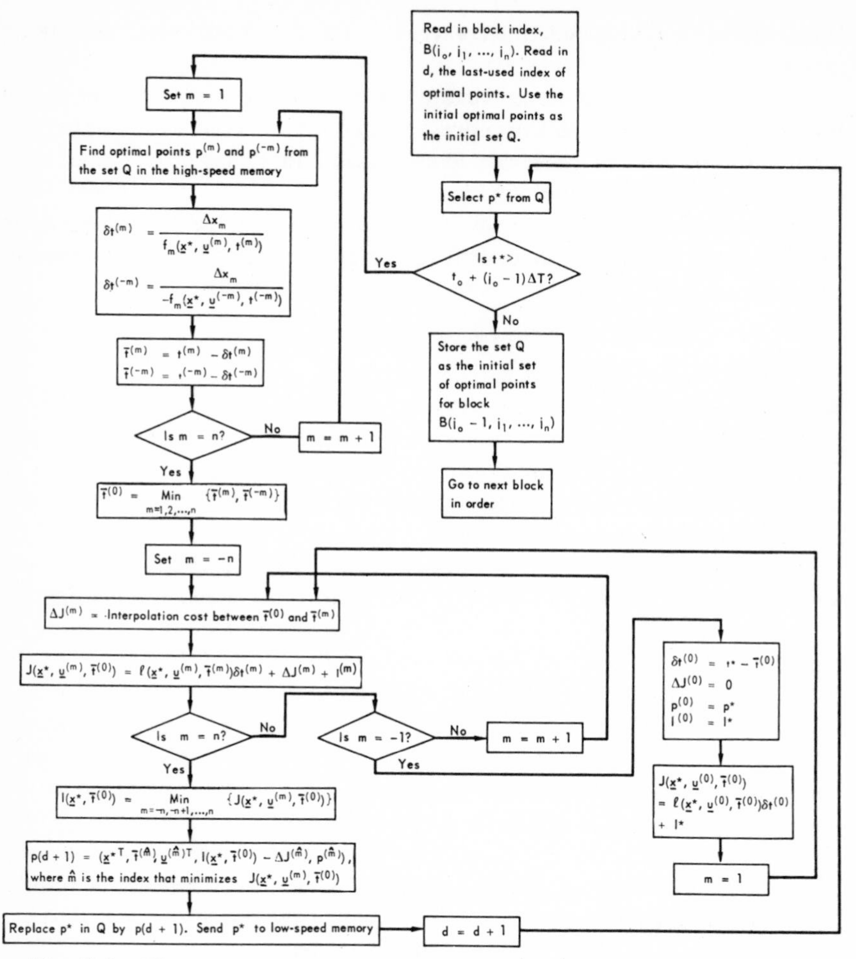

Fig. 5.6. Flow chart for computations within a block for case with complete state-transition control.

from Eqs. (5.6) and (5.10). The quantities $\bar{t}^{(m)}$ and $\bar{t}^{(-m)}$ are then determined from Eq. (5.9).

The computations for control $\mathbf{u}^{(0)}$ are accomplished by setting $\bar{t}^{(0)}$ equal to the minimum of the various $\bar{t}^{(m)}$ and $\bar{t}^{(-m)}$ and using $\mathbf{p}^*$ as the next state for this control.

At this time the controls $\mathbf{u} \in U^{(c)}$ have been applied at state $\mathbf{x}^*$, but at a number of different values of t. The next state for each of these controls is at an optimal point in Q; consequently, the only interpolation required is one-dimensional, in t. The functional equation based on the principle of optimality is then applied in Eq. (5.15), where $J(\mathbf{x}^*, \mathbf{u}^{(m)}, \bar{t}^{(0)}), m = -n, -n+1, \ldots, n,$

is defined in Eq. (5.13). The interpolations, if any, are subtracted out, and a new optimal point is established. The new optimal point, which is assigned the next index in sequence, replaces $\mathbf{p}^*$ in Q, and the old $\mathbf{p}^*$ is sent to the low-speed memory. The entire process is then repeated for the new Q.

This procedure continues until the t^* that is obtained is less than $t_0 + (j_0 - 1)\,\Delta T$, the least value of t in the block. At this time one optimal point has been computed for every quantized state in the block containing the same states and the next lower interval ΔT. These values are stored as the initial conditions for this block, and the next block is processed.

Chapter Six

PROCEDURES FOR PROCESSING BLOCKS

6.1 INTRODUCTION

In the previous two chapters procedures have been given for carrying out the computations within a given block. In this chapter procedures for processing the blocks are presented. The next three sections describe the calculations required to allow trajectories to pass from block to block. In the section after that the computations at $t = t_f$ needed to start the procedure are examined. Then, modifications to the procedure that are appropriate in on-line computations are indicated. Next, the computational requirements of the complete procedure are discussed. Finally, the complete procedure is summarized and a flow chart given.

6.2 TRANSITIONS TO PREVIOUSLY COMPUTED BLOCKS

As long as the computations take place on the interior of the block, the next state will lie in the same block, and there is no need to consider the passage of trajectories from one block to another. When computations are done on or near the boundary of a block, however, these interblock transitions must be explicitly taken into account.

The simplest case occurs at a boundary with a block that has already been computed. As noted in Chapter 3, this boundary is considered to be in both blocks. It is generally not necessary to recompute optimal control and minimum cost along this boundary.* However, the minimum costs at these states are retrieved and stored in the high-speed memory; they are then used in interpolation formulas for the minimum cost at next states corresponding to present states on the interior of the block for which the control takes the next state to this boundary.

The storage of these values allows transitions from the block presently being computed to the previously computed block; if a control which takes the state toward the boundary is selected as the optimal control, then the optimal trajectory will go toward the latter block. The optimal control at this next state

* The circumstances under which this might need to be done are discussed in Sec. 6.4.

can then take the trajectory further into the previously computed block. As long as values of minimum cost on such a boundary are available, transitions of this type to previously computed blocks can be made without restriction. Several such transitions can take place within one ΔT increment.

If the computations within a block are done according to the procedure of Chapter 4, then interblock transitions take place as in Figs. 6.1 and 6.2. The block sizes are taken to be $w_1 \Delta x_1 = 3 \Delta x$ by $\Delta T = S \Delta t = 10 \Delta t$. Also,

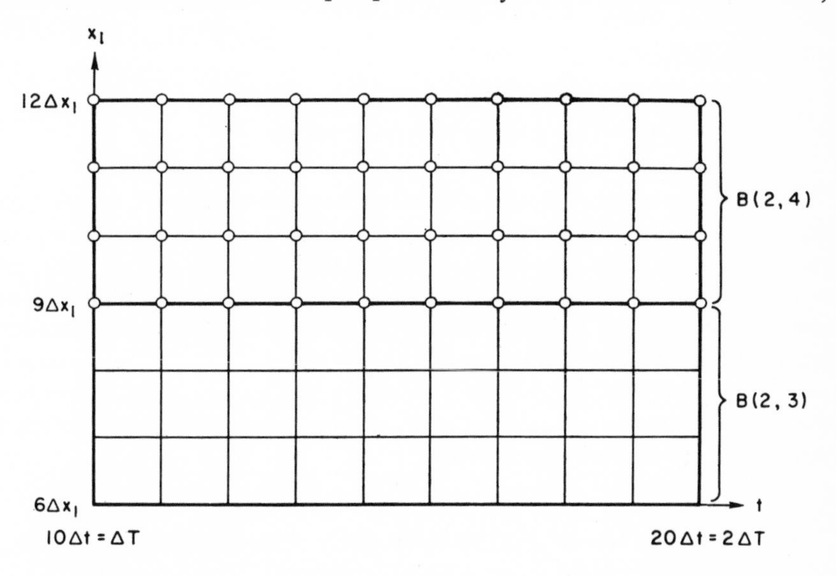

O POINTS AT WHICH OPTIMAL CONTROL HAS ALREADY BEEN COMPUTED

Fig. 6.1. Interblock transitions when block $B(2, 4)$ has been computed.

$\beta_i = 0$ and $t_0 = 0$, so that the region covered by block $B(j_0, j_1)$ contains the intervals $10(j_0 - 1) \Delta t \leq t \leq 10 j_0 \Delta t$ and $3(j_1 - 1) \Delta x_1 \leq x_1 \leq 3 j_1 \Delta x_1$.

The blocks $B(2, 3)$ and $B(2, 4)$ are shown in Fig. 6.1. It is assumed that block $B(2, 4)$ has already been processed. The values of x_1 and t in this block for which optimal control and minimum cost have already been computed are indicated in the figure.

It is clear from the figure that in block $B(2, 3)$ no computations need to be made at $x_1 = 9 \Delta x_1$; computations for this value of x_1 have already been done as part of the calculations for block $B(2, 4)$. However, at $x_1 = 8 \Delta x_1$ it is possible to consider controls for which the next state is $x_1 = 9 \Delta x_1$. Since $x_1 = 9 \Delta x_1$ lies in block $B(2, 4)$, a transition to this block from block $B(2, 3)$ is achieved by this operation. In general, this transition is implemented by using the values of minimum cost computed at $x_1 = 9 \Delta x_1$ for interpolation, i.e., when a computation is made at $x_1 = 8\ \Delta x_1$, $t = 10\ \Delta t + s\ \Delta t$, it is necessary to have available in the high-speed memory values of the minimum cost at $x_1 = 9\ \Delta x_1$, $t = 10\ \Delta t + s\ \Delta t$ and $x_1 = 9\ \Delta x_1$, $t = 10\ \Delta t + (s + 1)\ \Delta t$. Otherwise, the procedure is exactly as in Chapter 4.

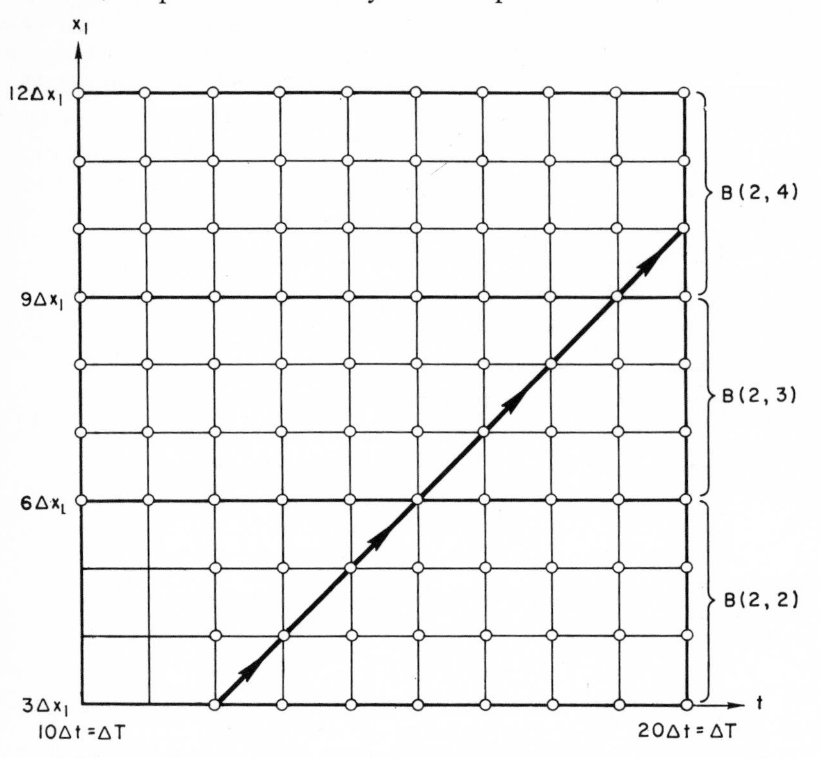

Fig. 6.2. Trajectory for multiple interblock transitions.*

Multiple interblock transitions of this type take place as shown in Fig. 6.2. In this figure an optimal trajectory is shown which begins in block $B(2, 2)$, passes thence to block $B(2, 3)$, and finally goes into block $B(2, 4)$. In this case block $B(2, 4)$ is processed before block $B(2, 3)$ and block $B(2, 3)$ is processed before $B(2, 2)$.

* For clarity the trajectory in this figure changes exactly one increment Δx_1 during each increment Δt. In general, the trajectory does not pass through quantized states and quantized values of t, so that the optimal control at nonquantized points must be obtained by interpolation. Nevertheless, the mechanism through which interblock transitions take place is the same as in this figure.

An analogous procedure can be followed for the case of complete state-transition control. In the one-dimensional case a transition to the boundary is achieved by adding an additional optimal point to the set Q. This point is never considered as a candidate for $\mathbf{p}^*$, but it is used as one of the next states in the computation of optimal control. The optimal point along the boundary which is selected is the one having the largest value of t which is still less than the current value of t^*. In Fig. 6.3 the point along the boundary which is placed in Q for the value of t^* indicated is point $\mathbf{p}(a_5)$. The computational procedure is exactly the same as in Chapter 5, except that an additional transition to the point on the boundary is considered as a possible optimal control.

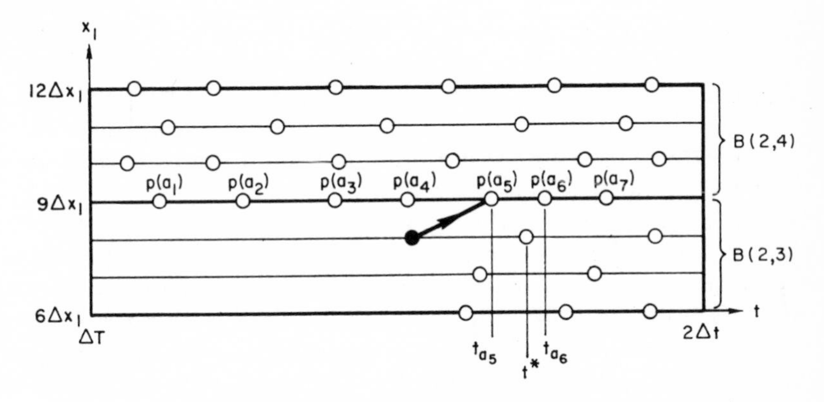

Fig. 6.3. Interblock transition for the case of complete state-transition control.

6.3 PREFERRED DIRECTION OF MOTION

In the previous section a straightforward procedure has been developed for implementing transitions from blocks not yet computed to blocks that have been previously processed. Although a number of methods are available for allowing transitions in the opposite direction, it will be shown in the next section that these methods are not as simple and accurate as the former procedure. It thus follows that certain interblock transitions are more desirable than others and that there might be some way to order the processing of blocks so as to take advantage of this fact. This idea is made explicit in the concept of a *preferred direction of motion*.

In a system to which state increment dynamic programming is applied there is generally some a priori knowledge of how the optimal trajectories behave. This information can be the result of operating experience or preliminary calculations. The amount of information can be quite limited or very large. In any case, if this a priori information indicates that in some region of state space the optimal trajectories tend to be in a certain direction, then this direction of motion is called the preferred direction of motion.

The preferred direction of motion can be well-defined or quite vague, depending on the amount and nature of information available. One example of a case where it is well-defined is the optimal control of a supersonic transport. It will be seen in Chapter 9 that as the aircraft consumes fuel and becomes lighter, the optimal trajectories are at higher altitudes and higher velocities. Thus, except for the terminal descent portion of the flight, the preferred direction of motion is towards increasing altitude and increasing velocity. Another class of problems where this direction is fairly well-defined is the regulator problem of control theory, where the object is to bring the system optimally to some specified state, usually the origin; the preferred direction of motion in this case is towards this state.

Knowledge of the preferred direction of motion can be used to order the blocks in such a way that in most of the cases where the optimal trajectory passes from one block to another, the transition is to a block that has been previously processed. This is implemented by processing the blocks in inverse order to the preferred direction of motion; for example, in the aircraft problem mentioned above the blocks at higher altitudes and velocities would be processed before those at the lower altitudes and velocities. If this is done, then the transitions which are likely to be portions of optimal trajectories are done efficiently and accurately according to the procedure of the previous section; only transitions which are unlikely to be portions of optimal trajectories need to use the more complex procedures of the next section. If the preferred direction of motion is known with a high degree of confidence, then these latter transitions can be computed very roughly, if at all, and a significant amount of computation time can be saved.

6.4 TRANSITIONS TO BLOCKS NOT PREVIOUSLY COMPUTED

Since the preferred direction of motion cannot always be determined exactly before the computations begin, it is necessary to have available techniques for allowing interblock transitions into a block that has not been previously computed. These techniques can be either very complex, if the preferred direction is not well-defined, or else quite simple, if it is known accurately.

The simplest technique is to exclude all controls which result in such a transition during the computation of a block, but to consider such a transition after both blocks have been computed. In the n-dimensional case, during the computations along a boundary, no controls are allowed which result in the next state lying in a block that has not yet been computed; however, when the latter block has been computed, such controls are applied at the boundary at the least value of t within the block. The costs for these controls, which can now be computed using values of the minimum cost inside this block, are

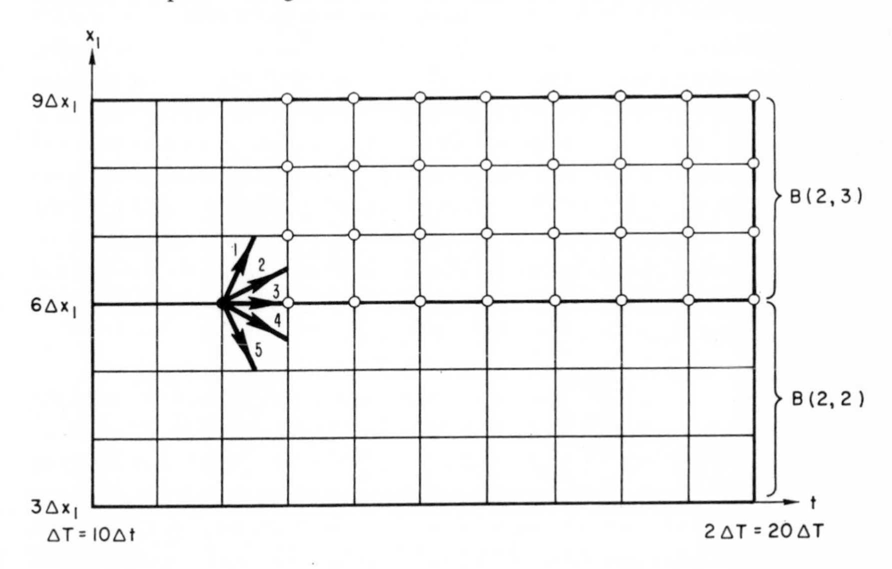

Fig. 6.4. Transition to a block not yet computed.

then compared with the minimum cost already computed at that point on the basis of controls that did not enter this block. If one of these costs is less than the existing minimum cost, then it replaces the existing minimum cost, and the corresponding control becomes the new optimal control. Otherwise, there is no change. The net effect of this procedure is that motion in the non-preferred direction along a given state variable x_i is constrained to proceed at a rate less than $w_i\, \Delta x_i/\Delta T$, where w_i is the number of increments Δx_i in a given block. Since $w_i\, \Delta x_i$ is determined from considerations of the high-speed memory requirements, ΔT can often be selected so that this constraint is not unduly restrictive.

This technique is illustrated for a one-dimensional example in Figs. 6.4 and 6.5. Again, it is assumed that the procedure for the general case is being used, that the block size is 3 Δx_1 by 10 Δt, and that $\beta_1^- = t_0 = 0$. In Fig. 6.4 the upper block, $B(2, 3)$, is currently being processed, while the lower block, $B(2, 2)$, has not yet been processed. At a typical point along the lower boundary, such as $x_1 = 6\ \Delta x_1$, $t = 12\ \Delta t$, the next states might appear as in the figure. Using this technique, only controls 1, 2, and 3 are considered as candidates for optimal control; the controls 4 and 5 are simply ignored. This

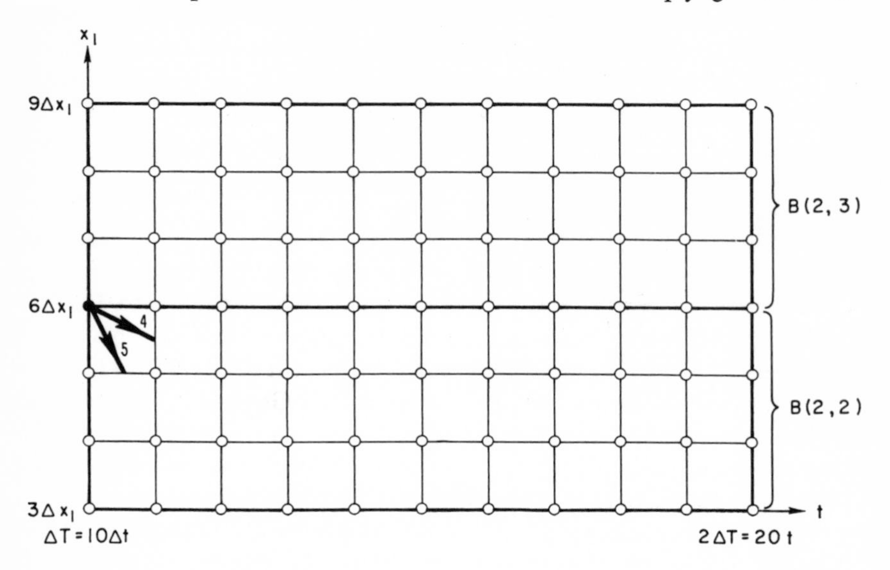

Fig. 6.5. Transition at $t = \Delta T$.

same procedure is followed for $x = 6\ \Delta x_1$ at every quantized value of t in the block.

After the lower block, $B(2, 2)$, has been computed, it is possible to use the values of minimum cost in that block to compute a transition from $x_1 = 6\ \Delta x_1$ into $B(2, 2)$ at $t = 10\ \Delta t$, the least value of t in the block. As shown in Fig. 6.5, this is implemented by applying the controls 4 and 5, for which the next state is $x_1 = 5\ \Delta x_1$. The results of these controls are compared with the value of minimum cost already computed at this point—it is recalled that this existing value of minimum cost was computed on the basis of controls for which the next state was in block $B(2, 3)$, i.e., either $x_1 = 6\ \Delta x_1$ or $x_1 = 7\ \Delta x_1$. A new value of minimum cost is computed at $x_1 = 6\ \Delta x_1$, $t = 10\ \Delta t$ by comparing the costs of the new controls 4 and 5 and the existing value of minimum cost. If the new value is the same as the existing value, then nothing is changed. If the new value corresponds to one of the new controls, then this control becomes the optimal control at that point.

In the case with complete state-transition control an analogous procedure can be applied with the same results.

A more accurate procedure for allowing these transitions is to extrapolate the minimum cost function into the not-yet-computed block. Generally this is accomplished by fitting a low-order polynomial in the state variables to values of the minimum cost at quantized states on or near the boundary of the present block and then evaluating this polynomial at points inside the not-yet-computed block. The extrapolation formulas can be obtained exactly as in Sec. 2.11 and Chapters 4 and 5.

In general, extrapolation procedures tend to be somewhat more inaccurate than interpolation formulas. However, because the extrapolation extends at most Δx_i along the x_i coordinate, the errors can be strictly bounded. Furthermore, by recomputing minimum cost at points along the boundary once every ΔT units in t, as described earlier in this section, it is possible to limit the accumulation of errors due to extrapolation. Generally, by using an appropriate degree of complexity in the extrapolation procedure and by holding ΔT at a reasonable value, sufficient accuracy can be obtained for transitions in the nonpreferred direction.

If the computations within a block are done according to the procedure of Chapter 4, then the extrapolation for the two blocks $B(2, 3)$ and $B(2, 2)$ discussed above takes place as shown in Fig. 6.6. A similar extrapolation can be made in conjunction with the procedure of Chapter 5.

If extrapolation does not provide adequate results, then recomputation of the results along the boundary can be done; in this case the optimal control and minimum cost along a boundary are recomputed when the block into which the extrapolation was made is processed. If necessary, the minimum cost can also be recomputed at points in the previously computed block where the optimal control results in a next state which lies on this boundary. In general, however, the increases in computing time and the high-speed memory requirement are not worth the slight increase in accuracy.

An alternative procedure if extrapolation is not sufficient is to do some precalculations in order to better define the preferred direction of motion. If this is done, then less of the optimal trajectories will go in the nonpreferred direction, and hence the results of the extrapolation procedure will be used in computing minimum cost less often. Consequently, the extrapolations need not always be done with the high accuracy needed to obtain correct values of the minimum cost, but only with sufficient accuracy to check whether or not the transition in the nonpreferred direction corresponds to the optimal control.

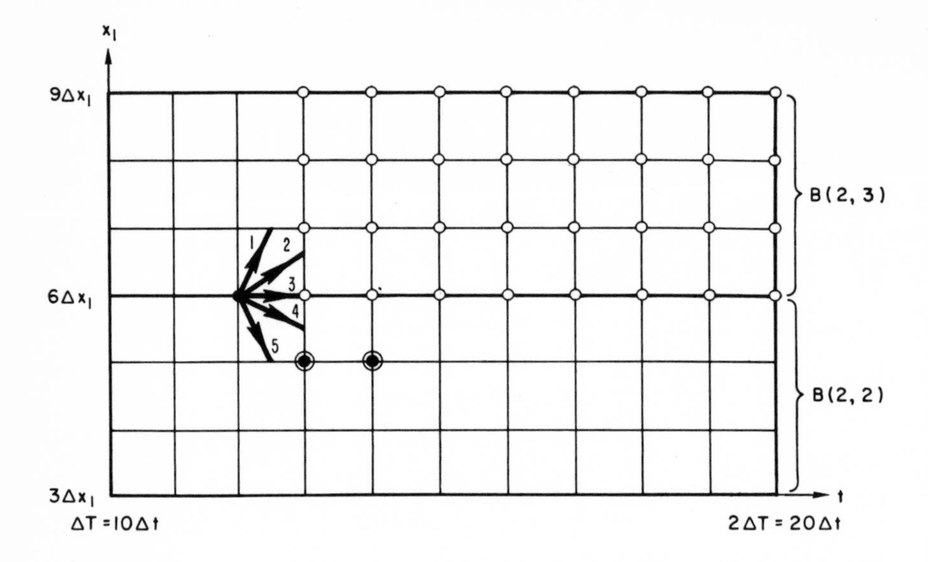

Fig. 6.6. Extrapolation of minimum cost function into not-yet-computed block.

6.5 INITIALIZATION PROCEDURES

Initialization procedures for state increment dynamic programming are capable of handling a wide variety of terminal cost functions, $\psi(\mathbf{x}, t_f)$, and constraints on $X(t_f)$, the set of allowable states at the final time.

If the computational procedure of Chapter 4 is used within the blocks, then values of minimum cost are required for both $t = t_f$ and $t = t_f - \Delta t$. For all admissible states at $t = t_f$, i.e., for all $\mathbf{x} \in X(t_f)$, it is seen immediately that

$$I(\mathbf{x}, t_f) = \psi(\mathbf{x}, t_f). \tag{6.1}$$

At $t = t_f - \Delta t$, however, the computations are slightly more complex. If $\psi(\mathbf{x}, t_f)$ is known as an explicit function, then the conventional dynamic programming computational procedure can be used for the calculations. The equation is

$$I(\mathbf{x}, t_f - \Delta t) = \min_{\mathbf{u} \in U} \{l(\mathbf{x}, \mathbf{u}, t)\, \Delta t + \psi(\mathbf{x} + \mathbf{f}(\mathbf{x}, \mathbf{u}, t)\, \Delta t, t_f)\} \tag{6.2}$$

where $\psi(\mathbf{x} + \mathbf{f}(\mathbf{x}, \mathbf{u}, t)\, \Delta t, t_f)$ is evaluated directly. Often, $\psi(\mathbf{x}, t_f) = 0$ for all $\mathbf{x} \in X(t_f)$, in which case this determination is particularly straightforward.

If ψ is given as a table at quantized values of $\mathbf{x} \in X(t_f)$, then the conventional procedure may be inapplicable because of too large a high-speed memory requirement. In this case, it is generally possible to use the second interpolation procedure described in Sec. 4.5, for which it is sufficient to store minimum costs at only *one* time increment Δt into the future. However, this procedure requires that minimum cost be computed for at least one quantized state at the present value of t. For this one state it is always possible to search for

$$\psi[\mathbf{x} + \mathbf{f}(\mathbf{x}, \mathbf{u}, t)\,\Delta t, t_f]$$

in low-speed storage. This is clearly undesirable, but since it needs to be done only once, the additional computational effort is tolerable.

The procedure is particularly straightforward for implementing terminal constraints on $\mathbf{x}(t_f)$. The function $\psi(\mathbf{x}, t_f)$ is evaluated only for $\mathbf{x} \in X(t_f)$, the set of allowable terminal states, and it is noted that all other states are inadmissible. If for a given state and control at $t = t_f - \Delta t$ the next state is found to be inadmissible, then the control, even though it is in the set U, is taken to be inadmissible. If at a state it is found that for all $\mathbf{u} \in U$ the next state is inadmissible, then this state—even though it is in the set $X(t_f - \Delta t)$—is taken to be inadmissible. Such a state is called nonterminable [Refs. 1, 2].

If the set $X(t_f)$ contains only one element, it may occur that there is *no* state $\mathbf{x} \in X(t_f - \Delta t)$ for which an admissible control $\mathbf{u} \in U(\mathbf{x}, t_f - \Delta t)$, results in *exactly* this state as the next state. It is then necessary to extend $X(t_f)$ to cover some small but finite region around the single state $\mathbf{x} \in X(t_f)$. Generally, this region covers less than half an increment along any state variable axis.

In the case of complete state-transition control, for which the computations within a block follow the procedure of Chapter 5, it is only necessary to obtain one optimal point for each quantized state at $t = t_f$. The terminal cost in Eq. (6.1) suffices for this purpose. In this case it is often possible to precompute the set of nonterminable states. Computational savings can then be made by establishing an optimal point for each quantized $\mathbf{x} \in X$ at the *largest* value of t for which an admissible control can be found that still results in an admissible next state. No computations at quantized states need be done for larger values of t, since the state is then nonterminable. This procedure is utilized very effectively in the supersonic transport (SST) trajectory optimization problem of Chapter 9.

6.6 MODIFICATIONS FOR ON-LINE CONTROL

The procedure described in the previous section computes optimal control at every admissible state for all values of t. These results are exactly what is obtained in the conventional dynamic programming procedure. As will be shown in the next section, the high-speed memory requirement for the state increment dynamic programming procedure is much less than that of conventional dynamic programming. However, because both procedures compute exactly as many values of optimal control and minimum cost, there is generally little difference in either computing time or the low-speed memory requirement.

In many optimization problems it is not necessary to compute optimal control at all possible values of $\mathbf{x}$ and t. For example, in on-line control applications the only calculation required is an optimal trajectory starting from the present state.

The block structure of state increment dynamic programming is ideally suited for restricting the computations so as to obtain only the results which are needed for the problem under consideration. The basic idea is to compute optimal control only in a region in which the optimal trajectories of interest are expected to lie. If such a region can be defined, then computations are performed only in those blocks contained in it. Clearly, if much information is available about these trajectories, then the region can be small, and hence a considerable reduction in both computing time and low-speed memory requirement can be obtained.

One method of determining the region in which computations take place is to use the results of either operating experience or previous computations on the system. For example, consider the application of real-time control to the illustrative example of Sec. 5.5, namely the optimal control of the altitude of an aircraft. Operating experience and previous computations indicate that the minimum-fuel trajectory is close to the trajectory defined by first climbing at a maximum rate to the cruise altitude, maintaining the cruise altitude until near the destination, and then diving at maximum rate. This trajectory is illustrated in Fig. 6.7, and the blocks which would then be processed are indicated by shading.

If there is not sufficient information to determine a reasonably small region, then an iterative computational procedure can be applied. One promising procedure begins by quantizing the state variables in very large increments. A computation is then made over the entire $X - t$ space using this quantization. On the basis of the results of this computation, an approximation to the optimal trajectory is obtained, and a region about it is determined. Computations are then repeated in this region only, but the quantization increments are reduced. Generally, the increments are reduced in proportion to the reduction of the region, so that the time required for a single iteration remains constant. This process is repeated until a sufficiently accurate trajectory is obtained. A study of this procedure, including a number of examples, appears in Ref. 3.

It is important to note that optimal control is computed everywhere in the region, not just along a single trajectory. Consequently, if in actual operation the system deviates slightly from the initially computed optimal trajectory, but these deviations do not take the trajectory outside the region, then no new calculations have to be made. In this manner the number of times that a recomputation has to be made in an on-line control system is considerably reduced.

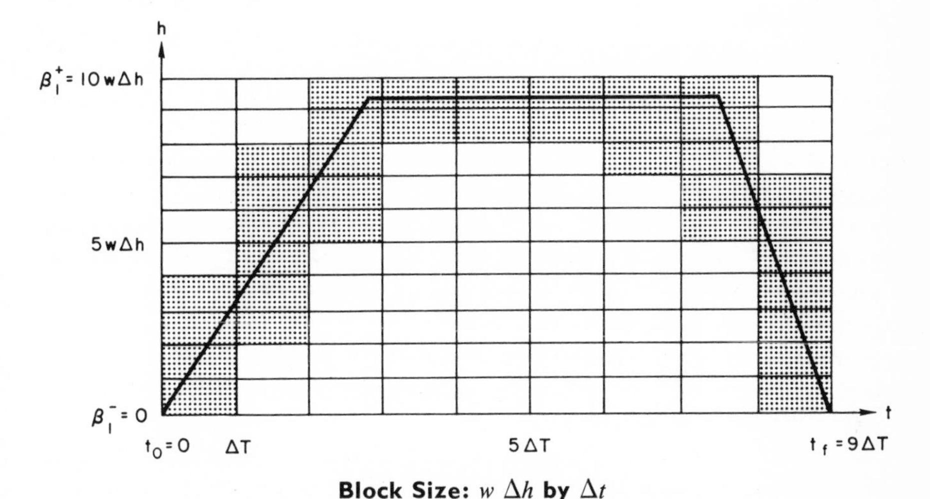

Block Size: $w\ \Delta h$ **by** Δt

Fig. 6.7. Determination of region from knowledge of an approximation to the optimal trajectory.

Another procedure for this case is Bellman's successive approximations [Refs. 4, 5, 6]. This extremely important technique is discussed in Chapter 12. Still another alternative is quasilinearization [Ref. 7]; this method is briefly discussed in Chapter 8.

6.7 COMPUTATIONAL REQUIREMENTS

It is now possible to state explicitly the computational requirements of state increment dynamic programming and to compare them with those of conventional dynamic programming.

As discussed in Chapter 2, the high-speed memory requirement is the amount of data which must be stored in the high-speed memory when a computation of optimal control is taking place. In the conventional dynamic programming procedure this requirement is one storage location for every quantized state. This is because the next state can occur anywhere in the space of admissible states; hence, if the present state is at t, values of minimum cost for every quantized state at $(t + \Delta t)$ are required in order to be able to determine the minimum cost at the next state. If there are n state variables and N_i quantization levels in the ith state variable, then the high-speed memory requirement is given by

$$N_h = \prod_{i=1}^{n} N_i. \tag{6.3}$$

For a system with three state variables and 100 quantization levels in each state variable,

$$N_h = (100)^3 = 10^6. \tag{6.4}$$

Because in the state increment dynamic programming procedure the next state is always in the same block as the present state (or at most one increment away from the block along any state variable axis), the high-speed memory requirement is considerably less than that of the conventional procedure. If in the general case of state increment dynamic programming linear interpolation in t is performed, then the basic high-speed memory requirement is two storage locations for each state in the block. These locations are necessary for storing values of the minimum cost at $(t + \Delta t)$ and $(t + 2\,\Delta t)$ when computations are taking place at t. For a block covering $w_i\,\Delta x_i$ along the x_i-axis, so that there are $(w_i + 1)$ quantization levels in each block, this requirement is

$$N_h' = 2 \prod_{i=1}^{n} (w_i + 1). \tag{6.5}$$

If the values of minimum cost along the boundaries with previously computed blocks are also stored in the high-speed memory, then the number of storage locations required may be increased. Sometimes it is possible to keep these values in a buffer storage and bring them into the high-speed memory only as required; in this case the high speed memory requirement is still N_h'. However, in general it is necessary to store all the values along these boundaries in the high-speed memory. If half the boundaries of the block have been previously computed and if the block covers $S\,\Delta t$ units in t, then the high-speed memory requirement is two locations for the states not on these boundaries and $(S + 1)$ locations for the states on them.

$$N_h'' = 2 \prod_{i=1}^{n} w_i + (S + 1)\left[\prod_{i=1}^{n} (w_i + 1) - \prod_{i=1}^{n} w_i\right] \tag{6.6}$$

for a case with $n = 3$ and with a block size containing $(w_i + 1) = 3$ quantized values of state variable i, $i = 1, 2, 3$ and $(S + 1) = 5$ quantized values of t, the basic requirement is

$$N_h' = 2(3^3) = 54. \tag{6.7}$$

If the values of minimum cost along the boundaries with previously computed blocks are stored, then according to Eq. (6.6),

$$N_h'' = 2(2^3) + (5)[(27) - (8)] = 111. \tag{6.8}$$

This value of N''_h is of the order of 10^2, a far smaller number than 10^6, the corresponding requirement for conventional dynamic programming. If the block size is increased to $(w_i + 1) = 5$, $(S + 1) = 11$, a rather large set of values, then

$$N''_h = 2(4^3) + (11)[(125) - (64)] = 799 \tag{6.9}$$

which is still far less than 10^6.

It should be pointed out that once the N''_h initial values of minimum cost are placed in the high-speed memory, the computation of optimal control throughout the block is performed without the need for further data from the low-speed memory. This is because the values of minimum cost that are currently being computed are used in the computations at values of t smaller than the present value. For example, in the procedure for the general case with linear interpolation in t, the computations at t make use of the values of minimum cost at $(t + \Delta t)$ and $(t + 2\,\Delta t)$; when the computations at $(t - \Delta t)$ are made, the values of $(t + 2\,\Delta t)$ are no longer needed, and the just-computed values at t replace them in the high-speed memory. Thus, using the N''_h initial values of minimum cost for a block, new values of optimal control and minimum cost are computed at S values of t for all quantized states in the block not on the boundaries with previously computed blocks, a total of $N_{c/b}$ computations per block, where

$$N_{c/b} = S \prod_{i=1}^{n} w_i. \tag{6.10}$$

Since the high-speed memory requirement is so small, it is often possible to store the data for several blocks in the high-speed memory. In this manner the boundary between two adjacent blocks can sometimes be retained in the high-speed memory after the first block is computed, so that when the second is to be computed it is not necessary to retrieve the data for this boundary from the low-speed memory.

In summary, all the computations for a single block, a total of $N_{c/b}$, can be performed on the basis of N''_h values of the minimum cost function, which can be retrieved from the low-speed memory in a single operation. If several blocks can be stored in the high-speed memory, careful programming can permit several blocks to be computed on the basis of a single transfer from low-speed to high-speed memory. If N''_h high-speed memory locations are not available, but N'_h are, then the state increment dynamic programming procedure can still be used, but values of the minimum cost along previously computed blocks must be retrieved from the low-speed memory after all the computations in a block have been performed for a given value of t. If the computational procedure within the blocks is other than the procedure for the general case with linear interpolation in t, then the formulas for N', N'', $N_{c/b}$, etc., are different, but the same basic conclusions apply.

If computation of optimal control and minimum cost is made for all admissible states at all possible values of t, then, provided that increment sizes Δt and Δx_i, the sets of admissible controls, and interpolation formulas are comparable, the computing time and low-speed memory requirement of state increment dynamic programming are about the same as those of conventional dynamic programming. In general, the total number of computations is about the same, but because state increment dynamic programming performs additional operations, such as processing of the blocks and computation of δt, it generally requires more computing time than the conventional procedure. However, clever programming can make the additional time required negligible. In any case, this increase is generally well worth the great reduction in high-speed memory requirement.

If the modified procedure for on-line control is used, both computing time and low-speed memory requirement are reduced by a large factor. This is because only a small percentage of the total number of blocks are actually processed. For the procedure in the general case with linear interpolation in t, the number of computations performed in a block is given by $N_{c/b}$ in Eq. (6.10). If the number of blocks processed is N_b, then the total number of computations performed, and hence the low-speed memory requirement, is given by

$$N_c = N_{c/b} N_b = N_b S \prod_{i=1}^{n} w_i. \tag{6.11}$$

If Δt_c seconds are required for each computation, then the total computing time is

$$T_c = N_c \, \Delta t_c = N_b S \, \Delta t_c \prod_{i=1}^{n} w_i. \tag{6.12}$$

For the example cited in Chapter 2 with $n = 3$, $N_i = 100$, $i = 1, 2, 3$, and with $K = 50$ discrete values of t, the low-speed memory requirement for conventional dynamic programming is

$$N_c = 5 \cdot 10^7. \tag{6.13}$$

If the time required for each computation is $\Delta t_c = 50 \cdot 10^{-6}$ sec, then the total computing time is

$$T_c = 2{,}500 \text{ seconds}. \tag{6.14}$$

On the other hand, in state increment dynamic programming, if the block size is $(w_i + 1) = 5$, $S + 1 = 11$, then the number of computations performed per block is

$$N_{c/b} = 640. \tag{6.15}$$

Each block covers $\Delta T = S \, \Delta t = 10 \, \Delta t$ in t by $w_i \, \Delta x_i = 4 \, \Delta x_i$ along each coordinate x_i. Thus, there are 5 ΔT intervals for the 50 Δt increments and $(25)^3 = 15{,}625$ blocks in each interval ΔT. Suppose that a priori information is used to reduce the region in which optimal control is computed to $12\,\Delta x_i$ along each coordinate; i.e., 12% of the total state space is covered along each coordinate by this region. The number of blocks covered by this region in each interval ΔT is clearly

$$\prod_{i=1}^{3} \frac{12\,\Delta x_i}{4\,\Delta x_i} = 3^3 = 27 \text{ blocks},$$

so that over the total range of t the number of blocks is $N_b = 27 \cdot 5 = 135$. The total number of computations, and hence the low-speed memory requirement, is exactly

$$N_c = (135)(640) = 86{,}400. \tag{6.16}$$

This number is less than 0.2% of $5 \cdot 10^7$, the low-speed memory requirement for conventional dynamic programming.

This result could have been arrived at directly by observing that computations are performed only over 12% of the range of each x_i, so that the new low-speed memory requirement must be $(.12)^3 = .001728$ of $5 \cdot 10^7$. The result is thus independent of the block size, and depends only on the amount of state space covered by the region.

If the time required for a single computation is $\Delta t_c = 35 \cdot 10^{-6}$ sec, a reasonable estimate if $25 \cdot 10^{-6}$ sec is the time required for one computation in conventional dynamic programming, then the total computing time becomes

$$T_c = (86{,}400)(35 \cdot 10^{-6}) = 3.024 \text{ seconds},$$

which is a reduction by a factor of over 800.

6.8 SUMMARY AND FLOW CHART OF PROCEDURE

The complete state increment dynamic programming computational procedure has now been described in detail. In this section the procedures for processing blocks are briefly reviewed, and a flow chart of the complete procedure is given.

First, an initialization procedure from Sec. 6.5 is applied. This provides sufficient data for the calculations in blocks covering the interval $t_f - \Delta T \leq t \leq t_f$ to be performed.

Next, all the calculations within blocks in this interval of t are carried out according to one of the procedures of Chapters 4 and 5. The blocks are ordered so that, whenever possible, interblock transitions that are portions of optimal trajectories are to blocks that have already been computed; the ordering makes use of the preferred direction of motion, as discussed in Sec. 6.3. Interblock transitions to previously computed blocks are done as in Sec. 6.2, while transitions to not-yet-computed blocks use one of the methods of Sec. 6.4.

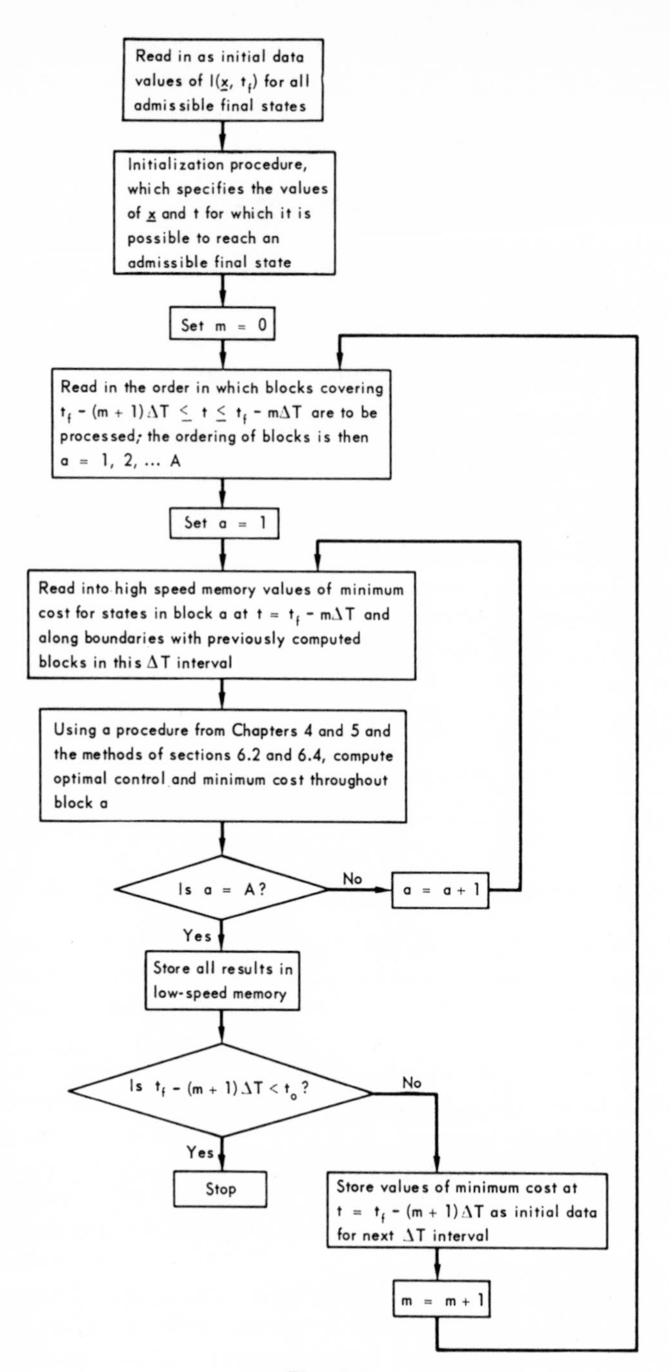
Read in as initial data values of $I(\underline{x}, t_f)$ for all admissible final states
Initialization procedure, which specifies the values of $\underline{x}$ and t for which it is possible to reach an admissible final state
Set m = 0
Read in the order in which blocks covering $t_f - (m+1)\Delta T \leq t \leq t_f - m\Delta T$ are to be processed; the ordering of blocks is then a = 1, 2, ... A
Set a = 1
Read into high speed memory values of minimum cost for states in block a at $t = t_f - m\Delta T$ and along boundaries with previously computed blocks in this ΔT interval
Using a procedure from Chapters 4 and 5 and the methods of sections 6.2 and 6.4, compute optimal control and minimum cost throughout block a
Is a = A?
No
a = a + 1
Yes
Store all results in low-speed memory
Is $t_f - (m+1)\Delta T < t_o$?
No
Yes
Stop
Store values of minimum cost at $t = t_f - (m+1)\Delta T$ as initial data for next ΔT interval
m = m + 1

Fig. 6.8

When all of the computations have been performed for blocks in $t_f - \Delta T \leq t \leq t_f$, the set of blocks in $t_f - 2\,\Delta T \leq t \leq t_f - \Delta T$ is processed. The same procedures are used, except that the values of minimum cost just computed at $t_f - \Delta T$ (and possibly at other nearby values of t) are used to initialize these blocks.

These computations continue until $t = t_0$ is reached. The overall procedure is illustrated in the flow chart in Fig. 6.8.

REFERENCES

1. Kahne, S. J., "Feasible Control Computations Using Dynamic Programming," *Proc. Third IFAC Congress*, London, June 1966, Paper No. 18G, Session 18.
2. Larson, R. E., "Comments on 'Feasible Control Computations Using Dynamic Programming'," *Proc. Third IFAC Congress*, London, June 1966.
3. Keckler, W. G., "Optimization About a Nominal Trajectory Via Dynamic Programming," Engineer's Thesis, Department of Electrical Engineering, Stanford, California, September 1967.
4. Bellman, R., *Dynamic Programming*, Princeton University Press, Princeton, N.J., 1957.
5. Bellman, R., *Adapative Control Processes*, Princeton University Press, Princeton, N.J., 1961.
6. Bellman, R, and Dreyfus, S., *Applied Dynamic Programming*, Princeton University Press, Princeton, N.J., 1962.
7. Bellman, R. E. and R. E. Kalaba, *Quasilinearization and Nonlinear Boundary-Value Problems*, American Elsevier Publishing Co., Inc., New York, N.Y., 1965.

Chapter Seven

AN ILLUSTRATIVE EXAMPLE

7.1 INTRODUCTION

In the preceding four chapters the complete state increment dynamic programming computational procedure has been presented. In this chapter the method is illustrated by working out an example in complete detail. The computations within a block use the procedure of Chapter 4. The problem is the same as that worked out in Sec. 2.9 by the conventional computational procedure; this provides an opportunity for comparing these methods with each other and with the exact solution.

7.2 FORMULATION OF THE PROBLEM

The problem of this chapter has a single state variable, x, and a single control variable, u. The system equation is

$$\dot{x} = f(x, u, t) = u. \tag{7.1}$$

The performance criterion, which is to be minimized, is

$$\begin{aligned} J &= \int_{t_0}^{t_f} l(x, u, \sigma)\, d\sigma + \psi[x(t_f), t_f] \\ &= \int_0^{10} [x^2(\sigma) + u^2(\sigma)]\, d\sigma + 2.5[x(10) - 2]^2. \end{aligned} \tag{7.2}$$

The control is bounded

$$-2 \leq u \leq 2. \tag{7.3}$$

The state variable obeys the constraint

$$0 \leq x \leq 8. \tag{7.4}$$

The final state of the system is constrained as

$$0 \leq x(10) \leq 2. \tag{7.5}$$

This problem is exactly the same as that described in Sec. 2.9. The objective is to bring the state of the system, described by Eq. (7.1), into the interval $0 \leq x \leq 2$ at the final time, $t = 10$, while minimizing the sum of an integral cost function [the first term in Eq. (7.2)], plus a penalty for not reaching $x = 2$

as the terminal state [the second term in Eq. (7.2)]. No terminal penalty is incurred if the final state is exactly at $x = 2$, while a cost proportional to the square of the distance from $x = 2$ is charged if the final state is not 2. The state variable and control variable are bounded by Eqs. (7.3) and (7.4).

7.3 CONSTRAINTS, QUANTIZATION, AND INTERPOLATION

The problem is first converted to a discrete form. The differential equation, Eq. (7.1), is replaced by a difference equation

$$x(t + \delta t) = x(t) + u(t)\,\delta t. \tag{7.6}$$

The iterative equation based on the principle of optimality is written as

$$I(x, t) = \underset{u \in U}{\text{Min}}\,\{(x^2 + u^2)\,\delta t + I(x + u\,\delta t, t + \delta t)\}. \tag{7.7}$$

The quantization is taken to be the same as in Sec. 2.9. The state variable is quantized in uniform increments of $\Delta x = 1$.

$$x = j \qquad (j = 0, 1, 2, \ldots, 8). \tag{7.8}$$

The control variable is also quantized in increments of 1. The set of admissible controls, U, is

$$U = \{-2, -1, 0, 1, 2\}. \tag{7.9}$$

The procedure of Chapter 4 is used for the computations within a block. The variable t thus must also be quantized. This quantization is taken to be uniform with an increment of $\Delta t = 1$. The quantized values of t are

$$t = k \qquad (k = 0, 1, 2, \ldots, 10). \tag{7.10}$$

Since $f(x, u, t)$ does not depend on either x or t, the increment δt can be pre-computed. The equation for δt becomes

$$\begin{aligned} \delta t &= \min\left\{\frac{\Delta x}{|f(x, u, t)|}, \Delta t\right\} \\ &= \min \frac{1}{|u|}, 1. \end{aligned} \tag{7.11}$$

For $u = \pm 1$,

$$\begin{aligned} \delta t^{(1)} &= 1 \\ \delta t^{(-1)} &= 1. \end{aligned} \tag{7.12}$$

For $u = 0$

$$\delta t^{(0)} = \Delta t = 1. \tag{7.13}$$

For $u = \pm 2$,

$$\begin{aligned} \delta t^{(2)} &= \tfrac{1}{2} \\ \delta t^{(-2)} &= \tfrac{1}{2}. \end{aligned} \tag{7.14}$$

From Eq. (7.2) it is seen that

$$x + u\,\delta t = \begin{cases} x+1, & u = 1, 2 \\ x-1, & u = -1, -2 \\ x, & u = 0. \end{cases} \tag{7.15}$$

Consequently, the next state is always at a quantized value of x. However, because δt is not always an integer, the next state is not necessarily at a quantized value of t. Therefore, in order to compute $I(x + u\,\delta t, t + \delta t)$, the minimum cost at the next state, an interpolation in time is required.

The interpolation of $I(x + u\,\delta t, t + \delta t)$ is taken to be *linear* in time. Except during the initialization procedure, this interpolation uses the two most recently computed values of minimum cost at $(x + u\,\delta t)$. If $I(x + u\,\delta t, t)$ has been computed (where t is the time of the present computation), then the two most recent values of minimum cost at $(x + u\,\delta t)$ are $I(x + u\,\delta t, t)$ and $I(x + u\,\delta t, t + I)$. The interpolation formula can be written as

$$\begin{aligned} I(x + u\,\delta t, t + \delta t) &= I(x + u\,\delta t, t + 1) \\ &\quad + [I(x + u\,\delta t, t + 1) - I(x + u\,\delta t, t)](\delta t - 1). \end{aligned} \tag{7.16}$$

If $I(x + u\,\delta t, t)$ has *not* been computed, the two most recent values are $I(x + u\,\delta t, t + 1)$ and $I(x + u\,\delta t, t + 2)$. The interpolation formula then becomes

$$\begin{aligned} I(x + u\,\delta t, t + \delta t) &= I(x + u\,\delta t, t + 1) \\ &\quad + [I(x + u\,\delta t, t + 2) - I(x + u\,\delta t, t + 1)](\delta t - 1). \end{aligned} \tag{7.17}$$

These two expressions can be combined as

$$I(x + u\,\delta t, t + \delta t) = I(x + u\,\delta t, t + 1) + \Delta J \tag{7.18}$$

where

$$\Delta J = \begin{cases} [I(x + u\,\delta t, t + 1) - I(x + u\,\delta t, t)](\delta t - 1) \\ \qquad\qquad \text{if } I(x + u\,\delta t, t) \text{ has been computed;} \\ [I(x + u\,\delta t, t + 2) - I(x + u\,\delta t, t + 1)](\delta t - 1) \\ \qquad\qquad \text{otherwise.} \end{cases} \tag{7.19}$$

Substituting Eq. (7.18) into Eq. (7.7), the basic iterative equation becomes

$$I(x, t) = \operatorname*{Min}_{u \in U} \{(x^2 + u^2)\,\delta t + I(x + u\,\delta t, t + 1) + \Delta J\}. \tag{7.20}$$

7.4 INITIALIZATION PROCEDURE

The initialization procedure for this problem is carried out in a sequence of three steps. The first step is to evaluate the final-value term in Eq. (7.2) for the admissible final states

$$I(x, 10) = \psi(x, 10) = 2.5(x - 2)^2. \tag{7.21}$$

This term is evaluated for the quantized admissible final states, $x = 0$, $x = 1$, and $x = 2$, and placed in Fig. 1. The calculations are shown in Table 7.1

Because there is one state variable and one control variable, the conditions for complete state-transition control are met. It is then possible to pre-compute the set of terminable states, i.e., the set of states for which it is possible to reach an admissible terminal state at the terminal time. For $0 \leq x \leq 2$, the quantized states are terminable for all t, $0 \leq t \leq 10$.

Table 7.1. Minimum Cost by Admissible Final States

x	$I(x, 10)$
0	10
1	2.5
2	0

For $x > 2$, the set of terminable states is determined by finding the trajectory along which control is fixed at its most negative value, $u = -2$, and for which the terminal state is $x(10) = 2$. The equation of this trajectory is

$$x = 22 - 2t, \qquad (0 \leq t \leq 10). \tag{7.22}$$

The portion of this trajectory for $2 \leq x \leq 8$ is shown as a dotted line in Fig. 7.1. The costs along the trajectory are computed from the iterative relation

$$I(x, t) = (x^2 + 4)\tfrac{1}{2} + I(x - 1, t + \tfrac{1}{2}) \tag{7.23}$$

where the computations have as a boundary condition

$$I(2, 10) = 0. \tag{7.24}$$

The calculations are summarized in Table 7.2. For the quantized values of x and t through which this trajectory passes, the minimum cost and optimal control (always $u = -2$) are entered in Fig. 7.1. As in Sec. 2.9 the minimum cost is above and to the right of the point, while the optimal control is below and to the right.

The set of admissible states for $2 \leq x \leq 8$ is all states to the left of the trajectory in Fig. 7.1; equivalently, any state x, $2 \leq x \leq 8$, is terminable for all values of t less than the value of t for which the trajectory of Eq. (7.22) passes through the state.

Table 7.2. Minimum Costs Along Trajectory for Determining Terminable States

x	u	δt	t	$[x^2 + u^2]\,\delta t$	$I(x + u\,\delta t, t + \delta t)$	$I(x, t)$
2	—	—	10	—	—	0
3	−2	0.5	9.5	6.5	0	6.5
4	−2	0.5	9	10	6.5	16.5
5	−2	0.5	8.5	14.5	16.5	31
6	−2	0.5	8	20	31	51
7	−2	0.5	7.5	26.5	51	77.5
8	−2	0.5	7	34	77.5	111.5

The final step is to evaluate $I(x, k)$ at the *two* largest quantized values of t for which x is a terminable state. The exact procedure followed depends on which block is being processed. In block $B(2, 1)$, $I(x, 10)$ is given by $\psi(x, 10)$ as in Eq. (7.21), and $I(x, 9)$ is computed using the conventional dynamic procedure and the closed-form expression for cost at the next state. In the other blocks the required values are computed on the basis of points along the trajectory of Table 7.2 and on the boundary with previously computed blocks. The details of the computations are discussed further in the next section.

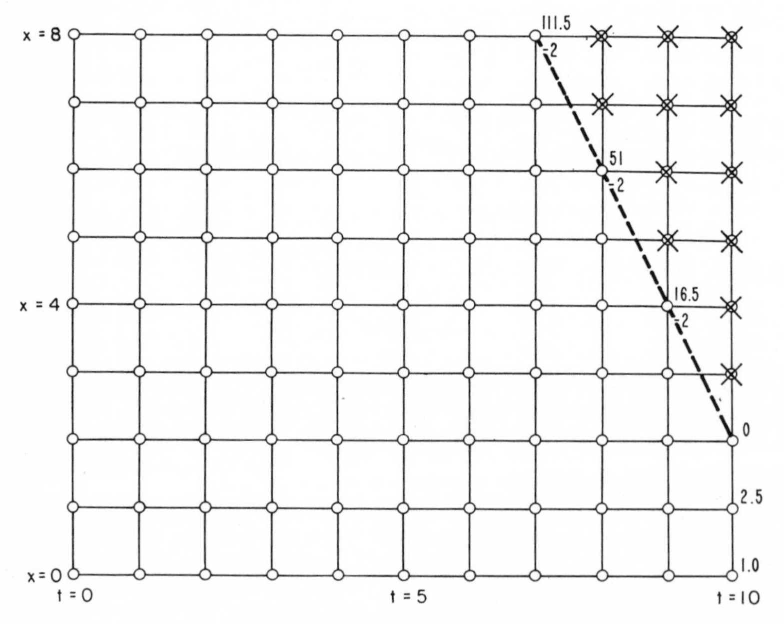

Fig. 7.1. Results of first two steps of the initialization procedure.

7.5 PROCESSING OF THE BLOCKS

The block size is taken to be $2\,\Delta x = 2$ by $5\,\Delta t = 5$. The blocks for the example are pictured in Fig. 7.2.

Since the cost function, $l(x, u, t) = x^2 + u^2$, has its minimum value for $x = 0, u = 0$, the preferred direction of motion is taken to be in the direction of decreasing x. The order in which blocks are to be processed is shown in Table 7.3. The computations within a block take place according to the

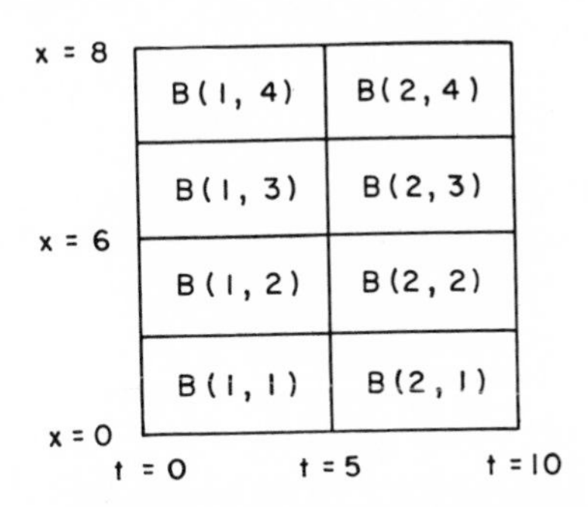

Fig. 7.2. Blocks for example.

procedures of Chapter 4. The initialization procedure has been described in the previous section. The interpolation procedure used if the next state is in the same block has already been described in Eqs. (7.17) to (7.19). If the next state lies in a not-yet-computed block, quadratic extrapolation of the minimum cost function is performed. If the present state and time are x^*, t^*, the *two* minimum costs at $x^* + 1, t^* + 1$ and $x^* + 1, t^* + 2$ are extrapolated by the quadratic formula

$$I(x^* + 1, t + j) = 3I(x^*, t + j) - 3I(x^* - 1, t + j) + I(x^* - 2, t + j) \qquad (j = 1, 2). \quad (7.25)$$

If $\delta t = \frac{1}{2}$, the minimum cost $I(x^* + 1, t + \frac{1}{2})$ is obtained by using Eqs. (7.17) to (7.19) and the two values in Eq. (7.25).

Table 7.3. Order of Processing Blocks

1. $B(2, 1)$	5. $B(1, 1)$
2. $B(2, 2)$	6. $B(1, 2)$
3. $B(2, 3)$	7. $B(1, 3)$
4. $B(2, 4)$	8. $B(1, 4)$

7.6 COMPUTATIONS WITHIN THE BLOCKS

From Table 7.3 the first block to be processed is Block $B(2, 1)$. The points at which minimum cost have already been computed, in this case from the terminal cost function, are shown in Fig. 7.3.

Since $I(x, 10) = \psi(x, 10)$ is an explicit function of x, it is possible to apply the conventional dynamic programming procedure at $t = 9$ without using

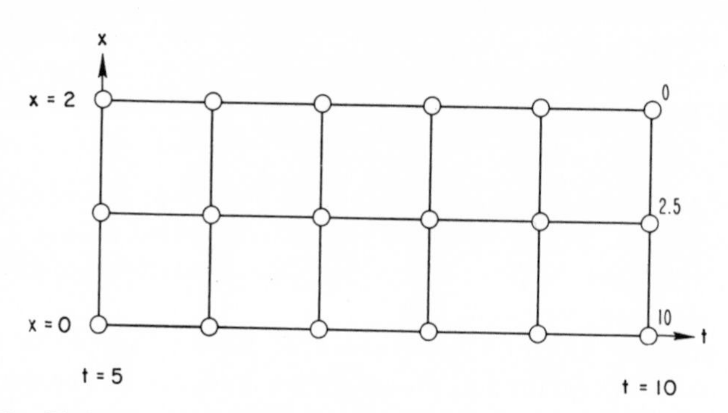

Fig. 7.3. Values of minimum cost known in block $B(2, 1)$ before computations begin.

the high-speed memory for storage of $I(x, 10)$. The calculations can be carried out as in Sec. 2.9. The computations for $x = 0$, $t = 9$ appear in Table 7.4. As in Sec. 2.9, an $\times$ is placed in the cost entries if the state $x + u$ is inadmissible. The minimum cost and optimal control, denoted by asterisks in the table, are $I(0, 9) = 3.5$, $\hat{u}(0, 9) = 1$. These values are entered on Fig. 7.4.

The calculations for $x = 1$, $t = 9$ are also done in this manner. The results appear in Table 7.5. The minimum cost and optimal control are $I(1, 9) = 2$, $\hat{u}(1, 9) = 1$. These values are also entered on Fig. 7.4.

Table 7.4. Initialization Procedure Computations at $x = 0$, $t = 9$

u	$x + u$	$x^2 + u^2$	$I(x + u, t + 1)$	*Total cost*
2	2	4	0	4
* 1	1	1	2.5	3.5 *
0	0	0	10	10
−1	−1	×	×	×
−2	−2	×	×	×

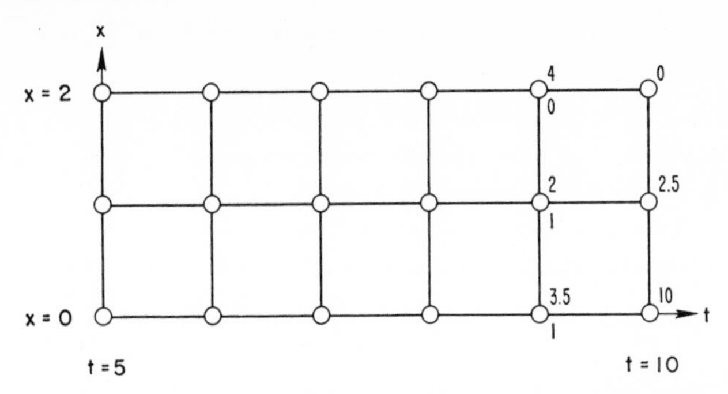

Fig. 7.4. Complete initial conditions for block $B(2, 1)$.

For $x = 2$, $t = 9$, the calculations are as in Table 7.6. The minimum cost and optimal control, $I(2, 9) = 4$, $\hat{u}(2, 9) = 0$, are entered on Fig. 7.4.

After these initialization procedure computations have been completed, the data for the block appears as in Fig. 7.4.

The first point at which computations take place according to the state increment dynamic programming procedure is $x = 0$, $t = 8$. The calculations are illustrated in Table 7.7. From Eq. (7.20), the total cost is the sum of the *three* quantities, $(x^2 + u^2)\,\delta t$, $I(x + u\,\delta t, t + 1)$, and ΔJ. The superscript (a) in the ΔJ column indicates that the first expression in Eq. (7.19) is used

Table 7.5. Initialization Procedure Computations at $x = 1$, $t = 9$

u	$x + u$	$x^2 + u^2$	$I(x + u, t + 1)$	*Total cost*
2	3	×	×	×
* 1	2	2	0	2 *
0	1	1	2.5	3.5
−1	0	2	10	12
−2	−1	×	×	×

Table 7.6. Initialization Procedure Computations at $x = 2$, $t = 9$

u	$x + u$	$x^2 + u^2$	$I(x + u, t + 1)$	*Total cost*
2	4	×	×	×
1	3	×	×	×
* 0	2	4	0	4 *
−1	1	5	2.5	7.5
−2	0	8	10	18

Table 7.7. Computations at $x = 0$, $t = 8$

u	δt	$x + u\,\delta t$	$(x^2 + u^2)\,\delta t$	$I(x + u\,\delta t, t + 1)$	ΔJ	*Total cost*
2	0.5	1	2	2	$-.25^{(b)}$	3.75
* 1	1	1	1	2	0	3 *
0	1	0	0	3.5	0	3.5
−1	1	−1	×	×	×	×
−2	0.5	−1	×	×	×	×

for the interpolation in t, while a superscript (b) indicates that the second expression is used. A double parentheses around the entry in the $I(x + u\,\delta t, t + 1)$ column signifies that the number is obtained by quadratic extrapolation as in Eq. (7.25). An × is entered in the remaining columns if the next state, $x + u\,\delta t$, is not admissible.

As indicated by the asterisks, the minimum cost and optimal control are $I(0, 8) = 3$ and $\hat{u}(0, 8) = 1$. The result of the calculation is shown in Fig. 7.5.

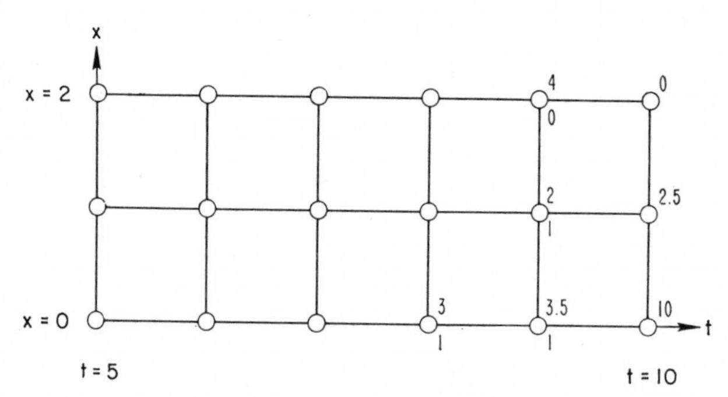

Fig. 7.5. Results after computations at $x = 0$, $t = 8$.

For $x = 1$, $t = 8$.

Table 7.8. Computations at $x = 1$, $t = 8$

u	δt	$x + u\,\delta t$	$(x^2 + u^2)\,\delta t$	$I(x + u\,\delta t, t + 1)$	ΔJ	*Total cost*
2	0.5	2	2.5	4	$2^{(b)}$	8.5
1	1	2	2	4	0	6
* 0	1	1	1	2	0	3 *
−1	1	0	2	3.5	0	5.5
−2	0.5	0	2.5	3.5	$-.25^{(a)}$	5.75

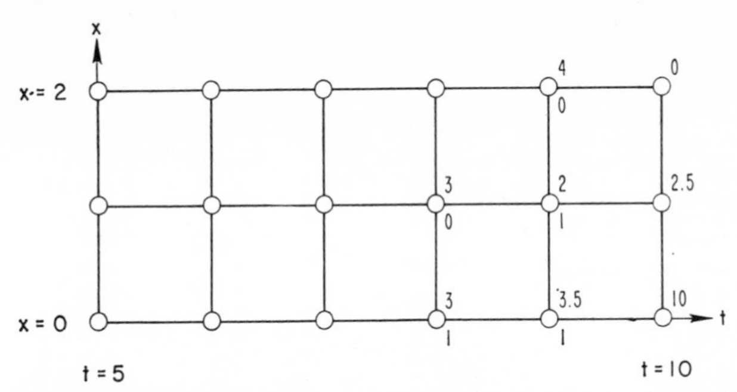

Fig. 7.6. Results after computations at $x = 1, t = 8$.

The minimum cost and the optimal control are $I(1, 8) = 3$, $\hat{u}(1, 8) = 0$. This result is shown in Fig. 7.6.

At $x = 2$, $t = 8$,

Table 7.9. Computations at $x = 2$, $t = 8$

	u	δt	$x + u\,\delta t$	$(x^2 + u^2)\,\delta t$	$I(x + u\,\delta t, t + 1)$	ΔJ	*Total cost*	
	2	0.5	3	4	((9.5))	3.5[b]	17	
	1	1	3	5	((9.5))	0	14.5	
	0	1	2	4	4	0	8	
	−1	1	1	5	2	0	7	
*	−2	0.5	1	4	2	0.5[a]	6.5	*

In the interpolation in t for $u = 2$, the two values of minimum cost at $x = 3$ are both obtained by quadratic extrapolation. One value is $I(x + u\,\delta t, t + 1) = I(3, 9) = 9.5$, and the other is $I(x + u\,\delta t, t + 2) = I(3, 10) = 2.5$. The second expression in the interpolation formula of Eq. (7.17) is used. The minimum cost is $I(2, 8) = 6.5$, corresponding to optimal control $\hat{u}(2, 8) = -2$. The results of all the computations at $t = 8$ are pictured in Fig. 7.7.

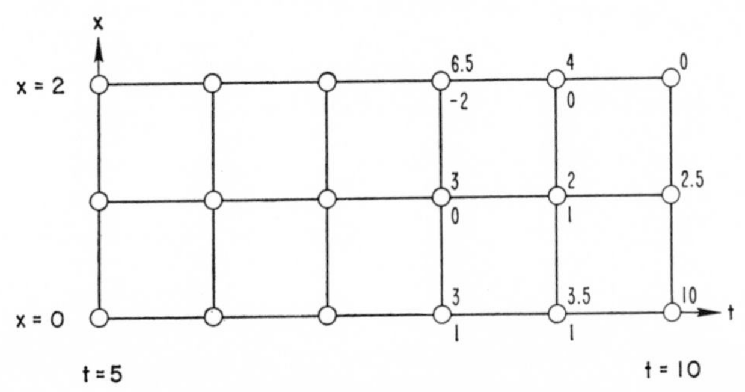

Fig. 7.7. Results after all computations at $t = 8$ in $B(2, 1)$.

Similar computations take place at $t = 7$. At $x = 0$, $t = 7$,

Table 7.10. Computations at $x = 0$, $t = 7$

u	δt	$x + u\,\delta t$	$(x^2 + u^2)\,\delta t$	$I(x + u\,\delta t, t + 1)$	ΔJ	*Total cost*
2	0.5	1	2	3	0.5[(b)]	5.5
1	1	1	1	3	0	4
* 0	1	0	0	3	0	3 *
1	1	−1	×	×	×	×
2	0.5	−1	×	×	×	×

The minimum cost is $I(0, 7) = 3$ and the optimal control is $\hat{u}(0, 7) = 0$.
At $x = 1$, $t = 7$,

Table 7.11. Computations at $x = 1$, $t = 7$

u	δt	$x + u\,\delta t$	$(x^2 + u^2)\,\delta t$	$I(x + u\,\delta t, t + 1)$	ΔJ	*Total cost*
2	0.5	2	2.5	6.5	1.25[(b)]	10.25
1	1	2	2	6.5	0	8.5
* 0	1	1	1	3	0	4 *
−1	1	0	2	3	0	5
−2	0.5	0	2.5	3	0[(a)]	5.5

The minimum cost is $I(1, 7) = 4$ and the optimal control is $\hat{u}(1, 7) = 0$.
At $x = 2$, $t = 7$,

Table 7.12. Computations at $x = 2$, $t = 7$

u	δt	$x + u\,\delta t$	$(x^2 + u^2)\,\delta t$	$I(x + u\,\delta t, t + 1)$	ΔJ	*Total cost*
2	0.5	3	4	((13.5))	2[(b)]	19.5
1	1	3	5	((13.5))	0	18.5
0	1	2	4	6.5	0	10.5
−1	1	1	5	3	0	8
* −2	0.5	1	4	3	0.5[(a)]	7.5 *

Again, the interpolation for $u = 2$ utilizes two values of extrapolated cost, $I(3, 8) = 13.5$ and $I(3, 9) = 9.5$. The minimum cost is $I(2, 7) = 7.5$ and the optimal control is $\hat{u}(2, 7) = -2$.

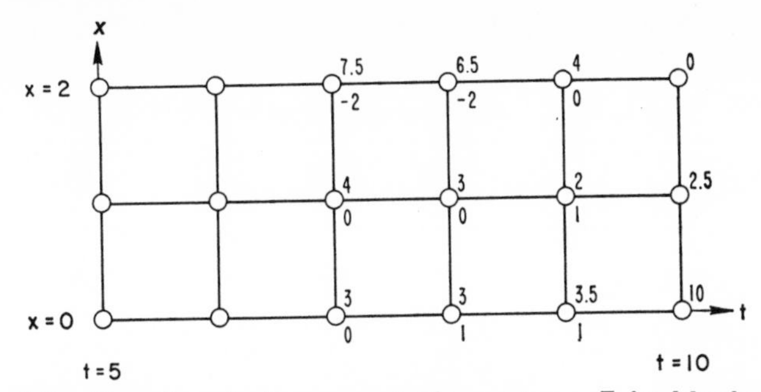

Fig. 7.8. Results of all computations at $t = 7$ in block $B(2, 1)$.

The results of all the computations at $t = 7$ are shown in Fig. 7.8.

The other computations are very similar. The results for the rest of block $B(2, 1)$ are shown in Fig. 7.9.

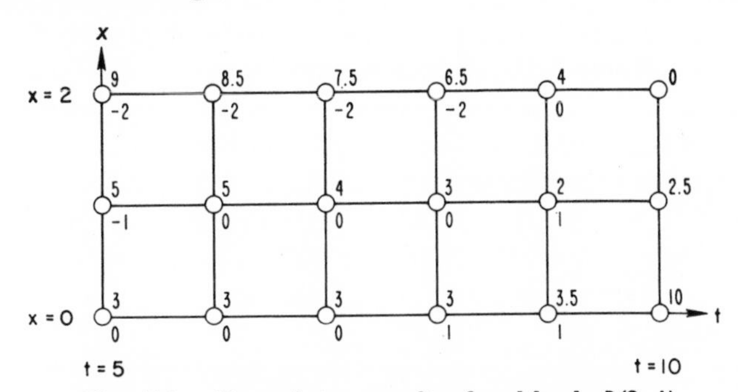

Fig. 7.9. Complete results for block $B(2, 1)$.

From Table 7.3 the next block to be processed is Block $B(2, 2)$. The values of minimum cost that have already been obtained are indicated in Fig. 7.10. Values are known along the boundary with Block $B(2, 1)$ ($x = 2, k = 5 - 10$) as a result of the calculations for this block. The value at $x = 4$, $k = 9$ was obtained as part of the second step in the initialization procedure.

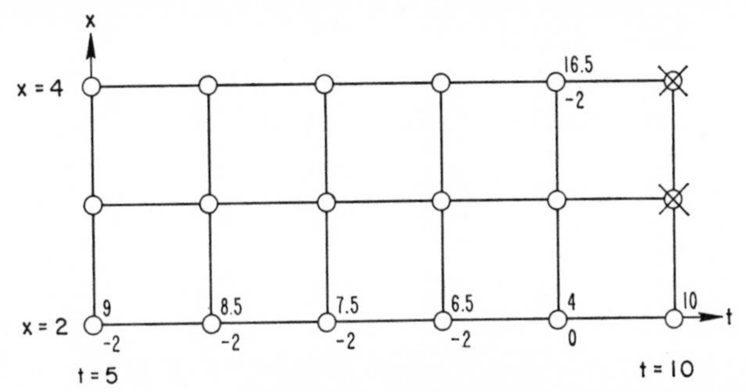

Fig. 7.10. Values of minimum cost known in block $B(2, 2)$ before computations begin.

Table 7.13. Computations at $x = 3$, $t = 9$

	u	δt	$x + u\,\delta t$	$(x^2 + u^2)\,\delta t$	$I(x + u\,\delta t, t + 1)$	ΔJ	Total cost	
	2	0.5	4	×	×	×	×	
	1	1	4	×	×	×	×	
	0	1	3	×	×	×	×	
	−1	1	2	10	0	0	10	
*	−2	0.5	2	6.5	0	2[a]	8.5	*

The remainder of the initialization procedure consists of specifying $I(3, 9)$, $I(3, 8)$, and $I(4, 8)$. The calculation of $I(3, 9)$ uses the normal procedure, except that for controls $u = 2, 1, 0$, the next state is inadmissible. The calculations are as in Table 7.13. The minimum cost is $I(3, 9) = 8.5$, and the corresponding optimal control is $\hat{u}(3, 9) = -2$. For the two values $I(3, 8)$ and $I(4, 8)$ it is necessary to use the second interpolation procedure of Sec. 4.5. It will be recalled that this procedure requires values of minimum cost at the next quantized t and one value of minimum cost at the present quantized t; it is seen from Fig. 7.10 (note that $I(3, 9)$ has just been computed) that these quantities are indeed available. The only change required in the procedure is that when $x^* + u\,\delta t = x^* - 1$, ΔJ is determined by using a linear fit in t based on the two values $I(x^* - 1, t)$ and $I(x^* - 1, t + 1)$. From Eqs. (4.13) and (4.14), the formula for ΔJ can be derived as

$$\Delta J = \left[\frac{I(x^* - 1, t^*)}{I(x^* - 1, t^* + 1)} - 1\right](1 - \delta t)I(x^* + 1, t^* + 1) \qquad (7.26)$$

$$0 \leq \delta t \leq 1$$

$$x^* = 3, 4$$

$$t^* = 8.$$

The calculations for $x = 3$, $t = 8$ are as in Table 7.14. The superscript (c) on ΔJ for $u = 2$ indicates that Eq. (7.26) was used. The minimum cost and optimal control are obtained as $I(3, 8) = 11.75$ and $\hat{u}(3, 8) = -2$.

Table 7.14. Calculations at $x = 3$, $t = 8$

	u	δt	$x + u\,\delta t$	$(x^2 + u^2)\,\delta t$	$I(x + u\,\delta t, t + 1)$	ΔJ	Total cost	
	2	0.5	4	6.5	16.5	5.16[c]	28.16	
	1	1	4	10	16.5	0	26.5	
	0	1	3	9	8.5	0	17.5	
	−1	1	2	10	4	0	14	
*	−2	0.5	2	6.5	4	1.25[a]	11.75	*

At $x = 4$, $t = 8$,

Table 7.15. Calculations at $x = 4$, $t = 8$

	u	δt	$x + u\,\delta t$	$(x^2 + u^2)\,\delta t$	$I(x + u\,\delta t, t + 1)$	ΔJ	*Total cost*	
	2	0.5	5	10	((28))	5.35[c]	43.35	
	1	1	5	17	((28))	0	45	
	0	1	4	16	16.5	0	32.5	
	−1	1	3	17	8.5	0	25.5	
*	−2	0.5	3	10	8.5	1.62[a]	20.12	*

The minimum cost and optimal control are $I(4, 8) = 20.12$ and $\hat{u}(4, 8) = -2$.
After these computations the initial conditions for Block $B(2, 2)$ are as shown in Fig. 7.11.

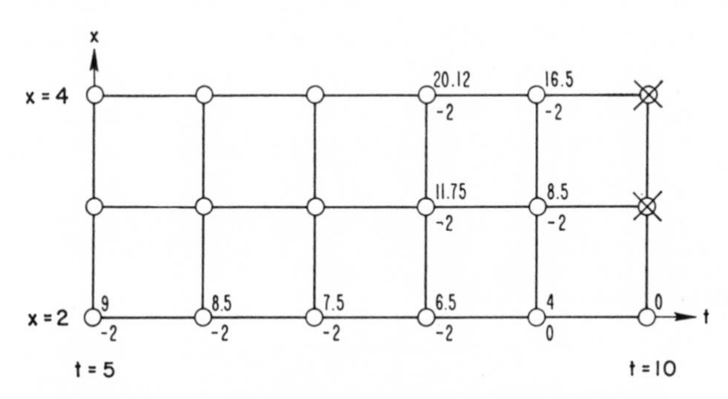

Fig. 7.11. Initial conditions for block (2, 2).

Continuing with this block, at $x = 3$, $t = 7$,

Table 7.16. Computations at $x = 3$, $t = 7$

	u	δt	$x + u\,\delta t$	$(x^2 + u^2)\,\delta t$	$I(x + u\,\delta t, t + 1)$	ΔJ	*Total cost*	
	2	0.5	4	6.5	20.12	2[b]	28.62	
	1	1	4	10	20.12	0	30.12	
	0	1	3	9	11.75	0	20.75	
	−1	1	2	10	6.5	0	16.5	
*	−2	0.5	2	6.5	6.5	0.5[a]	13.5	*

The minimum cost is $I(3, 7) = 13.5$, and the optimal control is $\hat{u}(3, 7) = -2$.

At $x = 4$, $t = 7$,

Table 7.17. Computations at $x = 4$, $t = 7$

	u	δt	$x + u\,\delta t$	$(x^2 + u^2)\,\delta t$	$I(x + u\,\delta t, t + 1)$	ΔJ	Total cost	
	2	0.5	5	10	31.38	1[b]	43.07	
	1	1	5	17	31.38	0	47.5	
	0	1	4	16	20.12	0	36.12	
	−1	1	3	17	11.75	0	28.75	
*	−2	0.5	3	10	11.75	0.87[a]	22.62	*

The minimum cost is $I(4, 7) = 22.62$, again corresponding to optimal control $\hat{u}(4, 7) = -2$.

The results of these computations are shown in Fig. 7.12.

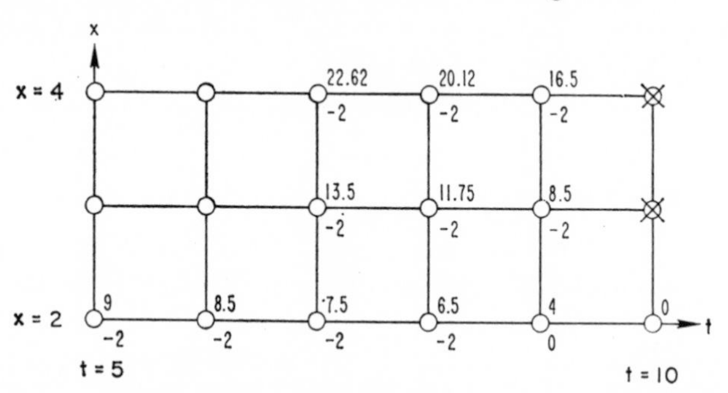

Fig. 7.12. Results of computations at $t = 7$ in block $B(2, 2)$.

The remainder of the computations in $B(2, 2)$ are carried out in exactly the same way. The results are shown in Fig. 7.13.

After this block has been processed, a recomputation at $x = 2$, $t = 5$ using the values of minimum cost just computed in $B(2, 2)$ can be made. However,

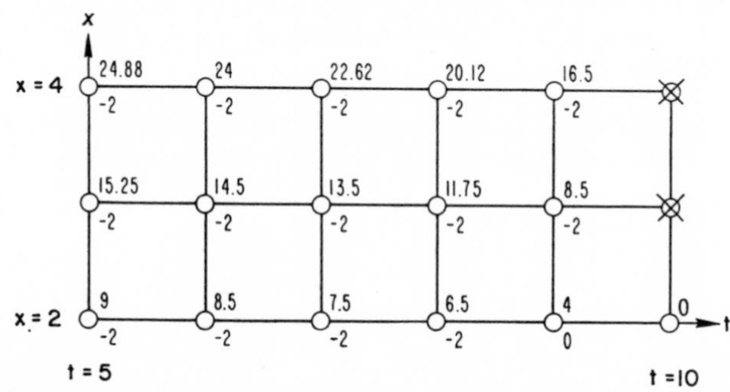

Fig. 7.13. Results of all computations in block $B(2, 2)$.

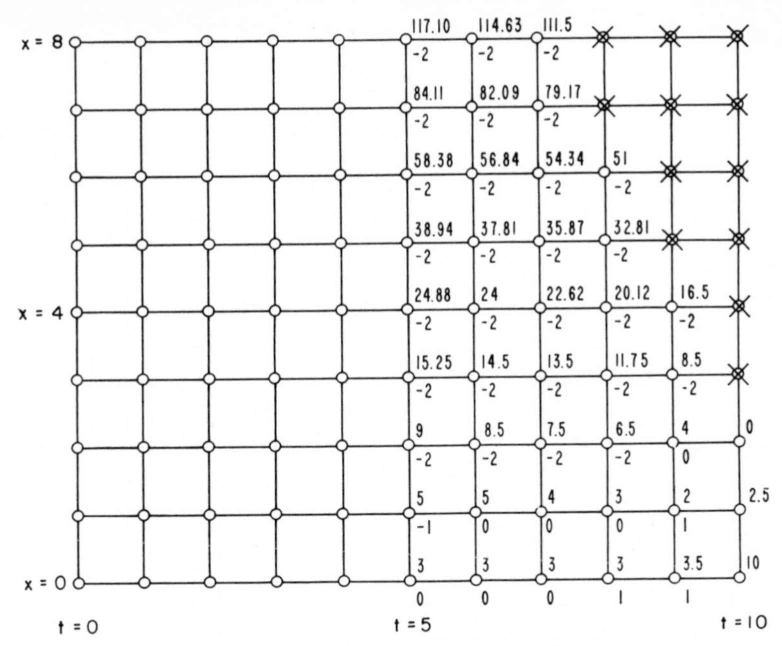

Fig. 7.14. Results for all blocks in $5 \leq t \leq 10$.

this recomputation changes neither the optimal control nor the minimum cost at this point.

The computations for the other blocks in this time interval are similar. The results are summarized in Fig. 7.14.

The processing of blocks in the interval $0 \leq t \leq 5$ follows essentially the same procedure. The main difference is that the two initial values of minimum cost for each state must be taken at $t = 5$, the largest value of t in the interval, and at $t = 6$, the smallest quantized value of t beyond the interval. The initial conditions for Block $B(1, 1)$ are thus as in Fig. 7.15. The complete results for this problem are shown in Fig. 7.16.

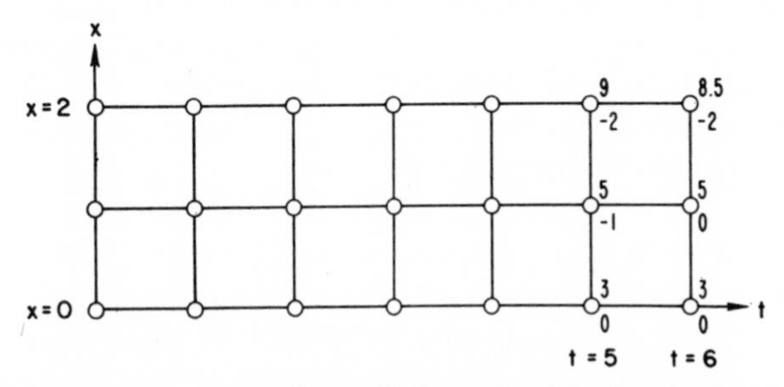

Fig. 7.15. Initial conditions for block $B(1, 1)$.

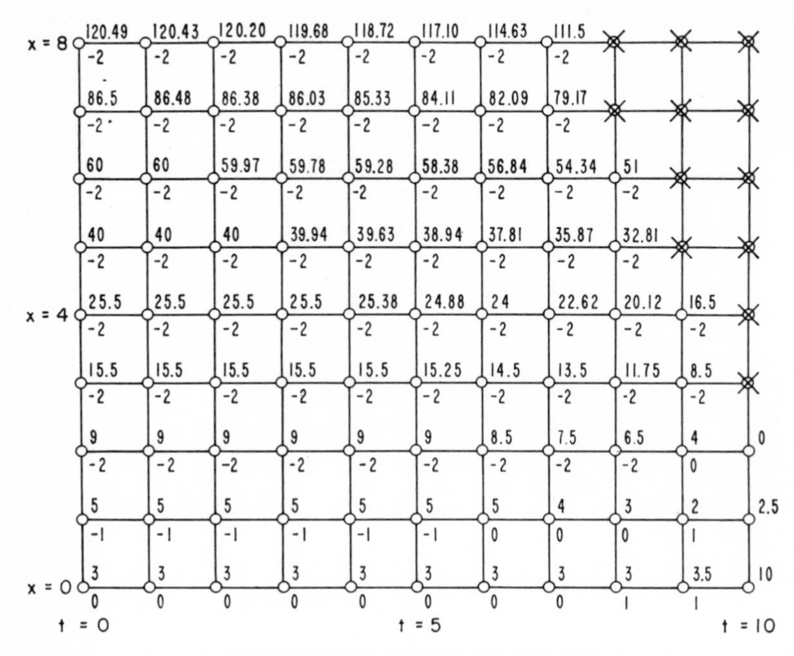

Fig. 7.16. Complete results.

The recomputations for points on the boundary once in each ΔT interval never changed either the optimal control or the minimum cost. This indicates that in this example the extrapolation procedure used is adequate. As a further check on these extrapolations, the average percentage difference between extrapolated cost and the cost actually calculated at quantized points was computed for both linear extrapolation and quadratic extrapolation. The average percentage error was 11.7% for linear extrapolation and 2.3% for quadratic extrapolation. The quadratic extrapolation procedure actually used is thus seen to be a good approximation.

It is of considerable interest to compare the results of Fig. 7.16 with those derived in Sec. 2.9, where the same problem was solved by the conventional dynamic programming computational procedure using the same increment sizes. The values of optimal control and minimum cost obtained by the conventional procedure are shown in Fig. 7.17. The optimal control computed by the two methods is essentially the same. A notable exception is that at $x = 2$ for $0 \leq t \leq 8$, optimal control is $u = -1$ for the conventional method and $u = -2$ for state increment dynamic programming. The same discrepancy occurs at $x = 3$ for $t = 8$ and $t = 9$. The minimum costs also are not exactly the same, except at $x = 0$ and $x = 1$. In general, the minimum cost computed by state increment dynamic programming is 80–90% of the minimum cost computed by conventional dynamic programming for $x \geq 2$.

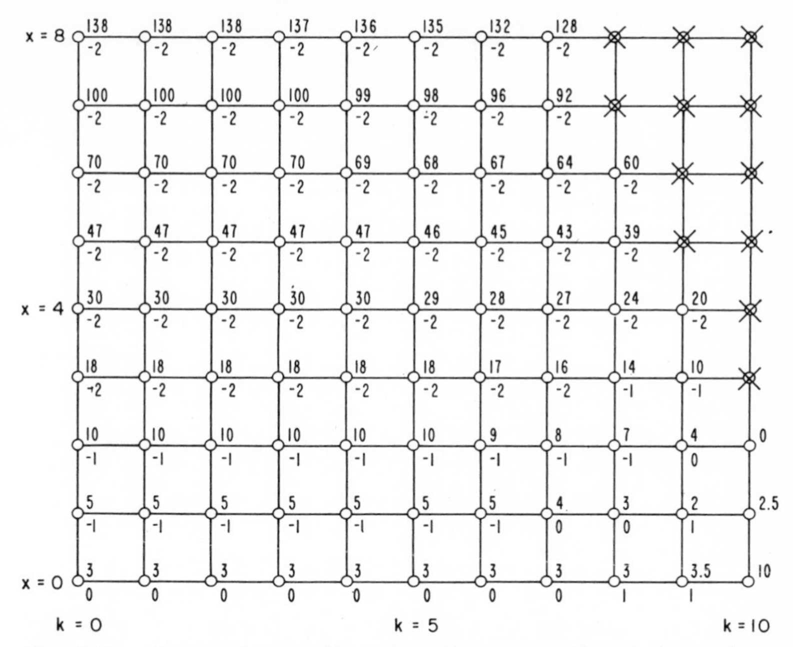

Fig. 7.17. Complete results using the conventional dynamic programming computational procedure.

Before any conclusions can be drawn, the exact analytical solution to the original continuous problem as expressed in Eqs. (7.1 to 7.5) must be computed. Because of the special properties of this problem [Ref. 1], it is possible (though laborious) to obtain it. The solution evaluated at the same quantized grid points as in the two dynamic programming procedures is shown in Fig. 7.18.

Because of the relatively coarse quantization used in the dynamic programming procedures, there is some difference between these optimal controls and minimum costs and the values from the dynamic programming procedures. However, in every case where the values from the two dynamic programming procedures differ, the state increment dynamic programming values are closer to the exact analytical relation. As an illustration of this, the minimum costs obtained by all these methods for $t = 0$ are plotted as a function of x in Fig. 7.19. In this figure, the state increment dynamic programming curve is about halfway between the two others. As another illustration of this point note that at $x = 2$ for $0 \leq t \leq 8$ the exact analytical value of optimal control is always closer to $u = -2$, the state increment dynamic programming value, than to $u = -1$, the conventional dynamic programming value.

Thus, in this example, in which the quantization increments are the same, the solution based on state increment dynamic programming is a better

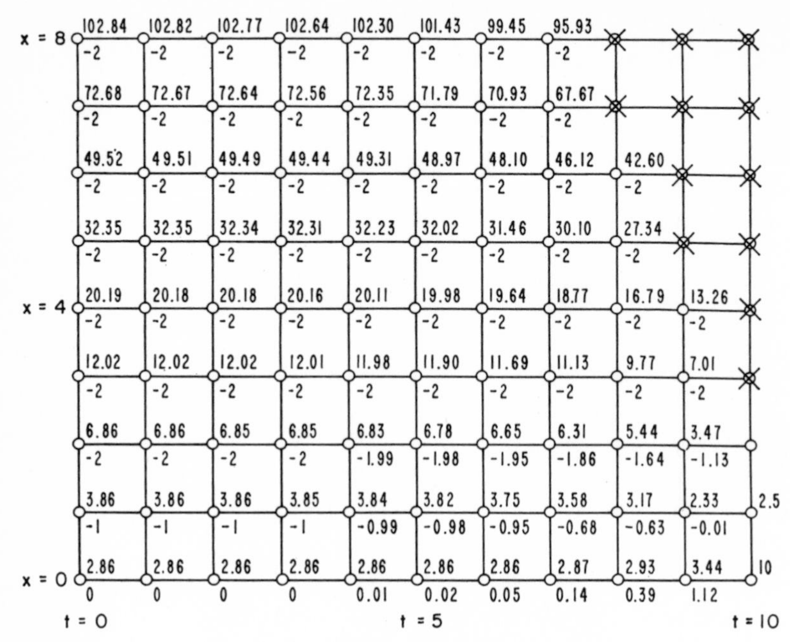

Fig. 7.18. Exact analytical solution.

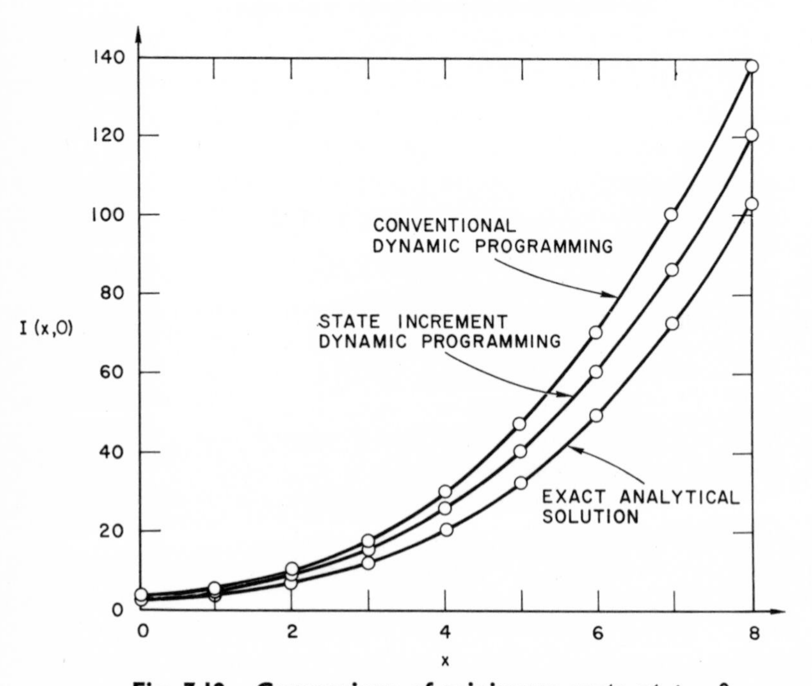

Fig. 7.19. Comparison of minimum costs at $t = 0$.

approximation to the solution of the original problem than is the solution obtained from the conventional dynamic programming computational procedure.

This result, which at first sight might seem surprising, has a simple explanation. It is true that there are interpolation and extrapolation errors in the state increment dynamic programming procedure, while there are no such errors in the conventional procedure. However, another source of error in both procedures is the approximations to the system differential equation and the integral in the performance criterion. Both methods use the first term of the Taylor series expansion in determining the next state. For the general system differential equation,

$$x(t + \delta t) = x(t) + f[x(t), u(t), t]\,\delta t. \tag{7.27}$$

Both methods also use the same approximation to the integral in the performance criterion. In general,

$$\int_t^{t+\delta t} l(x, u, \sigma)\,d\sigma = l[x(t), u(t), t]\,\delta t. \tag{7.28}$$

Clearly, if the change in x is great over the interval δt, these approximations are not good ones. This can happen with the conventional method, because δt is fixed at $\Delta t = 1$. However, with state increment dynamic programming, δt is determined so that the change in x is at most one increment, Δx. As a result, if u is very large, state increment dynamic programming produces better approximations in both cases.

7.7 CONCLUSIONS

In this chapter the computational procedure of state increment dynamic programming has been illustrated. Determination of a preferred direction was straightforward. Linear interpolation and quadratic extrapolation procedures proved to be quite accurate.

The same example was also solved by the conventional dynamic programming procedure in Chapter 2. A comparison of the two solutions showed that they were quite similar, but not identical. When the two were checked against the exact analytical solution, the state increment dynamic programming solution was more accurate. This was shown to be a consequence of the fact that δt as determined by state increment dynamic programming makes the simple integration formulas used in this example more accurate than would be the case for a fixed value of $\delta t = \Delta t$.

REFERENCE

1. L. S. Pontryagin et al., *The Mathematical Theory of Optimal Processes*, John Wiley & Sons, Inc., New York, 1962.

Chapter Eight

A COMPUTER PROGRAM FOR PROBLEMS WITH FOUR STATE VARIABLES

8.1 INTRODUCTION

In this chapter a computer program for implementing the state increment dynamic programming procedure for problems having four or less state variables is discussed. The class of problems to which the program applies is defined in the next section. In the section after that, the calculations at a given state and stage are described. In the following section the procedure for processing blocks is presented. In the next section the results of applying the program to a particular example are shown. The example is such that a closed-form solution can be found; this enables a comparison to be made between the true solution and the computed results. In the final section of this chapter another example, one for which a closed-form solution is not known, is solved.

8.2 PROBLEMS TO WHICH THE PROGRAM APPLIES

The program has been written for problems having four or less state variables and one control variable. It is not difficult to expand the program to handle more control variables, but because of the importance of this program as a building block in a successive approximations technique for even larger problems (see Chapter 12), attention has thus far been restricted to this case.

The system equations are assumed to be in the form

$$\dot{\mathbf{x}} = \mathbf{f}(\mathbf{x}, u, t). \tag{8.1}$$

The performance criterion takes the form

$$J = \int_{t_0}^{t_f} l(\mathbf{x}, u, \sigma)\, d\sigma + \psi[\mathbf{x}(t_f), t_f]. \tag{8.2}$$

The constraints are

$$\begin{aligned} &\mathbf{x} \in X(t) \\ &u \in U(\mathbf{x}, t). \end{aligned} \tag{8.3}$$

The basic iterative equation is thus

$$I(\mathbf{x}, t) = \min_{u \in U} \{l(\mathbf{x}, u, t)\,\delta t + I[\mathbf{x} + \mathbf{f}(\mathbf{x}, u, t)\,\delta t, t + \delta t]\}. \tag{8.4}$$

In this program it is assumed that the system equations are linear and the performance criterion quadratic, i.e., the system equation can be written as

$$\dot{\mathbf{x}} = \mathbf{f}(\mathbf{x}, u, t) = A(t)\mathbf{x} + \mathbf{b}(t)u \tag{8.5}$$

where $A(t)$ is a time varying, 4×4 matrix and $\mathbf{b}(t)$ is a time-varying 4-dimensional vector, and the functions in the performance criterion can be expressed as

$$\begin{aligned} l(\mathbf{x}, u, t) &= \mathbf{x}^T Q(t)\mathbf{x} + R(t)u^2(t) \\ \psi(\mathbf{x}, t_f) &= \mathbf{x}^T Q(t_f)\mathbf{x} \end{aligned} \tag{8.6}$$

where $Q(t)$ is a time-varying 4×4 matrix and $R(t)$ is a time-varying scalar. Without significant loss of generality, * the matrix $A(t)$ and the vector $b(t)$ can be written as

$$A(t) = \begin{bmatrix} 0 & 1 & 0 & 0 \\ 0 & 0 & 1 & 0 \\ 0 & 0 & 0 & 1 \\ a_1(t) & a_2(t) & a_3(t) & a_4(t) \end{bmatrix} \qquad b(t) \begin{bmatrix} 0 \\ 0 \\ 0 \\ b(t) \end{bmatrix}. \tag{8.7}$$

The system equations thus become

$$\begin{aligned} \dot{x}_1 &= f_1(\mathbf{x}, u, t) = x_2 \\ \dot{x}_2 &= f_2(\mathbf{x}, u, t) = x_3 \\ \dot{x}_3 &= f_3(\mathbf{x}, u, t) = x_4 \\ \dot{x}_4 &= f_4(\mathbf{x}, u, t) = \mathbf{a}^T(t)\mathbf{x} + b(t)u \end{aligned} \tag{8.8}$$

where $\mathbf{a}(t)$ is a 4-dimensional vector with components $a_i(t)$, $i = 1, 2, 3, 4$.

The expansion of nonlinear system equations and a nonquadratic performance criterion into this form has been extensively studied in connection with quasilinearization [Ref. 2] and second-order gradient methods [Ref. 3]. The interpretation of this expansion in the dynamic programming formulation is discussed in Ref. 2.

It should be pointed out that there is a considerable difference in the degree to which these linearizations are used in the present program and in these other methods. For the latter procedures the linearization is made over the entire space in which computations are made, and the closed-form solution of an unconstrained problem with linear system equations and quadratic performance criterion [Refs. 2, 4] is utilized. In this program, on the other hand,

* It is required that the system be controllable in Kalman's sense [Ref. 1].

the linearization is required to be valid only within one block, and no attempt is made to use the closed-form solution. The only purpose of the linearization is to expand the terms in brackets in Eq. (8.4) to the same order as the interpolation formulas for the minimum cost function at the next state; these latter formulas are quadratic in the state variables and linear in the stage variable. This makes possible the use of particularly efficient search procedures, in which constraints are implemented directly, for determining the optimal control and minimum cost at a given quantized state and stage.

As will be shown in Sec. 8.4, the block processing procedures are such that in every block there is at least one quantized state for which optimal control has already been computed; it is thus possible to repeat the linearization in every block at every quantized value of t. In fact, it is even possible to arrange the calculations within a block so that relinearizations can be made within the block, with the result that the linearization never extends more than one increment over any state variable.

8.3 COMPUTATIONS WITHIN A BLOCK

Once the linearizations described in the previous section have been made, the calculations in a given block follow the procedure outlined in this section. The basic equation that is to be solved is

$$I(\mathbf{x}^*, t^*) = \min_{u \in U} \{l(\mathbf{x}^*, u, t^*)\, \delta t + I[\mathbf{x}^* + \mathbf{f}(\mathbf{x}^*, u, t^*)\, \delta t, t^* + \delta t]\} \tag{8.9}$$

where $\mathbf{x}^*$ and t^* are particular quantized values of $\mathbf{x}$ and t and where l and $\mathbf{f}$ are as in Eqs. (8.6) and (8.7), respectively.

The function $I[\mathbf{x}^* + \mathbf{f}(\mathbf{x}^*, u, t^*)\, \delta t, t^* + \delta t]$ is evaluated by performing a quadratic interpolation in $\mathbf{x}$ and a linear interpolation in t based on values at quantized values of $\mathbf{x}$ and t. The quadratic interpolation in $\mathbf{x}$ is made by a least-squares fit over all quantized points in the block at time $(t^* + \Delta t)$: the minimum cost function at $(\mathbf{x}, t^* + \Delta t)$ is assumed to have the form

$$I(\mathbf{x}, t^* + \Delta t) = \mathbf{x}^T P(t^* + \Delta t)\mathbf{x} \tag{8.10}$$

where $P(t^* + \Delta t)$ is a symmetric 4×4 matrix; and the components of this matrix are selected by minimizing the sum of the squares of the errors between the values of the function in Eq. (8.10) and the actual minimum costs at the quantized states. Within a block, this calculation needs to be made only once for each quantized value of t.

The first step in the linear interpolation in t is to select the nearest quantized state in the block for which computations have already been made at time t^*. The computations are ordered so that it is always possible to find one such state that differs from $\mathbf{x}^*$ only in one component and by only one increment in this component. If this state is denoted as $\mathbf{x}^+$, the next step is to locate the minimum cost values $I(\mathbf{x}^+, t^*)$ and $I(\mathbf{x}^+, t^* + \Delta t)$. A linear function of time, $I(\mathbf{x}^+, t)$, $t^* \leq t \leq t^* + \Delta t$, is then defined such that it agrees with $I(\mathbf{x}^+, t^*)$ at $t = t^*$ and with $I(\mathbf{x}^+, t^* + \Delta t)$ at $t = t^* + \Delta t$. This function is written in the form

$$I(x^+, t) = [c_1 + c_2(t - t^*)]I(\mathbf{x}^+, t^* + \Delta t) \tag{8.11}$$

$$c_1 = \frac{I(\mathbf{x}^+, t^*)}{I(\mathbf{x}^+, t^* + \Delta t)}$$

$$c_2 = \frac{1 - c_1}{\Delta t}.$$

Finally, it is assumed that $I[\mathbf{x}^* + \mathbf{f}(\mathbf{x}^*, u, t)\,\delta t, t^* + \delta t]$, $t^* \leq t^* + \delta t \leq t^* + \Delta t$, also has this dependence on time. The complete interpolation formula is thus

$$I[\mathbf{x}^* + \mathbf{f}(\mathbf{x}^*, u, t^*)\,\delta t, t^* + \delta t]$$
$$= [c_1 + c_2\,\delta t]\{(\mathbf{x}^* + \mathbf{f}(\mathbf{x}^*, u, t^*)\,\delta t]^T P(t^* + \Delta t)[\mathbf{x}^* + \mathbf{f}(\mathbf{x}^*, u, t^*)\,\delta t]\} \tag{8.12}$$

The application of this formula to a problem with two state variables is illustrated in Fig. 8.1.

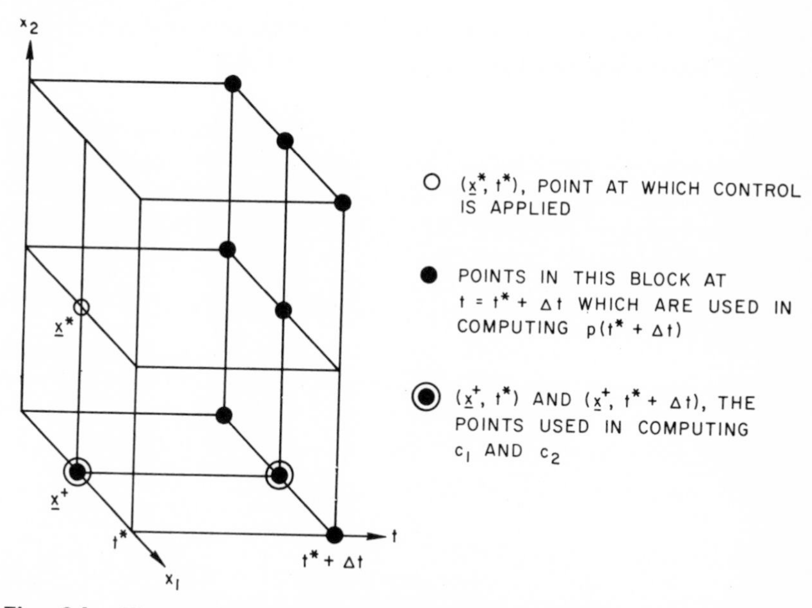

Fig. 8.1. Illustration of the interpolation formula for a two-dimensional example.

Substituting Eqs. (8.6), (8.8), and (8.11) into Eq. (8.9), the basic iterative equation now becomes

$$I(\mathbf{x}^*, t^*) = \min_{u\in U} \{[\mathbf{x}^{*T} Q(t^*)\mathbf{x}^* + R(t^*)u^2]\delta t + [c_1 + c_2\,\delta t][(\mathbf{x}^* + [A(t^*)\mathbf{x}^* + \mathbf{b}(t^*)u]\,\delta t)^T \times P(t^* + \Delta t)(\mathbf{x}^* + [A(t^*)\mathbf{x}^* + \mathbf{b}(t^*)u]\,\delta t)]\}. \quad (8.13)$$

Using the basic formula of Chapter 3, the increment δt is determined as

$$\delta t = \min\left\{\left|\frac{\Delta x_1}{x_2^*}\right|, \left|\frac{\Delta x_2}{x_3^*}\right|, \left|\frac{\Delta x_3}{x_4^*}\right|, \left|\frac{\Delta x_4}{\mathbf{a}^T(t^*)\mathbf{x}^* + b(t^*)u}\right|, \Delta t\right\} \quad (8.14)$$

The only one of these terms that varies with u is the fourth one. The minimization in Eq. (8.14) can thus be re-written as

$$\delta t = \min\left\{\left|\frac{\Delta x_4}{\mathbf{a}^T(t^*)\mathbf{x}^* + b(t^*)u}\right|, dt\right\} \quad (8.15)$$

where

$$dt = \min\left\{\left|\frac{\Delta x_1}{x_2^*}\right|, \left|\frac{\Delta x_2}{x_3^*}\right|, \left|\frac{\Delta x_3}{x_4^*}\right|, \Delta t\right\}$$

As u is varied from $-\infty$ to $+\infty$, there will be three different functions for δt. If $b(t^*) > 0$, these functions are

$$\begin{aligned}\delta t &= \frac{\Delta x_4}{-[\mathbf{a}^T(t^*)\mathbf{x}^* + b(t^*)u]}, -\infty < u \le \frac{1}{b(t^*)}\left[-\frac{\Delta x_4}{dt} - \mathbf{a}^T(t^*)\mathbf{x}^*\right] \\ &= dt, \frac{1}{b(t^*)}\left[-\frac{\Delta x_4}{dt} - \mathbf{a}^T(t^*)\mathbf{x}^*\right] \le u \le \frac{1}{b(t^*)}\left[\frac{\Delta x_4}{dt} - \mathbf{a}^T(t^*)\mathbf{x}^*\right] \\ &= \frac{\Delta x_4}{[\mathbf{a}^T(t^*)\mathbf{x}^* + b(t^*)u]}, \frac{1}{b(t^*)}\left[\frac{\Delta x_4}{dt} - \mathbf{a}^T(t^*)\mathbf{x}^*\right] \le u < \infty \end{aligned} \quad (8.16)$$

If $b(t^*) < 0$, then the functions are

$$\begin{aligned}\delta t &= \frac{\Delta x_4}{-[\mathbf{a}^T(t^*)\mathbf{x}^* + b(t^*)u]}, \frac{1}{b(t^*)}\left[-\frac{\Delta x_4}{dt} - \mathbf{a}^T(t^*)\mathbf{x}\right] \le u < \infty \\ &= dt, \frac{1}{b(t^*)}\left[\frac{\Delta x_4}{dt} - \mathbf{a}^T(t^*)\mathbf{x}^*\right] \le u \le \frac{1}{b(t^*)}\left[-\frac{\Delta x_4}{dt} - \mathbf{a}^T(t^*)\mathbf{x}^*\right] \\ &= \frac{\Delta x_4}{[\mathbf{a}^T(t^*)\mathbf{x}^* + b(t^*)u]}, -\infty \le u \le \frac{1}{b(t^*)}\left[\frac{\Delta x_4}{dt} - \mathbf{a}^T(t^*)\mathbf{x}^*\right]. \end{aligned} \quad (8.17)$$

In Fig. 8.2 the use of these formulas is illustrated for a two-dimensional example; the locus of the next states in (x_1, x_2, t) space traced out as u varies from $-\infty$ to $+\infty$ is shown, both for $dt = \Delta t$ and for $dt = \Delta x_1/x_2^*$.

Constraints on u are implemented directly by restricting the range of u that is searched over. Constraints on $\mathbf{x}$ are applied by eliminating values of u for which the next state, $\mathbf{x}^* + [A(t^*)\mathbf{x}^* + b(t^*)u]\,\delta t$, is inadmissible.* A search is then made over the ranges of admissible values of u corresponding to each function δt.

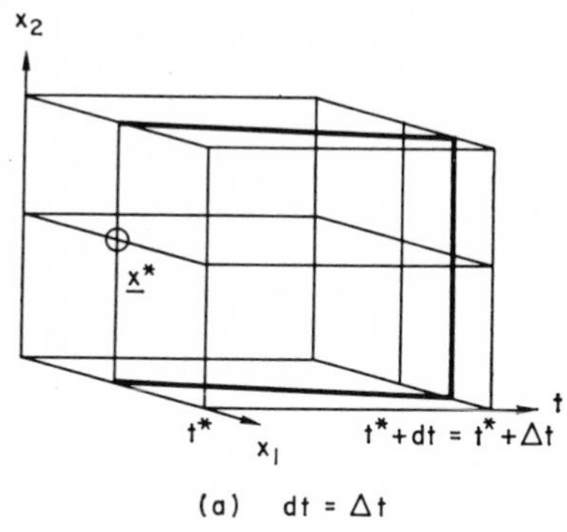

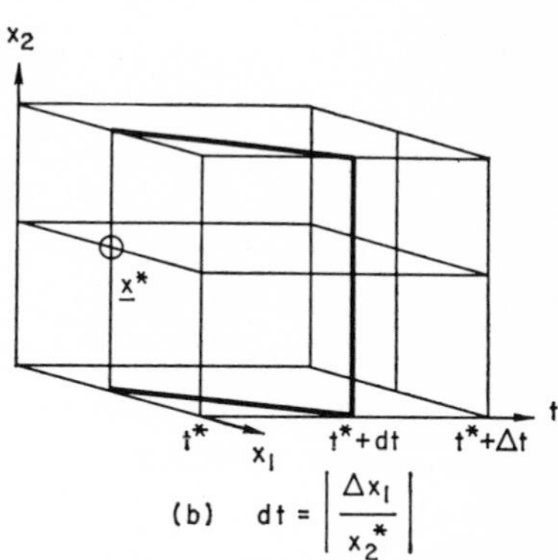

○ = $\underline{x}^*$, POINT AT WHICH CONTROL IS APPLIED

— = LOCUS OF NEXT STATES AS u VARIES FROM $-\infty$ TO $+\infty$; SHOWN FOR BOTH $dt = \Delta t$ AND FOR $dt = \left|\frac{\Delta x_1}{x_2^*}\right|$

Fig. 8.2. Illustration of the formulas for δt.

If $\delta t = dt$, the expression in brackets in Eq. (8.13) is a quadratic function of u. A simple search procedure suffices to find the optimum value in this range of u. This procedure evaluates the cost at all local optima (values of u where the derivative with respect to u vanishes), at the end points of the ranges of admissible values of u, and at the end points of the interval of u over which $\delta t = dt$; the minimum cost of those evaluated is then retained along with the corresponding value of u.

If $\delta t < dt$, the expression in brackets in Eq. (8.13) is a more complicated function of u. In this case it is more convenient to write u as a function of δt using the appropriate part of either Eq. (8.16) or (8.17) and to optimize with respect to δt. The expression in brackets in Eq. (8.13) then has the form

$$I(\mathbf{x}^*, t^*) = \min_{\delta t} \left\{ \frac{d_0 + d_1(\delta t) + d_2(\delta t)^2 + d_3(\delta t)^3}{\delta t} \right\} \tag{8.18}$$

where the admissible values of δt are found by an appropriate transformation

* If $\mathbf{x}^*$ is itself inadmissible for $t = t^*$, no calculations will be performed at this point.

of the admissible values of u. The expression in Eq. (8.18) can be minimized by a procedure analogous to that for $\delta t = dt$; the optimum value of u is obtained from the optimum value of δt by using the equation for u as a function of δt derived from Eq. (8.16) or (8.17).

When the optimum admissible value of u and the corresponding cost in each of the three intervals of u has been obtained, a comparison is made of the three costs. The minimum cost of the three is stored as $I(\mathbf{x}^*, t^*)$, and the corresponding value of u is stored as $\hat{u}(\mathbf{x}^*, t^*)$, the optimal control at $\mathbf{x}^*, t^*$.

8.4 THE PROCESSING OF BLOCKS

In this program a block covers two increments along each state variable axis ($w_i\,\Delta x_i = 2\,\Delta x_i$, $i = 1, 2, 3, 4$) and five increments in $t(S\,\Delta t = 5\,\Delta t)$; each block thus contains $81 \cdot 6 = 486$ quantized values of $(\mathbf{x}, t)$ at which optimal control and minimum cost are computed. In the time interval $t_f - 5\,\Delta t \le t \le t_f$, the first block processed is centered at the origin of $\mathbf{x}$-space ($x_1 = 0, x_2 = 0, x_3 = 0, x_4 = 0$). The computations in this block are done by an initialization procedure that is specialized to each particular problem.

The next set of blocks processed is the layer of blocks surrounding this first block. The $\mathbf{x}$-coordinates of the center of these blocks are of the form

$$\begin{aligned} x_1 &= 2h_1\,\Delta x_1 \\ x_2 &= 2h_2\,\Delta x_2 \\ x_3 &= 2h_3\,\Delta x_3 \\ x_4 &= 2h_4\,\Delta x_4 \end{aligned} \tag{8.19}$$

where each $h_i = 1, 0$ or -1, $i = 1, 2, 3, 4$. The block $h_i = 0$, $i = 1, 2, 3, 4$, is of course, the one block that has already been processed. The remaining blocks within this layer are taken in order of the ternary number defined as

$$h = h_1'3^3 + h_2'3^2 + h_3'3^1 + h_4'3^0 \tag{8.20}$$

where

$$\begin{aligned} h_i' &= h_i, \qquad & h_i &= 0 \quad \text{or} \quad 1 \\ h_i' &= 2, & h_i &= -1. \end{aligned}$$

The first block processed in this layer is thus ($h_1 = 0, h_2 = 0, h_3 = 0, h_4 = 1$), the second block is ($h_1 = 0, h_2 = 0, h_3 = 0, h_4 = -1$), the third block is ($h_1 = 0, h_2 = 0, h_3 = 1, h_4 = 0$), etc. The last block is ($h_1 = -1, h_2 = -1, h_3 = -1, h_4 = -1$). There are thus a total of 80 blocks in this layer.

The equivalent layer for a two-dimensional problem is shown as the shaded area nearest the origin in Fig. 8.3. The blocks are numbered in the order they are processed, where the initial block centered at the origin is numbered 0. In this case there are 8 blocks in this layer.

The next layer of blocks has center coordinates of the same form as Eq. (8.19), except that now each $h_i = -2, -1, 0, +1$, or $+2$, $i = 1, 2, 3, 4$.

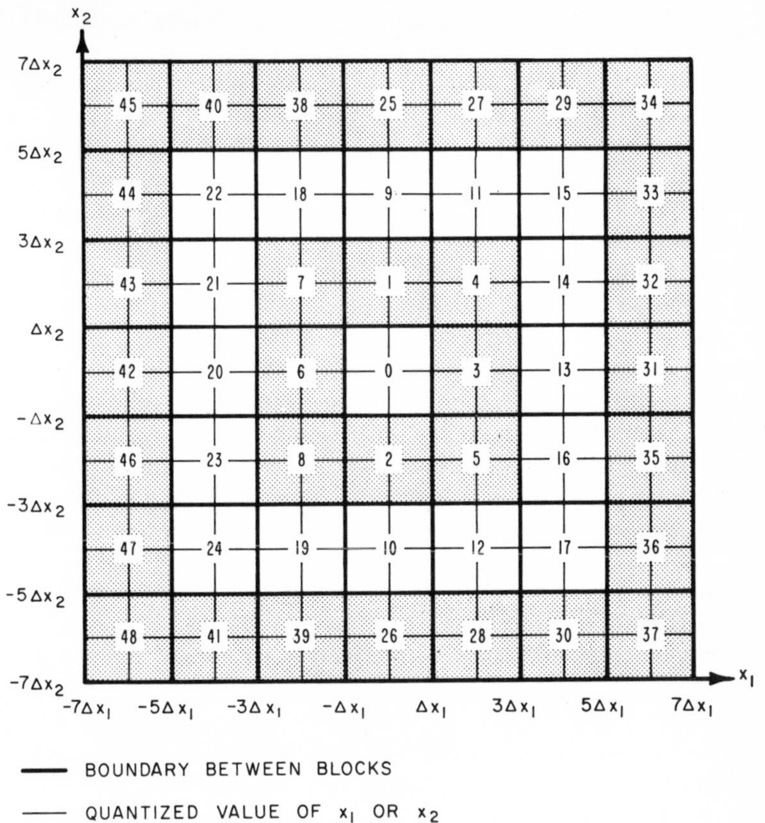

Fig. 8.3. Processing of blocks in two-dimensional case.

Blocks with all $h_i = -1$, 0, or $+1$, $i = 1, 2, 3, 4$, have already been processed, and they need not be considered again. The remaining blocks are taken in order of the base 5 number

$$h = h_1'5^3 + h_2'5^2 + h_3'5^1 + h_4'5^0 \tag{8.21}$$

where

$$\begin{aligned} h_i' &= h_i, & h_i &= 0, 1, \text{ or } 2 \\ h_i' &= 3, & h_i &= -1 \\ h_i' &= 4, & h_i &= -2. \end{aligned}$$

In this case the number h is *not* necessarily the numerical order in which the block is processed; however, the *relative* order is determined by the relative values of h, excluding expansions with all $h_i = -1, 0$, or 1. There are 544 blocks in this layer.

For $n = 2$ this layer is the unshaded area surrounding the shaded area nearest the origin in Fig. 8.3. Again the ordering of the blocks in this layer is listed on the blocks. There are 16 blocks in this layer for this case.

The processing of the next layer follows a similar rule using expansions of numbers to the base 7. There are 1776 blocks in this layer. In the two-dimensional case these blocks are shown as the outer shaded area in Fig. 8.3. The ordering within this layer is again specified on the blocks. For $n = 2$ there are 24 blocks in this layer.

The program has been written to accommodate only four such layers. The total numbers of quantized states at a given value of t is $15^4 = 50{,}625$; the high-speed memory requirement for such a problem using conventional dynamic programming is thus in excess of 50,000 storage locations. For the block size specified here, the total number of points in an entire block is 486, which is an upper bound on the high-speed memory requirement. The program could be expanded to cover more layers without much difficulty; if this were done, the high-speed memory requirement for the conventional procedure would increase dramatically, while for this program it would remain at or below 486 locations.

Each time a particular block is processed, the values of $\mathbf{x}$ for which minimum costs have already been computed during the processing of neighboring blocks are determined. These costs are used in the interpolation formulas as in the previous section, and no new computations are performed at these states. As a boundary condition for these calculations $I(\mathbf{x}, t_f)$ is specified; in this program it is taken to be $x^T Q(t_f)\mathbf{x}$, where $Q(t_f) = 0$ is an admissible choice.

After all the computations in the interval $t_f - 5\,\Delta t \leq t \leq t_f$ have been made, the minimum costs at $t = t_f - 5\,\Delta t$ are stored in low-speed memory for retrieval as the boundary conditions for blocks in the interval $t_f - 10\,\Delta t \leq t \leq t_f - 5\,\Delta t$. The block with $(h_1 = 0, h_2 = 0, h_3 = 0, h_4 = 0)$ is then computed according to the specified initialization procedure, and all calculations take place just as in the previous interval of t. Computations continue in this manner until $t = t_0$ is reached.

8.5 A LINEAR EXAMPLE

One problem to which the program has been applied is the unconstrained linear example with system equation

$$\begin{aligned} \dot{x}_1 &= x_2 \\ \dot{x}_2 &= x_3 \\ \dot{x}_3 &= x_4 \\ \dot{x}_4 &= u \end{aligned} \tag{8.22}$$

and performance criterion

$$J = \int_0^1 (\mathbf{x}^T\mathbf{x} + u^2)\, d\sigma. \tag{8.23}$$

This problem can be written directly in the form of Eqs. (8.6) and (8.8), where

$$\begin{aligned} Q(t) &= I \\ R(t) &= 1 \\ \mathbf{a}^T(t) &= 0 \\ b(t) &= 1. \end{aligned} \tag{8.24}$$

Thus, no linearization as in Sec. 8.2 is required.

This problem is particularly worthy of study because the system equation in Eq. (8.22) is unstable—i.e., with a bounded input function it is possible to have an unbounded function for one or more state variables.* The numerical difficulties in an unstable system are more pronounced than in a stable system [Ref. 6], and thus the degree of accuracy attained in this problem is pessimistic relative to what can be expected in the latter case.

As mentioned before, there is a closed-form expression for the solution to this problem in the unconstrained case, namely, the well-known Riccati equation [Refs. 2, 4]. This equation can be derived in a straightforward manner. If the increment δt is set equal to a fixed value Δt, the iterative equation for the general linear case, with an m-dimensional control vector u and an n-dimensional state vector $\mathbf{x}$, becomes

$$I(\mathbf{x}, t) = \min_{\mathbf{u}} \{[\mathbf{x}^T Q(t)\mathbf{x} + \mathbf{u}^T R(t)\mathbf{u}]\, \Delta t + I(\mathbf{x} + (A(t)\mathbf{x} + B(t)\mathbf{u})\, \Delta t, t + \Delta t]\} \tag{8.25}$$

where all quantities are as in Eqs. (8.5) and (8.6), except that $\mathbf{x}(t)$ and $\mathbf{u}(t)$ are now n-dimensional and m-dimensional vectors, respectively, $Q(t)$ and $A(t)$ are time-varying $n \times n$ matrices, $R(t)$ is a time-varying $m \times m$ matrix, and $B(t)$ is a time-varying $n \times m$ matrix. As a boundary condition for this relation,

$$I(\mathbf{x}, t_f) = \mathbf{x}^T Q(t_f)\mathbf{x}. \tag{8.26}$$

The closed-form solution results from showing by induction that the form of $I(\mathbf{x}, t)$ is

$$I(\mathbf{x}, t) = \mathbf{x}^T P(t)\mathbf{x} \tag{8.27}$$

where $P(t)$ is a time-varying $n \times n$ matrix. This is clearly true for $t = t_f$, since

$$P(t_f) = Q(t_f). \tag{8.28}$$

* This is only one possible definition of stability. However, this system is unstable under almost all other definitions. The reader interested in further study of stability is referred to Ref. 5.

It then suffices to show that if Eq. (8.27) is true for $t + \Delta t$, then it is true for t. This is accomplished by re-writing Eq. (8.25) as

$$I(\mathbf{x}, t) = \min_{\mathbf{u}} \{[\mathbf{x}^T Q(t)\mathbf{x} + \mathbf{u}^T R(t)\mathbf{u}]\,\Delta t$$
$$+ [\mathbf{x} + (A(t)\mathbf{x} + B(t)\mathbf{u})\,\Delta t]^T P(t + \Delta t)[\mathbf{x} + (A(t)\mathbf{x} + B(t)\mathbf{u})\,\Delta t]\} \quad (8.29)$$

The value of u that accomplishes the minimization is obtained as

$$\mathbf{u}(\mathbf{x}, t) = -G(t)\mathbf{x} \quad (8.30)$$

where

$$G(t) = [R(t)\,\Delta t + B^T(t)P(t + \Delta t)B(t)(\Delta t)^2]^{-1}$$
$$\times\ [B^T(t)P(t + \Delta t)(I + A(t)\,\Delta t)](\Delta t).$$

The minimum cost is then found to be

$$I(\mathbf{x}, t) = \mathbf{x}^T P(t)\mathbf{x} \quad (8.31)$$

where

$$P(t) = Q(t)\,\Delta t + [I + A(t)\,\Delta t]^T P(t + \Delta t)[I + A(t)\,\Delta t - B(t)G(t)\,\Delta t].$$

The proof by induction is thus completed.

The iterative use of Eqs. (8.30) and (8.31) with Eq. (8.28) as a boundary condition is clearly far simpler than the application of any dynamic programming procedure that involves quantization of **x** and **u** and a direct search. This is clearly an important computational result, and extensions of this approach to nonlinear problems, notably quasilinearization [Ref. 2] and second-order gradient methods [Ref. 3], have been made. However, further consideration of this topic is beyond the scope of this book, and the interested reader is referred to Refs. 2 and 4 for more details on the proof, such as conditions for its applicability,* and on the related computational procedures. The only use of this result in this chapter is to obtain the exact solution of the unconstrained problem defined in Eqs. (8.22) and (8.23) for comparison with the state increment dynamic programming solution.

The program is run exactly as in Secs. 8.2 to 8.4. The increment sizes used in the program are

$$\begin{aligned} \Delta x_1 &= 0.2 \\ \Delta x_2 &= 0.2 \\ \Delta x_3 &= 0.2 \\ \Delta x_4 &= 0.2 \\ \Delta t &= 0.2. \end{aligned} \quad (8.32)$$

* Notably, the matrices $Q(t)$ and $R(t)$ are required to be positive semi-definite and definite respectively.

The boundary condition is taken to be $Q(t_f) = 0$. The initialization in the first block ($h_1 = 0, h_2 = 0, h_3 = 0, h_4 = 0$) is done by using the Riccati equation directly.

The computer time required for processing the first 81 blocks (two layers) is approximately 10 minutes on an IBM 360/65 system. The time required for the first 625 blocks (three layers) is about 90 minutes on an IBM 360/65.

The program was actually run for the first three layers and for certain regions in outer layers. In the remainder of this section a number of complete

Table 8.1. State Increment Dynamic Programming Trajectory from $\mathbf{x}^T(0) = (0.4, 0.4, 0.4, 0.4)$

t	x_1	x_2	x_3	x_4	$\hat{u}$	Accum. cost
0.0000	0.4000	0.4000	0.4000	0.4000	−0.3912	0.0000
0.2000	0.4800	0.4800	0.4800	0.3128	−0.2353	0.1586
0.4000	0.5760	0.5760	0.5444	0.2747	−0.1240	0.3286
0.4417	0.6000	0.5987	0.5558	0.2695	−0.1101	0.3724
0.4440	0.6014	0.6000	0.5564	0.2693	−0.1093	0.3750
0.6000	0.6950	0.6868	0.5984	0.2522	−0.0485	0.5491
0.6062	0.6992	0.6905	0.6000	0.2519	−0.0472	0.5576
0.7521	0.8000	0.7780	0.6367	0.2450	−0.0092	0.7606
0.7866	0.8268	0.8000	0.6452	0.2447	−0.0026	0.8196
0.8000	0.8376	0.8086	0.6485	0.2447	0.0000	0.8438
1.0000	0.9993	0.9383	0.6974	0.2447	0.0000	1.2109

trajectories recovered from these results are presented and compared with the true solution.

The first trajectory has initial state $x_1(0) = 0.4$, $x_2(0) = 0.4$, $x_3(0) = 0.4$, $x_4(0) = 0.4$. This initial state is interesting because all the state variables have the same sign, and hence in the absence of control the variables x_1, x_2, and x_3 will grow rapidly with t. The effect of control is to retard this growth somewhat by decreasing x_4. The cost of the computed trajectory is very slightly greater than the cost of the true optimal trajectory; this difference can be entirely accounted for by the interpolation and extrapolation errors.

The computed trajectory and the true optimal trajectory from this initial state are shown in Tables 8.1 and 8.2. The first column contains t; in the true solution t is always a multiple of $\Delta t = 0.2$ sec, while in the state increment dynamic programming solution other values of t are obtained from use of Eq. (8.14). The next four columns contain the values of x_1, x_2, x_3, and x_4 at these values of t. The sixth column is the optimal control for this value of t. The seventh column is the accumulated cost; the last entry in this column is thus the total cost of the trajectory.

Table 8.2. True Optimal Trajectory from $\mathbf{x}^T(0) = (0.4, 0.4, 0.4, 0.4)$

t	x_1	x_2	x_3	x_4	$\hat{u}$	Accum. cost
0.0000	0.4000	0.4000	0.4000	0.4000	−0.3911	0.0000
0.2000	0.4800	0.4800	0.4800	0.3218	−0.2353	0.1586
0.4000	0.5760	0.5760	0.5444	0.2747	−0.1240	0.3286
0.6000	0.6912	0.6849	0.5993	0.2499	−0.0481	0.5387
0.8000	0.8282	0.8047	0.6493	0.2403	0.0000	0.8129
1.0000	0.9891	0.9346	0.6973	0.2403	0.0000	1.1755

Table 8.3. State Increment Dynamic Programming Trajectory from $\mathbf{x}^T(0) = (0.2, 0.2, 0.2, 0.4)$

t	x_1	x_2	x_3	x_4	$\hat{u}$	Accum. Cost
0.0000	0.2000	0.2000	0.2000	0.4000	−0.3426	0.0000
0.2000	0.2400	0.2400	0.2800	0.3315	−0.2173	0.0795
0.4000	0.2880	0.2960	0.3463	0.2880	−0.1217	0.1496
0.5864	0.3432	0.3606	0.4000	0.2653	−0.0545	0.2220
0.6000	0.3481	0.3660	0.4036	0.2646	−0.0509	0.2285
0.6842	0.3789	0.4000	0.4259	0.2603	−0.0339	0.2698
0.7369	0.4000	0.4224	0.4396	0.2585	−0.0121	0.2990
0.8000	0.4266	0.4501	0.4559	0.2578	0.0000	0.3367
1.0000	0.5167	0.5413	0.5075	0.2578	0.0000	0.4685

Table 8.4. True Optimal Trajectory from $\mathbf{x}^T(0) = (0.2, 0.2, 0.2, 0.4)$

t	x_1	x_2	x_3	x_4	$\hat{u}$	Accum. Cost
0.0000	0.2000	0.2000	0.2000	0.4000	−0.3426	0.0000
0.2000	0.2400	0.2400	0.2800	0.3314	−0.2173	0.0795
0.4000	0.2880	0.2960	0.3463	0.2880	−0.1217	0.1497
0.6000	0.3472	0.3652	0.4039	0.2636	−0.0507	0.2273
0.8000	0.4202	0.4460	0.4566	0.2535	0.0000	0.3251
1.0000	0.5094	0.5373	0.5073	0.2535	0.0000	0.4548

The second trajectory has similar properties. The initial state is $x_1(0) = 0.2$, $x_2(0) = 0.2$, $x_3(0) = 0.2$, $x_4(0) = 0.4$. The computed trajectory and the corresponding true optimal trajectory are shown in Tables 8.3 and 8.4.

The initial state for the third trajectory is $x_1(0) = 1.2$, $x_2(0) = -0.4$, $x_3(0) = -0.4$, $x_4(0) = -0.4$. The computed trajectory and the corresponding true optimal trajectory are shown in Tables 8.5 and 8.6.

Table 8.5. State Increment Dynamic Programming Trajectory from $\mathbf{x}^T(0) = (1.2, -0.4, -0.4, -0.4)$

t	x_1	x_2	x_3	x_4	$\hat{u}$	Accum. cost
0.0000	1.2000	−0.4000	−0.4000	−0.4000	0.3888	0.0000
0.2000	1.1200	−0.4800	−0.4800	−0.3222	0.2356	0.4142
0.4000	1.0240	−0.5760	−0.5444	−0.2752	0.1242	0.7892
0.4417	1.0000	−0.5987	−0.5559	−0.2699	0.1103	0.8628
0.4440	0.9986	−0.6000	−0.5565	−0.2697	0.1095	0.8670
0.6000	0.9050	−0.6868	−0.5986	−0.2526	0.0486	1.1402
0.6055	0.9012	−0.6901	−0.6000	−0.2523	0.0475	1.1496
0.7522	0.8000	−0.7781	−0.6370	−0.2454	0.0092	1.0411
0.7866	0.7733	−0.8000	−0.6454	−0.2451	0.0026	1.4599
0.8000	0.7625	−0.8087	−0.6487	−0.2450	0.0000	1.4829
1.0000	0.6008	−0.9384	−0.6977	−0.2450	0.0000	1.8262

The fourth example is somewhat different from the previous three trajectories. In this case the state variables have alternating signs, so that with no control the system will tend to drive states x_1, x_2, and x_3 to zero. It would be expected that the magnitude of the control is somewhat less than for the previous three cases. For values of $\mathbf{x}$ that are at comparable distances from the origin, this is in fact observed.

The fourth trajectory has initial state $x_1(0) = 0.6$, $x_2(0) = -0.6$, $x_3(0) = 0.6$, $x_4(0) = -0.6$. The computed trajectory and the corresponding true optimal trajectory are shown in Tables 8.7 and 8.8. In this case the cost of the computed trajectory is slightly lower than the cost of the true optimal trajectory; this is readily accounted for by interpolation errors.

The fifth trajectory is also from an initial state vector whose components have alternating signs, namely $x_1(0) = 1.6$, $x_2(0) = -1.6$, $x_3(0) = 1.6$, $x_4(0) = -1.6$. The behavior of this trajectory is similar to that of the previous case. The computed trajectory and the corresponding true optimal trajectory appear in Tables 8.9 and 8.10.

Table 8.6. True Optimal Trajectory from $\mathbf{x}^T(0) = (1.2, -0.4, -0.4, -0.4)$

t	x_1	x_2	x_3	x_4	u	Accum. cost
0.0000	1.2000	−0.4000	−0.4000	−0.4000	0.3889	0.0000
0.2000	1.1200	−0.4800	−0.4800	−0.3222	0.2356	0.4144
0.4000	1.0240	−0.5760	−0.5444	−0.2751	0.1242	0.7892
0.6000	0.9088	−0.6849	−0.5995	−0.2503	0.0481	1.1428
0.8000	0.7718	−0.8048	−0.6495	−0.2407	0.0000	1.4868
1.0000	0.6109	−0.9347	−0.6977	−0.2407	0.0000	1.8312

Table 8.7. State Increment Dynamic Programming Trajectory from $\mathbf{x}^T(0) = (0.6, -0.6, 0.6, -0.6)$

t	x_1	x_2	x_3	x_4	$\hat{u}$	Accum. cost
0.0000	0.6000	−0.6000	0.6000	−0.6000	0.3292	0.0000
0.2000	0.4800	−0.4800	0.4800	−0.5342	0.2496	0.3096
0.3497	0.4081	−0.4081	0.4000	−0.4968	0.1846	0.4652
0.3696	0.4000	−0.4002	0.3901	−0.4931	0.1790	0.4806
0.3700	0.3998	−0.4000	0.3899	−0.4930	0.1789	0.4809
0.4000	0.3879	−0.3883	0.3752	−0.4877	0.1705	0.5033
0.6000	0.3102	−0.3133	0.2776	−0.4536	0.0872	0.6450
0.7711	0.2566	−0.2658	0.2000	−0.4387	0.0111	0.7280
0.8000	0.2489	−0.2600	0.1873	−0.4383	0.0000	0.7387
0.9880	0.2000	−0.2248	0.1049	−0.4383	0.0000	0.8058
1.0000	0.1973	−0.2235	0.0007	−0.4383	0.0000	0.8093

Table 8.8. True Optimal Trajectory from $\mathbf{x}^T(0) = (0.6, -0.6, 0.6, -0.6)$

t	x_1	x_2	x_3	x_4	$\hat{u}$	Accum. cost
0.0000	0.6000	−0.6000	0.6000	−0.6000	0.3292	0.0000
0.2000	0.4800	−0.4800	0.4800	−0.5341	0.2496	0.3097
0.4000	0.3840	−0.3840	0.3731	−0.4842	0.1692	0.5174
0.6000	0.3072	−0.3093	0.2763	−0.4503	0.0866	0.6569
0.8000	0.2453	−0.2541	0.1862	−0.4330	0.0000	0.7523
1.0000	0.1945	−0.2168	0.0996	−0.4330	0.0000	0.8217

Table 8.9. State Increment Dynamic Programming Trajectory from $\mathbf{x}^T(0) = (1.6, -1.6, 1.6, -1.6)$

t	x_1	x_2	x_3	x_4	$\hat{u}$	Acum. cost
0.0000	1.6000	−1.6000	1.6000	−1.6000	0.8209	0.0000
0.1250	1.4000	−1.4000	1.4000	−1.4974	0.7011	1.3642
0.2000	1.2950	−1.2950	1.2877	−1.4448	0.6326	2.0103
0.2607	1.2164	−1.2168	1.2000	−1.4064	0.5289	2.4655
0.2728	1.2017	−1.2023	1.1830	−1.4000	0.5153	2.5462
0.4000	1.0490	−1.0522	1.0050	−1.3346	0.4607	3.3720
0.4467	1.0000	−1.0054	0.9427	−1.3132	0.4274	3.6142
0.5556	0.8911	−0.9036	0.8000	−1.2671	0.3063	4.1327
0.6000	0.8510	−0.8681	0.7438	−1.2535	0.2652	4.3080
0.6587	0.8000	−0.8244	0.6701	−1.2379	0.1859	4.5237
0.6951	0.7700	−0.8000	0.6251	−1.2311	0.1430	4.6450
0.7155	0.7537	−0.7872	0.6000	−1.2282	0.1099	4.7095
0.8000	0.6872	−0.7366	0.4962	−1.2189	0.0000	4.9687
0.8789	0.6290	−0.6974	0.4000	−1.2189	0.0000	5.1855
0.9205	0.6000	−0.6807	0.3493	−1.2189	0.0000	5.2906
1.0000	0.5459	−0.6530	0.2524	−1.2189	0.0000	5.4838

Table 8.10. True Optimal Trajectory from $\mathbf{x}^T(0) = (1.6, -1.6, 1.6, -1.6)$

t	x_1	x_2	x_3	x_4	$\hat{u}$	*Accum. cost*
0.0000	1.6000	−1.6000	1.6000	−1.6000	0.8678	0.0000
0.2000	1.2800	−1.2800	1.2800	−1.4244	0.6657	2.2021
0.4000	1.0240	−1.0240	0.9951	−1.2913	0.4513	3.6796
0.6000	0.8192	−0.8250	0.7368	−1.2010	0.2310	4.6713
0.8000	0.6542	−0.6776	0.4966	−1.1548	0.0000	5.3494
1.0000	0.5187	−0.5783	0.2657	−1.1548	0.0000	5.8429

Table 8.11. State Increment Dynamic Programming Trajectory from $\mathbf{x}^T(0) = (1.6, -1.6, 1.6, -0.8)$

t	x_1	x_2	x_3	x_4	$\hat{u}$	*Accum. cost*
0.0000	1.6000	−1.6000	1.6000	−0.8000	0.3062	0.0000
0.1250	1.4000	−1.4000	1.5000	−0.7617	0.3049	1.0517
0.2000	1.2950	−1.2875	1.4429	−0.7389	0.2884	1.5650
0.2580	1.2203	−1.2038	1.4000	−0.7221	0.2702	1.9157
0.2749	1.2000	−1.1801	1.3878	−0.7176	0.2634	2.0082
0.4000	1.0524	−1.0066	1.2981	−0.6846	0.2281	2.6765
0.4524	1.0000	−0.9388	1.2623	−0.6728	0.2042	2.9014
0.5449	0.9131	−0.8220	1.2000	−0.6539	0.1585	3.2687
0.5631	0.8981	−0.8000	1.1880	−0.6510	0.1511	3.3309
0.6000	0.8686	−0.7563	1.1641	−0.6454	0.1325	3.4525
0.6908	0.8000	−0.6506	1.1055	−0.6334	0.0715	3.7353
0.7366	0.7702	−0.6000	1.0765	−0.6301	0.0366	3.8586
0.8000	0.7321	−0.5317	1.0365	−0.6278	0.0000	4.0179
0.8582	0.7012	−0.4714	1.0000	−0.6278	0.0000	4.1509
0.9296	0.6676	−0.4000	0.9552	−0.6278	0.0000	4.3014
1.0000	0.6394	−0.3327	0.9110	−0.6278	0.0000	4.4361

Table 8.12. True Optimal Trajectory from $\mathbf{x}^T(0) = (1.6, -1.6, 1.6, -0.8)$

t	x_1	x_2	x_3	x_4	$\hat{u}$	*Accum. cost*
0.0000	1.6000	−1.6000	1.6000	−0.8000	0.2896	0.0000
0.2000	1.2800	−1.2800	1.4400	−0.7421	0.2671	1.6808
0.4000	1.0240	−0.9920	1.2916	−0.6887	0.2125	2.8753
0.6000	0.8256	−0.7337	1.1538	−0.6462	0.1243	3.7193
0.8000	0.6789	−0.5029	1.0246	−0.6213	0.0000	4.3162
1.0000	0.5782	−0.2980	0.9004	−0.6213	0.0000	4.7461

The sixth trajectory has initial state $x_1(0) = 1.6$, $x_2(0) = -1.6$, $x_3(0) = 1.6$, $x_4(0) = -0.8$. The computed trajectory and the corresponding true optimal trajectory appear in Tables 8.11 and 8.12.

The seventh trajectory has initial state $x_1(0) = 1.2$, $x_2(0) = -0.8$, $x_3(0) = -0.4$, $x_4(0) = 0.4$. The computed trajectory and the corresponding true optimal trajectory are shown in Tables 8.13 and 8.14.

Table 8.13. State Increment Dynamic Programming Trajectory from $\mathbf{x}^T(0) = (1.2, -0.8, -0.4, 0.4)$

t	x_1	x_2	x_3	x_4	$\hat{u}$	Accum. cost
0.0000	1.2000	−0.8000	−0.4000	0.4000	−0.1881	0.0000
0.2000	1.0400	−0.8800	−0.3200	0.3624	−0.1613	0.4871
0.2455	1.0000	−0.8945	−0.3035	0.3550	−0.1503	0.5833
0.4000	0.8618	−0.9415	−0.2487	0.3318	−0.1162	0.8987
0.4656	0.8000	−0.9578	−0.2269	0.3242	−0.0922	1.0177
0.5486	0.7205	−0.9766	−0.2000	0.3165	−0.0781	1.1606
0.6000	0.6703	−0.9869	−0.1837	0.3125	−0.0601	1.2439
0.6712	0.6000	−1.0000	−0.1615	0.3082	−0.0319	1.3549
0.8000	0.4712	−1.0208	−0.1218	0.3041	0.0000	1.5457
0.8698	0.4000	−1.0292	−0.1005	0.3041	0.0000	1.6414
1.0000	0.2660	−1.0423	−0.0609	0.3041	0.0000	1.8135

Table 8.14. True Optimal Trajectory from $\mathbf{x}^T(0) = (1.2, -0.8, -0.4, 0.4)$

t	x_1	x_2	x_3	x_4	$\hat{u}$	Accum. cost
0.0000	1.2000	−0.8000	−0.4000	0.4000	−0.1882	0.0000
0.2000	1.0400	−0.8800	−0.3200	0.3624	−0.1613	0.4872
0.4000	0.8640	−0.9440	−0.2475	0.3301	−0.1156	0.9104
0.6000	0.6752	−0.9935	−0.1815	0.3070	−0.0590	1.2746
0.8000	0.4765	−1.0298	−0.1201	0.2952	0.0000	1.5893
1.0000	0.2705	−1.0538	−0.0610	0.2952	0.0000	1.8670

Table 8.15. State Increment Dynamic Programming Trajectory from $\mathbf{x}^T(0) = (-0.8, 0.8, -0.8, -0.4)$

t	x_1	x_2	x_3	x_4	$\hat{u}$	Accum. cost
0.0000	−0.8000	0.8000	−0.8000	−0.4000	0.4434	0.0000
0.2000	−0.6400	0.6400	−0.8800	−0.3113	0.2650	0.4553
0.2455	−0.6109	0.6000	−0.8941	−0.2993	0.2398	0.5354
0.2636	−0.6000	0.5837	−0.8996	−0.2949	0.2300	0.5659
0.4000	−0.5204	0.4611	−0.9398	−0.2635	0.1344	0.7909
0.4650	−0.4904	0.4000	−0.9569	−0.2548	0.1077	0.8854
0.6000	−0.4364	0.2708	−0.9913	−0.2403	0.0462	1.0734
0.6361	−0.4267	0.2351	−1.0000	−0.2386	0.0390	1.1205
0.6711	−0.4184	0.2000	−1.0084	−0.2372	0.0319	1.1660
0.7632	−0.4000	0.1071	−1.0302	−0.2343	0.0071	1.2847
0.8000	−0.3961	0.0692	−1.0388	−0.2340	0.0000	1.3321
0.8666	−0.3914	0.0000	−1.0544	−0.2340	0.0000	1.4184
1.0000	−0.3914	−0.1406	−1.0856	−0.2340	0.0000	1.5944

The eighth and final trajectory obtained from this program has initial state $x_1(0) = -0.8$, $x_2(0) = 0.8$, $x_3(0) = -0.8$, $x_4(0) = -0.4$. The computed trajectory and the corresponding true optimal trajectory are shown in Tables 8.15 and 8.16.

Table 8.16. True Optimal Trajectory from $\mathbf{x}^T(0) = (-0.8, 0.8, -0.8, -0.4)$

t	x_1	x_2	x_3	x_4	$\hat{u}$	*Accum. cost*
0.0000	−0.8000	0.8000	−0.8000	−0.4000	0.4434	0.0000
0.2000	−0.6400	0.6400	−0.8800	−0.3113	0.2650	0.4554
0.4000	−0.5120	0.4640	−0.9423	−0.2583	0.1326	0.8075
0.6000	−0.4192	0.2756	−0.9939	−0.2318	0.0446	1.0974
0.8000	−0.3641	0.0768	−1.0403	−0.2229	0.0000	1.3564
1.0000	−0.3487	−0.1313	−1.0849	−0.2229	0.0000	1.6105

8.6 A CONSTRAINED EXAMPLE

Another problem to which the program has been applied is that in which the system equations and performance criterion are the same as in the previous section, but constraints are present. These system equations are

$$\begin{aligned} \dot{x}_1 &= x_2 \\ \dot{x}_2 &= x_3 \\ \dot{x}_3 &= x_4 \\ \dot{x}_4 &= u. \end{aligned} \tag{8.33}$$

The performance criterion is

$$J = \int_0^1 (\mathbf{x}^T\mathbf{x} + u^2)\, d\sigma. \tag{8.34}$$

The constraint on the control variable is

$$-0.2 \le u \le 0.2. \tag{8.35}$$

The constraints on the state variables are

$$\begin{aligned} -0.6 &\le x_1 \le 0.6 \\ -0.6 &\le x_2 \le 0.6 \\ -0.6 &\le x_3 \le 0.6 \\ -0.6 &\le x_4 \le 0.6. \end{aligned} \tag{8.36}$$

This problem has all the numerical difficulties cited on the previous section. However, it is not possible to write down a closed-form solution as in the previous section.

The increment sizes used in the program were the same as in the previous section, namely

$$\begin{aligned} \Delta x_1 &= 0.2 \\ \Delta x_2 &= 0.2 \\ \Delta x_3 &= 0.2 \\ \Delta x_4 &= 0.2 \\ \Delta t &= 0.2. \end{aligned} \tag{8.37}$$

Initialization was handled as in the previous section.

Table 8.17. State Increment Dynamic Programming Solution for Constrained Problem with Initial State $\mathbf{x}^T(0) = (0.2, 0.2, 0.2, 0.4)$

t	x_1	x_2	x_3	x_4	$\hat{u}$	*Accum. cost*
0.0000	0.2000	0.2000	0.2000	0.4000	−0.2000	0.0000
0.2000	0.2400	0.2400	0.2800	0.3600	−0.2000	0.0640
0.4000	0.2880	0.2960	0.3520	0.3200	−0.1340	0.1366
0.5500	0.3324	0.3488	0.4000	0.2999	−0.0707	0.1988
0.6000	0.3498	0.3688	0.4150	0.2964	−0.0570	0.2232
0.6751	0.3776	0.4000	0.4373	0.2921	−0.0417	0.2624
0.7312	0.4000	0.4245	0.4537	0.2897	−0.0132	0.2950
0.8000	0.4292	0.4557	0.4736	0.2888	0.0000	0.3383
1.0000	0.5203	0.5504	0.5313	0.2888	0.0000	0.4782

The time required to carry out the calculation of a single optimal control was increased only slightly over the time for the unconstrained case.

The optimal trajectories for two of the initial states used in the previous section are shown in Tables 8.17 and 8.18. The trajectory for initial state $x_1(0) = 0.2$, $x_2(0) = 0.2$, $x_3(0) = 0.2$ $x_4(0) = 0.4$, is shown in Table 8.17.

The control is at the constrained value $\hat{u} = -0.2$ for the first 0.4 seconds. The corresponding trajectory for the unconstrained case appears in Table 8.3. It can be seen that cost is increased over the unconstrained case from 0.4686 to 0.4782.

The trajectory for initial state $x_1(0) = 0.6$, $x_2(0) = -0.6$, $x_3(0) = 0.6$, $x_4(0) = -0.6$ is shown in Table 8.18. Again, the control is at the constrained value $\hat{u} = 0.2$, for the first 0.4 seconds. The corresponding trajectory for the unconstrained case appears in Table 8.7. The cost over the unconstrained problem is increased from 0.8093 to 0.8108.

Table 8.18. State Increment Dynamic Programming Solution for Constrained Problem with Initial State $\mathbf{x}^T(0) = (0.6, -0.6, 0.6, -0.6)$

t	x_1	x_2	x_3	x_4	$\hat{u}$	Accum. cost
0.0000	0.6000	−0.6000	0.6000	−0.6000	0.2000	0.0000
0.2000	0.4800	−0.4800	0.4800	−0.5600	0.2000	0.2960
0.3429	0.4114	−0.4114	0.4000	−0.5314	0.2000	0.4453
0.3706	0.4000	−0.4003	0.3852	−0.5259	0.2000	0.4681
0.3714	0.3997	−0.4000	0.3848	−0.5257	0.2000	0.4687
0.4000	0.3883	−0.3890	0.3698	−0.5200	0.1829	0.4911
0.6000	0.3105	−0.3151	0.2658	−0.4834	0.0930	0.6396
0.7361	0.2676	−0.2789	0.2000	−0.4708	0.0246	0.7088
0.8000	0.2498	−0.2661	0.1699	−0.4692	0.0000	0.7352
0.9869	0.2000	−0.2343	0.0822	−0.4692	0.0000	0.8066
1.0000	0.1969	−0.2333	0.0761	−0.4692	0.0000	0.8108

8.7 CONCLUSIONS

The results presented in this chapter demonstrate the feasibility of applying state increment dynamic programming to high-dimensional problems. The program described here is capable of obtaining very accurate solutions to a significant class of optimization problems having four or less state variables; these solutions are obtained in a reasonable amount of computing time on a large, high-speed computer (IBM 360/65). At the present time this program is being run on other examples with four state variables; some results for these cases are reported in Ref. 7. Also, the expansion of the program to handle a larger number of state variables is planned for the near future.

REFERENCES

1. R. E. Kalman, "On the General Theory of Control Systems," *Proc. 1st IFAC Congress*, Moscow, USSR, pp. 481–92 (Butterworths Press, New York, N.Y., 1960).
2. R. E. Bellman and R. E. Kalaba, *Quasilinearization and Nonlinear Boundary-Value Problems* (American Elesevier Publishing Co., Inc., New York, N.Y., 1965).
3. J. V. Breakwell, J. L. Speyer, and A. E. Bryson, "Optimization and Control of Nonlinear Systems Using the Second Variation," *J. SIAM Control*, ser. A. vol. 1, No. 2, March, 1963, pp. 193–223.
4. R. E. Kalman and R. W. Koepcke, "Optimal Synthesis of Linear Sampling Control Systems Using Generalized Performance Indexes," *Trans. ASME*, vol. 80, No. 6, November, 1958, pp. 1820–38.
5. J. LaSalle and S. Lefschetz, *Stability by Liapunov's Direct Method with Applications* (Academic Press, New York, N.Y., 1961).
6. R. E. Bellman, *Dynamic Programming* (Princeton University Press, Princeton, N.J., 1957).
7. R. E. Larson and W. G. Keckler, "Computation of Optimum Feedback Control Using State Increment Dynamic Programming," submitted to *J. SIAM Control*.

Chapter Nine

APPLICATIONS TO GUIDANCE AND CONTROL PROBLEMS

9.1 INTRODUCTION

One problem area to which state increment dynamic programming has been extensively applied is the optimum guidance and control of aerospace vehicles. These problems have stimulated much work in the area of system optimization, and they continue to be of considerable interest today.

Guidance and control problems can generally be put into the formulation of Sec. 3.2 without much difficulty. Typically, the state variables are position and velocity coordinates, the control variables are propulsion and aerodynamic forces, and the system equations are the differential equations obtained by summing forces on the body being controlled. The performance criterion is generally either some measure of how much cost is involved in carrying out a particular mission or how successful a particular mission is. Constraints are usually related to physical limitations on the system.

The distinction between the functions of guidance and control is largely a matter of determining on what level the optimization is performed. Guidance is commonly associated with the determination of optimal trajectories over the entire mission, while control is more concerned with maintaining the system on a given trajectory.

In the next three sections two guidance applications are discussed. The first application is the computation of minimum-fuel trajectories for the supersonic transport (SST). Attention is given both to evaluation of a given aircraft design and to on-line computation of a single trajectory. The second application treated is minimum-time-to-intercept trajectories for an anti-missile missile.

In the following section the general control problem is considered in more detail. In the section after that several control examples are worked out.

9.2 MINIMUM-FUEL TRAJECTORIES FOR THE SUPERSONIC TRANSPORT

In this section the application of state increment dynamic programming to the computation of minimum fuel trajectories for the supersonic transport (SST) [Refs. 1, 2, 3] is considered. This problem is of considerable interest because the cost of fuel is estimated to be as much as 60 percent of the total cost of a flight [Ref. 1]; hence, knowledge of minimum fuel trajectories is an important factor in determining the economic feasibility of these aircraft. In order to evaluate different configurations, a computer program was developed to compute the optimal trajectory from any possible initial condition using any airframe/engine combination [Refs. 1, 2]. In order to study a particular configuration, it is necessary only to insert tabular data for the airframe and the engine into the program. The program can be adapted to treat more conventional aircraft, such as the jets currently in use.

The equations of motion used to represent the aircraft are derived from quasi steady-state assumptions. These assumptions have led to good results in applications very similar to this problem [Ref. 4 Chapter 6; Ref. 5]. Under these assumptions, the forces on an aircraft with an angle of attack α flying at an angle γ to the horizontal, as in Fig. 9.1, can be resolved into components along and perpendicular to the direction of flight, as follows:

$$F = \frac{W}{g}\frac{dv}{dt} = T\cos\alpha - D - W\sin\gamma \tag{9.1}$$

$$0 = L - T\sin\alpha - W\cos\gamma \tag{9.2}$$

where W = weight of aircraft, including fuel
g = acceleration due to gravity
v = velocity of aircraft
h = altitude
T = engine thrust
D = drag
L = lift
γ = flight path angle with respect to horizontal
α = angle of attack
r = range.

The following differential equations can be written to describe the motion of the aircraft:

$$\frac{dv}{dt} = \frac{g}{W}(T\cos\alpha - D - W\sin\gamma) \tag{9.3}$$

$$\frac{dh}{dt} = v\sin\gamma \tag{9.4}$$

$$\frac{dr}{dt} = v\cos\gamma \tag{9.5}$$

$$\frac{dW}{dt} = W_F \tag{9.6}$$

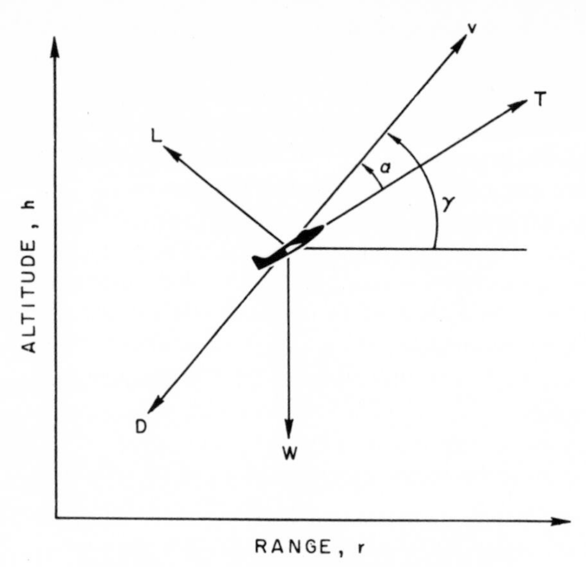

Fig. 9.1. Force diagram for aircraft.

where W_F is the fuel flow of the engines. The fuel flow is a function of thrust, velocity, altitude, and weight of the aircraft.

$$W_F = W_F(T, v, h, W). \tag{9.7}$$

This dependence for a given engine is completely specified by the tabular data for that engine. Similarly, drag can be written as a function of weight, velocity, altitude, and flight-path angle.

$$D = D(W, v, h, \gamma). \tag{9.8}$$

This equation is obtained by noting that lift is a function of velocity, altitude, and angle of attack,

$$L = L(v, h, \alpha) \tag{9.9}$$

and then setting angle of attack to satisfy Eq. (9.2).

The performance criterion is the total amount of fuel consumed during the flight, which is to be minimized. This criterion can be expressed as

$$J = \int_{t_0}^{t_f} W_F(T, v, h, W)\, d\sigma. \tag{9.10}$$

The fuel consumed, J, is to be minimized subject to a number of constraints. The first set of constraints determines the maximum and minimum allowed values of thrust and flight-path angle:

$$0 \le \alpha_1^-(v, h) \le T \le \alpha_1^+(v, h), \tag{9.11}$$

$$\alpha_2^-(v, h) \le \gamma \le \alpha_2^+(v, h). \tag{9.12}$$

The functions $\alpha_1^-(v, h)$ and $\alpha_1^+(v, h)$ are specified by the engine data, and the functions $\alpha_2^-(v, h)$ and $\alpha_2^+(v, h)$ are specified by the airframe data.

Another set of constraints limits the region of v-h space in which the aircraft can fly. First, there is a maximum operational altitude and maximum operational speed. Second, for a given altitude, if the speed is too low, drag will be greater than maximum thrust; and if speed is too high, aerodynamic heating of the aircraft becomes too great. These constraints can be summarized as:

$$0 \leq h \leq \beta_2^+, \tag{9.13}$$

$$\beta_1^-(h) \leq v \leq \beta_1^+(h). \tag{9.14}$$

The functions $\beta_1^-(h)$ and $\beta_1^+(h)$ are determined from considering both the engine and airframe data. The region of the v-h plane which satisfies both Eq. (9.13) and (9.14) is called the flight envelope. With some airframe/engine combinations, the flight envelope varies with time, and the boundary in Eq. (9.14) must then be constantly recomputed throughout the process.

As part of the study it was desired to insert still other constraints, such as maximum sonic boom and air traffic control restrictions. It was also desired to consider flights under variable weather conditions, including temperature and wind variations; emergency conditions, including loss of one or more engines, loss of cabin pressurization, or loss of trim drag control; and other conditions.

The aircraft is further constrained to fly to some specified set of terminal values of range, altitude, and velocity. These values can correspond to the entrance to the air traffic control zone at the destination airport, to the aircraft landed on the ground at the destination, or to any other desired condition.

The formulation of the problem according to state increment dynamic programming proceeds as follows. The system equations, Eqs. (9.3) through (9.6) are reduced to two equations by considering the range, r, to be the independent variable rather than time, t. However, since it is desirable to know the time to the end of the flight, as well as the distance to the end, the differential equations, Eqs. (9.3) and (9.4), are used as the system equations, and the range covered by a given flight is related to the time consumed by Eq. (5). The two-dimensional state vector thus consists of velocity and altitude.

$$\mathbf{x} = \begin{pmatrix} v \\ h \end{pmatrix}. \tag{9.15}$$

The control variables are taken to be thrust and flight path angle:

$$\mathbf{u} = \begin{pmatrix} T \\ \gamma \end{pmatrix}. \tag{9.16}$$

The state transformation equation is taken to be as follows:

$$\mathbf{x}(t + \delta t) = \mathbf{x}(t) + \mathbf{f}(\mathbf{x}, \mathbf{u}, t)\,\delta t \tag{9.17}$$

where

$$\mathbf{f}(\mathbf{x}, \mathbf{u}, t) = \begin{pmatrix} \dfrac{dv}{dt} \\ \\ \dfrac{dh}{dt} \end{pmatrix} = \begin{pmatrix} \dfrac{g}{W}(T\cos\alpha - D) - g\sin\gamma \\ \\ v\sin\gamma \end{pmatrix}. \tag{9.18}$$

The last of Eqs. (9.3)–(9.6) is used in the performance criterion.

$$J = \int_{t_0}^{t_f} l(\mathbf{x}, \mathbf{u}, \sigma)\,d\sigma = \int_{t_0}^{t_f} \frac{dW}{dt}\,dt = \int_{t_0}^{t_f} W_f\,dt. \tag{9.19}$$

The boundaries of the allowed state space, X, are determined by taking the maximum altitude as 90,000 ft, and the maximum velocity as Mach 3.00. The increment sizes are as follows:

$$\begin{aligned} \Delta x_1 &= \Delta v = \text{Mach } 0.02 \\ \Delta x_2 &= \Delta h = 1{,}000 \text{ ft.} \end{aligned} \tag{9.20}$$

The blocks are taken to be two increments in velocity by two increments in altitude by 10 nautical miles (nm) in range.

$$\begin{aligned} w_1\,\Delta x_1 &= 2\,\Delta v = \text{Mach } 0.04 \\ w_2\,\Delta x_2 &= 2\,\Delta h = 2{,}000 \text{ ft.} \\ \Delta R &= 10 \text{ nm.} \end{aligned} \tag{9.21}$$

Within this range, the flight envelope is determined by Eqs. (9.13) and (9.14). A typical flight envelope is shown as the shaded area of Fig. 9.2.

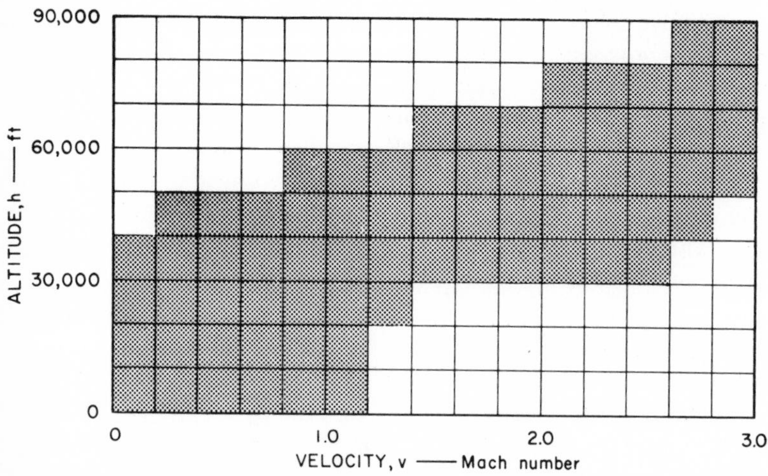

Fig. 9.2. Shape of flight envelope for a typical Mach 3 aircraft.

Since there are two control variables and two state variables, the conditions for complete state-transition control are met (see Chapter 5). Furthermore, a choice of U, the set of admissible controls, which satisfies the conditions for complete state-transition control, has been used in previous aircraft studies with good results [Ref. 4, Chapter 6; Ref. 5]. Consequently, this procedure can be used here.

The set of admissible controls consists of the following: climb at constant velocity, dive at constant velocity, accelerate at constant altitude, decelerate at constant altitude, and cruise at constant altitude and velocity. The maximum thrust, $T^+ = \beta^+$ from Eq. (9.13), is used for the controls climb and accelerate; minimum thrust, $T^- = \beta^-$ from Eq. (9.13), is used for the controls dive and decelerate; and thrust is set equal to drag for the control cruise. The flight path angle, γ, is chosen as $\gamma = 0$ for accelerate, decelerate, and cruise; and it is set at that value which keeps $dv/dt = 0$ in Eq. (9.3) for climb and dive. The five admissible control functions are then as follows:

$$U = [\mathbf{u}^{(-2)}, \mathbf{u}^{(-1)}, \mathbf{u}^{(0)}, \mathbf{u}^{(1)}, \mathbf{u}^{(2)}]$$
$$= \left\{ \begin{bmatrix} T^- \\ \sin^{-1}\left(\dfrac{T^-\cos\alpha - D}{W}\right) \end{bmatrix}, \begin{bmatrix} T^- \\ 0 \end{bmatrix}, \begin{bmatrix} D \\ 0 \end{bmatrix}, \begin{bmatrix} T^+ \\ 0 \end{bmatrix}, \begin{bmatrix} T^+ \\ \sin^{-1}\left(\dfrac{T^+\cos\alpha - D}{W}\right) \end{bmatrix} \right\} \tag{9.22}$$

where the five controls correspond to dive, decelerate, cruise, accelerate, and climb, respectively.

The set of admissible final states contains one element that specifies the terminal altitude and the terminal velocity. The weight of the aircraft at the final time, W_0, is also specified. The terminal state can be chosen to correspond to a landed position, the entrance to a holding pattern, or any other desired condition.

The initial set of optimal points uses a technique which has been extensively investigated [Refs. 4, 5]. In this technique, if the terminal state is $v(t_f) = 0$, $h(t_f) = 0$, then the optimal trajectory of the aircraft from any state to the terminal state is calculated under the restriction that the only admissible controls are a maximum-rate dive and a maximum-rate deceleration. The optimal points so generated are optimal in the sense that at the state where the optimal point in this set is generated, any other admissible control applied at $t \geq \bar{t}$, where $\bar{t}$ is the time of the optimal point, either fails to get the aircraft to the terminal state at $t = t_f$ or has a higher cost than the optimal control. A slight modification is necessary if the terminal state is not the origin.

An optimal point consists of the normal data for an optimal point listed in Chapter 5, plus the time to the end of flight. The index of an optimal point is determined by numbering the points consecutively as they are generated. Since in the case of complete state-transition control the next state is always an existing optimal point, the number of the next optimal point is also listed. In addition, the total weight, W, of the aircraft plus fuel is listed in the optimal point rather than the fuel consumed. The fuel consumed can be found by subtracting from W the quantity W_0, the weight at the final time.

The information at an optimal point thus has the following form:

$$\mathbf{p}(d) = (a, b, r, t, W, \hat{k}, c), \tag{9.23}$$

where $v = a\,\Delta v =$ velocity at point
$h = b\,\Delta h =$ altitude at point
$r =$ range at point
$t =$ time to the end of flight of point
$W =$ weight of aircraft plus fuel
$= W_0$ plus minimum fuel burned in reaching terminal state from point
$\hat{k} =$ index of optimal control; one of $-2, -1, 0, 1, 2$
$c =$ number of next optimal point
$d =$ number of this optimal point.

It is known from previous studies [Refs. 1, 2, 4] that as fuel is consumed and the weight of the aircraft plus fuel decreases, the optimal altitude and velocity at which the aircraft should fly both increase. Thus, except for the terminal descent portion of the flight, optimal trajectories tend to be in the direction of increasing altitude and increasing velocity. As a result, a preferred direction of motion for the problem is well defined.

The blocks in a given time interval are processed in an order such that interblock transitions in the preferred directions of motion take place without constraint. Specifically, all the blocks at a given altitude are processed consecutively in order of decreasing velocity, and the altitudes are taken in order from the highest to the lowest. Only blocks in the flight envelope are processed. In Fig. 9.2 the top row of blocks is computed first, then successively lower rows until the row at $h = 0$ is reached. Each row is processed from right to left, and only shaded blocks are computed.

Within each block, the computations follow the procedure described in Chapter 5. The change in range for control k, $\delta r^{(k)}$, is related to the corresponding change in time, $\delta t^{(k)}$, by

$$\begin{aligned} \delta r^{(k)} &= v\,\delta t^{(k)}, \qquad (k = 0, -2, 2) \\ \delta r^{(1)} &= (v + \tfrac{1}{2}\,\Delta v)\,\delta t^{(1)} \\ \delta r^{(-1)} &= (v - \tfrac{1}{2}\,\Delta v)\,\delta t^{(-1)}. \end{aligned} \tag{9.24}$$

The interpolation in t is performed by computing the cost of holding altitude and velocity constant over the interval. This amounts to introducing an "incremental cruise" into the flight. This incremental cruise is used for interpolation purposes only, and it is subtracted out from the actual optimal point.

The high-speed memory requirement for a single block is, on the average, 10 locations for each state on the boundary with a previously computed block plus one location for each of the other states in the block, a total of 54 locations. Actually, since this number is so small, all the optimal points along the boundary at the top altitude of a row of blocks are stored in the high-speed memory, so that if the initial points for the blocks in this row are read into the high-speed memory, then the entire row of blocks can be computed without further reference to the slow memory. At the same time optimal points along the lowest altitude of the row are retained in the high-speed memory to serve as the points on the boundary at the upper altitude of the row of blocks immediately below. Thus, referral to the high-speed memory takes place only once for each row of blocks, and the transfer of relatively few points from the slow memory to the high-speed memory makes possible the computation of a large number of optimal points. The amount of high-speed memory needed to implement this procedure is only about 1,100 locations.

A computer program to carry out these computations was written [Refs. 1, 2]. Data for a specific aircraft* were used in the program, and a portion of the program output is shown in Table 9.1. Part of the optimal trajectory from a

Table 9.1. Page of Print-Out from Program with $\Delta h = 2{,}000$ ft, $\Delta M = 0.04\,M$, $0 \leq h \leq 70{,}000$ ft, $0 \leq M \leq 3.00$, Weight of Empty Aircraft $= 80{,}000$ lb

Alt.	*Mach* $\times 10^{-2}$	*Dist.*	*Time*	*Weight*	*PPT*	*NPT*	*Opt. cont.*
60	288	482.4	1218.7	89178.1	9104	9057	Climb
58	288	484.0	1222.3	89256.6	9108	9104	Climb
60	260	514.2	1306.2	90329.6	9112	9065	Climb
60	256	522.0	1325.4	90564.2	9116	9075	Climb
50	252	502.1	1285.2	90605.9	9122	8418	Cruise
54	252	503.0	1287.3	90471.5	9124	8958	Dive
54	256	499.4	1276.2	90328.0	9125	8950	Dive
52	260	500.2	1278.2	90343.1	9126	8950	Decel.
50	260	501.3	1280.8	90510.7	9127	8954	Decel.
68	292	486.3	1226.5	88973.9	9135	8825	Cruise
68	288	496.2	1247.6	89264.3	9158	9019	Climb
70	284	501.8	1260.0	89402.8	9162	9019	Accel.
66	280	490.3	1236.3	89339.2	9164	9022	Climb
68	276	494.3	1245.5	89445.9	9169	9022	Accel.
70	268	496.9	1256.0	89595.8	9172	8854	Accel.
68	268	502.1	1268.2	89746.0	9174	9172	Climb
64	292	492.3	1237.3	89160.4	9187	9041	Cruise
62	296	484.8	1223.0	88991.4	9193	9103	Cruise
62	284	488.0	1231.1	89313.7	9202	9061	Climb
64	280	493.1	1242.9	89441.9	9206	9164	Climb
62	280	495.5	1248.2	89535.7	9210	9206	Climb
64	276	498.0	1253.8	89611.3	9215	9206	Accel.
62	276	500.4	1259.4	89706.1	9218	9215	Climb
64	252	524.2	1331.0	90580.6	9222	9079	Climb

* A proposed high-performance Mach 3 fighter.

Table 9.1. *Continued*

Alt.	*Mach* × 10^{-2}	*Dist.*	*Time*	*Weight*	*PPT*	*NPT*	*Opt. cont.*
62	252	527.4	1339.2	90700.7	9226	9222	Climb
60	296	486.6	1227.0	89073.8	9232	9193	Climb
58	296	488.2	1228.8	89139.9	9241	8899	Cruise
60	284	489.9	1235.5	89399.7	9245	9202	Climb
58	284	491.6	1239.3	89479.8	9249	9245	Climb
60	280	497.6	1252.9	89623.5	9253	9210	Climb
58	280	499.3	1256.7	89705.3	9257	9253	Climb
60	252	530.2	1346.2	90811.3	9261	9226	Climb
56	284	493.0	1242.6	89557.8	9266	9249	Climb
54	284	494.3	1251.9	89716.9	9271	8646	Cruise
68	296	487.8	1228.4	88957.8	9276	9144	Accel.
70	296	484.0	1220.9	88827.8	9281	9147	Cruise
68	300	485.1	1221.5	88861.8	9282	9156	Dive
70	292	493.5	1241.6	89085.5	9286	8995	Cruise
70	300	487.5	1228.2	88845.4	9294	9281	Decel.
66	288	499.6	1255.1	89376.4	9296	9158	Climb
68	284	506.2	1269.8	89534.2	9298	9162	Climb
66	284	509.7	1277.6	89648.6	9302	9298	Climb
66	276	498.0	1254.0	89565.4	9306	9169	Climb
70	276	493.6	1245.0	89394.7	9310	9025	Accel.
70	280	512.0	1282.7	89674.6	9312	9162	Accel.
66	268	506.1	1277.7	89874.5	9314	9174	Climb
68	264	510.6	1288.3	89988.7	9319	9174	Accel.
70	264	511.1	1289.6	89963.9	9322	9172	Accel.
66	264	514.9	1298.4	90122.6	9234	9319	Climb
62	300	487.8	1225.9	89015.6	9330	9037	Cruise
62	292	494.6	1245.0	89250.2	9334	9244	Cruise
64	300	489.6	1230.1	88947.5	9338	9047	Cruise
64	296	495.6	1243.6	89139.4	9343	9181	Cruise

specific initial condition is shown in Table 9.2. In both of these figures, the first column is the altitude in thousands of feet; the second column is Mach number in hundredths; the third column is the distance to end of flight in nautical miles; the fourth column is time to end of flight in seconds; the fifth column is weight in pounds of aircraft plus fuel, from which the total fuel consumed can be computed by subtracting 80,000 lb, the weight of the aircraft with no fuel on board; the sixth column is the index of the present point (not all consecutive numbers appear because this particular program assigns numbers to each possible control, not just the control that results in an optimum point); the seventh column is the index of the next point; and the last column is the optimal control that is to be performed.

On the basis of the running time for this program, a complete specification of all possible optimal trajectories would take about 1 hour of computing time for each 1,000 nm in range on the IBM 7090.

Table 9.2. Portion of an Optimal Trajectory as Determined by Program in Table 9.1

Alt.	*Mach* × 10^{-2}	*Dist.*	*Time*	*Weight*	*PPT*	*NPT*	*Opt. cont.*
56	272	508.6	1278.3	90089.2	9443	9410	Climb
58	272	507.1	1274.7	90008.9	9410	9405	Accel.
58	276	504.2	1268.1	89876.7	9405	9401	Climb
60	276	502.5	1264.1	89794.0	9401	9218	Climb
62	276	500.4	1259.4	89706.1	9218	9215	Climb
64	276	498.0	1253.8	89611.3	9215	9206	Accel.
64	280	493.1	1242.9	89441.9	9206	9164	Climb
66	280	490.3	1236.3	89339.2	9164	9022	Climb
68	280	486.7	1228.4	89224.0	9022	9012	Accel.
68	284	479.4	1212.6	89007.9	9012	8847	Accel.
68	288	472.1	1196.1	88793.9	8847	8825	Accel.
68	292	465.0	1180.6	88582.1	8825	8821	Accel.
68	296	458.0	1165.7	88372.8	8821	8682	Cruise
68	296	448.6	1145.8	88201.4	8682	8667	Cruise

9.3 MODIFICATIONS FOR ON-LINE CONTROL OF THE SUPERSONIC TRANSPORT

In connection with the study of the previous section, it was decided to develop a computer program for on-line control of the supersonic transport. Because of the large computing time, it was not possible to use the existing program directly.

One method of utilizing the existing program is to store all the results from it in a low-speed memory and to look up the optimal control for the present state during the flight. However, this procedure has the disadvantage that the constraints and conditions under which the computations were made, such as air traffic control constraints and weather conditions, may have changed considerably by the time it is desired to use the results.

An alternative procedure has been indicated in Sec. 6.6. In this procedure, computations are performed only in those blocks that are located in a region in which the optimal trajectory is expected to lie. By restricting the computations to such a region, the number of calculations can be decreased to the extent that the computations are feasible in real time.

In the aircraft problem the region was chosen to cover 10,000 ft. in altitude by 0.3 Mach. This area is large enough so that the region need not be determined accurately in advance, but small enough so that control can be computed in real time. A computing rate of 1,000 nm of trajectory per minute on the IBM 7090 was achieved for this procedure.

The region can be determined either from the results of previous off-line computations, from operating experience, or iteratively. One iterative approach that has been used in similar applications is to use coarse quantization in an initial computation covering the entire admissible state space and then to repeat the calculations in successively smaller regions using finer quantizations. This and other procedures are discussed in Ref. 6.

The region is completely specified by simply listing the blocks which lie in it; consequently, by processing only these blocks, it is easy to restrict the computations to the region. Furthermore, these computations yield the optimal control everywhere in the region, not just along a single trajectory. Thus, as long as the aircraft stays within the bounds of this region, and as long as the constraints and conditions under which the computations were made do not change very much, the results can be stored and used as needed. As a result, the computations usually need to be repeated only a few times during the entire flight.

9.4 MINIMUM-TIME-TO-INTERCEPT TRAJECTORIES FOR AN ANTI-MISSILE MISSILE

A program similar to that described in the previous two sections has been written for computing minimum-time-to-intercept trajectories for an anti-missile missile [Refs. 6, 7]. The problem is shown pictorially in Fig. 9.3. The objective is to find the minimum-time trajectory from the missile site, located at the origin of the coordinate system, to any point in three-dimensional

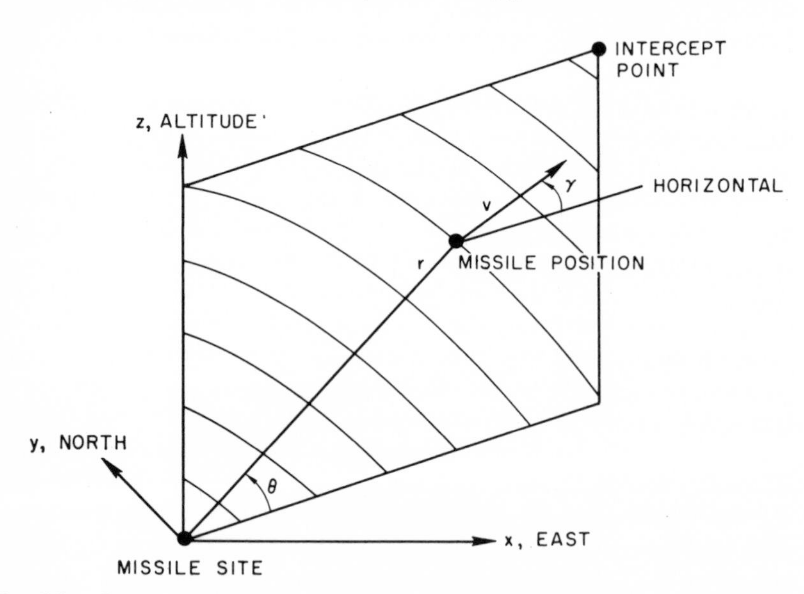

Fig. 9.3. Coordinates r, θ, v and γ in minimum-time-to-intercept problem.

space. It can be shown that, for the dynamic equations assumed in this problem, the minimum-time trajectory always lies in the plane containing the missile site and the intercept point. It is thus sufficient to solve the problem in the two spatial dimensions determined by this plane.

It is convenient to define the position and velocity coordinates of the missile as shown in Fig. 9.3. These coordinates are the distance along the vector from the missile site to the missile position (r), the angle of this vector with respect to horizontal (θ), the magnitude of missile velocity (v), and the angle of the velocity vector with respect to horizontal (γ). The distance r

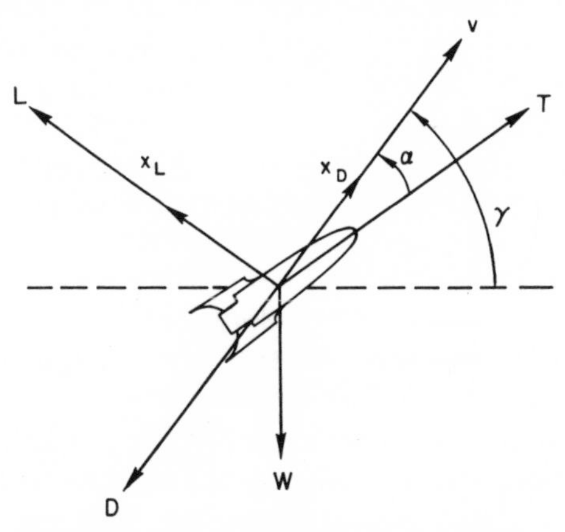

Fig. 9.4. Force diagram on missile.

is taken to be the stage variable, while the three remaining coordinates are the state variables. The single control variable is angle of attack, α; the thrust in this problem is taken to be a given function of time.

The system equations are derived from Fig. 9.4. These equations are written in a cartesian coordinate system, defined by displacement along the velocity vector (x_D) and the perpendicular to it (x_L). The system differential equations are thus

$$F_D = \frac{W}{g}\ddot{x}_D = T\cos\alpha - D - W\sin\gamma \tag{9.25}$$

$$F_L = \frac{W}{g}\ddot{x}_L = L - T\sin\alpha - W\cos\gamma \tag{9.26}$$

where the quantities W, D, L, and T are defined as in Sec. 9.2. The thrust and weight are given functions of time.

$$W = W(t) \tag{9.27}$$

$$T = T(t). \tag{9.28}$$

The lift is a function of altitude (z), velocity, and angle of attack.

$$L = L(v, z, \alpha) \tag{9.29}$$

where

$$z = r \sin \theta. \tag{9.30}$$

The drag is written as a function of weight, velocity, altitude, and angle of attack.

$$D = D(W, v, z, \alpha). \tag{9.31}$$

The system equations for a given control, α; stage, r; and state, θ, v, and γ, are updated by converting the quantities (r, θ, v, γ) to $(x_D, \dot{x}_D, x_L, \dot{x}_L)$, integrating Eqs. (9.25) and (9.26), and then converting the resulting values of $(x_D, \dot{x}_D, x_L, \dot{x}_L)$ to (r, θ, v, γ).

The performance criterion is the time required to reach a given value of r and θ. Formally,

$$J = \int_{t_0}^{t_f} d\sigma = t_f - t_0 \tag{9.32}$$

where t_f is the time that the value of r and θ is reached and t_0 is the time that the missile is launched.

Physical constraints on the missile are present, just as in Sec. 9.2.

Since there are three variables and only one control variable, the conditions for complete state-transition control are not met, and the procedure of Chapter 4 must be used. A program has been written to carry out the optimization calculations by this method [Refs. 6, 7]. The program has been used in two modes. In the first mode it was used to generate the minimum-time trajectory for taking the missile from the launch site to any reachable intercept point with any feasible terminal velocity vector. Because minimum-time trajectories from a single *initial* state to all possible *final* states are desired, it is advantageous to use the technique of *forward dynamic programming*. This technique is explained in Sec. 10.2. A detailed discussion of the program and a number of results from it appear in Ref. 7.

In the second mode the program was used to generate the minimum-time trajectory from a given initial point, not necessarily at the missile site, to a single intercept point. Applications of the program in this mode are treated in Ref. 6.

9.5 CONTROL OF TIME-INVARIANT SYSTEMS

The preceding three sections have discussed optimum guidance applications. In this section a class of control applications is defined. A number of examples of this type will be worked out in the next section.

The class of problems discussed in this section is characterized by the fact that the optimal control and minimum cost do not depend on the present time, t, but on the present state, $\mathbf{x}$. A necessary condition for this to occur is that the system equations, the integrand in the cost function, and the constraints do not depend explicitly on t. However, additional restrictions must be placed on the performance criterion for the minimum cost and optimal control to be time-invariant. Two performance criteria that meet that restriction are minimum time to drive the state to the origin and minimization of a general time-invariant integrand over an infinite time interval.

The minimum-time-to-origin problem occurs in many control applications. The interpretation of the problem is that the origin corresponds to the desired system configuration, as obtained from a guidance solution, and the objective is to return to this configuration as fast as possible. The origin is thus an equilibrium point of the system equation. If the system equation is

$$\dot{\mathbf{x}} = \mathbf{f}(\mathbf{x}, \mathbf{u}) \tag{9.33}$$

then the condition that the origin is an equilibrium point is

$$\mathbf{f}(\mathbf{0}, \mathbf{0}) = \mathbf{0} \tag{9.34}$$

where $\mathbf{0}$ = a vector with all components equal to zero. Often, the system equations are linearized about this point so that

$$\mathbf{f}(\mathbf{x}, \mathbf{u}) = A\mathbf{x} + B\mathbf{u} \tag{9.35}$$

where A and B are constant matrices.

The performance criterion is the time taken to reach the origin. If t_0 is the present time and if t_f is the time at which the state reaches the origin, then the cost function becomes

$$J = \int_{t_0}^{t_f} l(\mathbf{x}, \mathbf{u})\, d\sigma = \int_{t_0}^{t_f} d\sigma = t_f - t_0. \tag{9.36}$$

The constraints can have the form

$$\mathbf{x} \in X \tag{9.37}$$

$$\mathbf{u} \in U. \tag{9.38}$$

The minimization of a time-invariant integrand over an infinite time-interval also occurs frequently in control applications. Again, it is usually assumed that the origin corresponds to the solution of an optimum guidance problem. The use of an infinite time-interval of integration is justified if the response of the controlled system is significantly faster than the rate of change of the complete system. Generally this is insured by an appropriate choice of the integrand in the performance criterion.

For this problem the system equations are as in Eq. (9.33). Again, Eq. (9.34) is satisfied, and the system equations often take the form of Eq. (9.35).

The performance criterion is written as

$$J = \int_{t_0}^{\infty} l(\mathbf{x}, \mathbf{u})\, d\sigma. \tag{9.39}$$

The constraints are again as in Eqs. (9.37) and (9.38).

In these two cases the minimum cost function, I, depends only on the present state, $\mathbf{x}$.

$$I(\mathbf{x}) = \underset{\substack{\mathbf{u}(\sigma) \\ t \leq \sigma \leq t_f}}{\text{Min}} \left\{ \int_{t}^{t_f} l(\mathbf{x}, \mathbf{u})\, d\sigma \right\} \tag{9.40}$$

The principle of optimality then takes the form

$$I(\mathbf{x}) = \underset{\mathbf{u} \in U}{\text{Min}} \{ l(\mathbf{x}, \mathbf{u})\, \delta t + I(\mathbf{x} + \mathbf{f}(\mathbf{x}, \mathbf{u})\, \delta t) \}. \tag{9.41}$$

This problem requires less computations than does the general problem treated thus far because there is only one value of optimal control and minimum cost per state, rather than one value at each time increment per state. However, because the stage variable t has been suppressed, the computational procedures described in Chapters 2–6 cannot be applied directly.

One approach to solving this problem is Bellman's approximation in function space [Ref. 4]. In this method the procedure is to make an appropriate guess of $I(\mathbf{x})$, say $I^{(0)}(\mathbf{x})$, and solve for a sequence of minimum cost functions according to the relation

$$I^{(j+1)}(\mathbf{x}) = \min_{\mathbf{u}} \{ l(\mathbf{x}, \mathbf{u})\, \delta t + I^{(j)}[\mathbf{x} + \mathbf{f}(\mathbf{x}, \mathbf{u})\, \delta t] \}. \tag{9.42}$$

If $I^{(0)}(\mathbf{x})$ is a close approximation to the true $I(\mathbf{x})$, this procedure may be expected to converge to the proper function.

An alternative procedure is Bellman's approximation in policy space [Ref. 4]. This method, which has better convergence properties, is to guess an optimal policy, $\hat{\mathbf{u}}^{(0)}(\mathbf{x})$. The corresponding minimum cost function, $I^{(0)}(\mathbf{x})$, is then computed by a direct iteration according to the relation

$$I^{(0,i+1)}(\mathbf{x}) = l[\mathbf{x}, \hat{\mathbf{u}}^{(0)}(\mathbf{x})]\, \delta t + I^{(0,i)}\{\mathbf{x} + \mathbf{f}[\mathbf{x}, \hat{\mathbf{u}}^{(0)}(\mathbf{x})]\, \delta t\}. \tag{9.43}$$

The initial guess, $I^{(0,0)}(\mathbf{x})$, is usually

$$I^{(0,0)}(\mathbf{x}) = 0. \tag{9.44}$$

When $I^{(0)}(\mathbf{x})$ has been found from iteration of Eq. (9.43), a new policy, $\hat{\mathbf{u}}^{(1)}(\mathbf{x})$, is found by solving

$$I^{*}(\mathbf{x}) = \min_{\mathbf{u}} \{ l(\mathbf{x}, \mathbf{u})\, \delta t + I^{(0)}(\mathbf{x} + \mathbf{f}(\mathbf{x}, \mathbf{u})\, \delta t) \}. \tag{9.45}$$

The policy $\hat{\mathbf{u}}^{(1)}(\mathbf{x})$ for a given value of $\mathbf{x}$ is determined as the value of $\mathbf{u}$ for which the minimum is attained in Eq. (9.45). However, $I^{*}(\mathbf{x})$ is *not* $I^{(1)}(\mathbf{x})$, the minimum cost function corresponding to policy $\hat{\mathbf{u}}^{(1)}(\mathbf{x})$, because $I^{(0)}[\mathbf{x} + \mathbf{f}(\mathbf{x}, \mathbf{u})\,\delta t]$ appears inside the brackets in Eq. (9.45). Another direct iteration as in Eq. (9.43) is thus necessary.

In general, the minimum cost function $I^{(j)}(\mathbf{x})$ corresponding to the policy $\hat{\mathbf{u}}^{(j)}(\mathbf{x})$ is found by iterations of

$$I^{(j,i+1)}(\mathbf{x}) = l[\mathbf{x}, \hat{\mathbf{u}}^{(j)}(\mathbf{x})]\,\delta t + I^{(j,i)}\{\mathbf{x} + \mathbf{f}[\mathbf{x}, \hat{\mathbf{u}}^{(j)}(\mathbf{x})]\,\delta t\} \tag{9.46}$$

with initial guess

$$I^{(j,0)}(\mathbf{x}) = 0. \tag{9.47}$$

A new policy, $\hat{\mathbf{u}}^{(j+1)}(\mathbf{x})$, is then formed from

$$I^*(\mathbf{x}) = \min_{\mathbf{u}} \{l(\mathbf{x}, \mathbf{u})\,\delta t + I^{(j)}[\mathbf{x} + \mathbf{f}(\mathbf{x}, \mathbf{u})\,\delta t]\}. \tag{9.48}$$

Convergence to the true optimum can be proved for many important cases [Ref. 4].

Although the above procedures work well in many cases, they do retain some of the computational difficulties of other dynamic programming methods. In the first place, the high-speed memory requirement is just as severe a problem as in the conventional dynamic programming procedure, because $I^{(n-1)}(\mathbf{x})$ must be stored for all quantized values of $\mathbf{x}$ during the computation of $I^{(n)}(\mathbf{x})$. Also, if several iterations are required, the computing time may become a difficulty as well.

It is possible to bring state increment dynamic programming to bear in these problems in such a way that not only is the high-speed memory requirement reduced, but also the need for iterations is in many cases eliminated. The procedure will be described in detail in the next section in connection with a number of examples.

The basic idea is simply to construct blocks exactly as before, except that they are n-dimensional, rather than $(n + 1)$-dimensional, because there is no dependence on t in the problem. In the initialization procedure, however, minimum cost is specified only at some subset of the total state space. These initial computations can be done using one of the above iterative procedures or by any other suitable method. If the origin is an equilibrium point at which zero cost is incurred, as in the minimum-time-to-origin problem, then the subset is some small region about the origin. The blocks are then processed in the order opposite to the preferred direction, starting at this subset and radiating outwards until the entire state space is filled. In this manner, and provided that the extrapolations are made with suitable accuracy, optimal control and minimum cost are computed in a single pass through the state space. The high-speed memory requirement can be reduced to the size of a few or even just one block.

If the preferred direction of motion is not well known, it may be necessary to repeat the procedure in order to obtain sufficiently accurate results.

However, as will be shown in the next section, this is not usually necessary; even when it is, less iterations are required than if iteration in policy space is used. Of course, the saving in high-speed memory requirements, which is the most important consideration, is always obtained.

9.6 APPLICATIONS TO THE CONTROL OF TIME-INVARIANT SYSTEMS

The first problem to be discussed in this section is the problem of bringing the $1/s^2$ plant to the origin in minimum time with a bounded control.

The system equation is written as

$$\ddot{x} = u \tag{9.49}$$

which is converted to the standard form

$$\begin{aligned} \dot{x}_1 &= x_2 \\ \dot{x}_2 &= u. \end{aligned} \tag{9.50}$$

The performance criterion is minimum time to reach the origin.

$$J = \int_{t_0}^{t_f} d\sigma = t_f - t_0 \tag{9.51}$$

where t_f is determined by $x_1(t_f) = 0$, $x_2(t_f) = 0$.

The bounded control restriction is expressed as

$$|u| \leq 1. \tag{9.52}$$

The iterative equation from the principle of optimality can be written as

$$I(x_1, x_2) = \min_{u \in U} \{\delta t + I(x_1 + x_2\, \delta t, x_2 + u\, \delta t)\} \tag{9.53}$$

The quantization in x_1 and x_2 is done uniformly in increments of Δx_1 and Δx_2 respectively. The block size is chosen to be $2\,\Delta x_1$ by $2\,\Delta x_2$. The blocks themselves are arranged in hexagonal layers centered at the origin. These layers are shown in Fig. 9.5, where alternate layers are shaded.

The processing of blocks begins with the block centered at the origin. This block is taken first because the objective is to drive the system to this point, and hence the preferred direction of motion must be toward this block.

The next set of blocks processed is the second layer, which is not shaded in the figure. The first block taken within the layer is the upper right block. The remaining blocks in the layer are processed in counter-clockwise order. This order within a layer is chosen opposite to the preferred direction of motion; according to the first of the equations of motion, the change in x_1 is always in the direction of the sign of x_2, and hence *all* trajectories, including optimal trajectories, will tend to be in the clockwise direction.

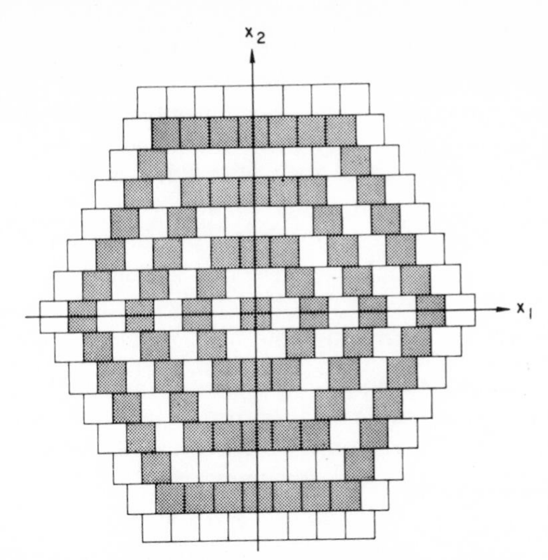

Fig. 9.5. Blocks for time-invariant systems.

If the blocks are processed one at a time, then the high-speed memory requirement is just 5, the points on the boundaries with blocks that have been already processed. (See Fig. 9.6.) However, this would mean that referral would have to be made to the low-speed memory after the computation of each block, which might be as few as 4 new points; clearly, this is not an efficient procedure.

A more reasonable procedure is to store the outer surface of the last layer computed; if this is done, the only points needed in computing a block are 2 from the block just computed plus 3 from this outer surface (see Fig. 9.6). Thus, with the points on this outer layer plus two points, there is never a need to refer to the low-speed memory. Furthermore, the points on the outer surface of the layer currently being computed can replace points from the

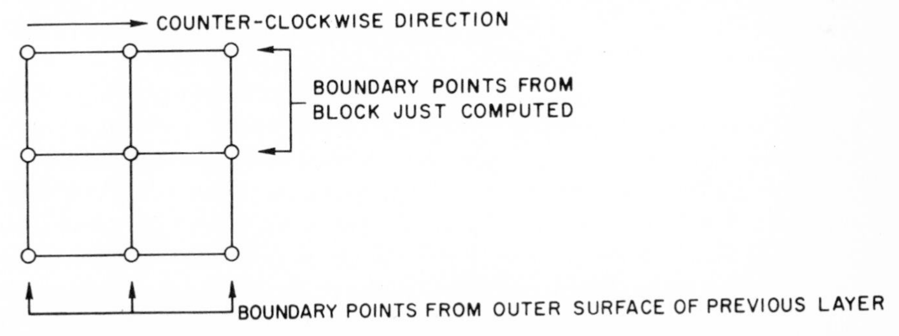

Fig. 9.6. High-speed memory requirement for computation of one block.

previous layer which are no longer needed in such a way that the outer points needed for the computation of the next layer are directly retained in the high-speed memory rather than being sent to the low-speed memory and then recalled.

In Table 9.3, a comparison is made of the high-speed memory requirement for the state increment dynamic programming procedure and for conventional dynamic programming. The comparison is made for several different values of the number of hexagonal layers (in Fig. 9.5 there are 8 layers, counting the block at the origin as the first layer). Two numbers are given for conventional dynamic programming: the first, which is the number of points

Table 9.3. Comparison of High-Speed Memory Requirement for State Increment Dynamic Programming (S.I.D.P.) and Conventional Dynamic Programming (C.D.P.) in First Example.

Number of layers	*Requirement for S.I.D.P.*	*Requirement for C.D.P. (referral to low-speed memory required once per layer)*	*Requirement for C.D.P. (no referral to low-speed memory required)*
7	106	737	1,474
8	122	937	1,874
9	138	1,161	2,322
10	154	1,409	2,818
11	170	1,681	3,362
12	186	1,977	3,954

contained in the complete hexagon, is the amount of storage necessary for a single iteration of the procedure; while the second, which is twice the first value, is the number of locations needed to store the results of the present iteration along with those of the previous iteration so that no referral has to be made to the low-speed memory in order to perform the next iteration. It is fair to compare the state increment dynamic programming number with this latter number, because in the state increment dynamic programming procedure there is never any need to refer to the low-speed memory during the computations. The improvement in high-speed memory requirements is thus seen to range from a factor of 14 up to 21 in this example. If referral to low-speed memory is allowed in the state increment dynamic programming procedure, it is possible to reduce the requirement even further by storing only a portion of the surface of the outer layer rather than the entire surface.

A computer program has been written for the SDS-910, a small computer, which implements the state increment dynamic programming procedure. This computer has a high-speed (core) storage of 8,000 words. However, if FORTRAN is used, 3,300 locations are reserved for it and its necessary subroutines. In addition, about 1,500 words are required to store a program which would implement an iterative conventional dynamic programming procedure. Consequently, if the number of points desired covers an area larger than that included in 11 layers of hexagons, then the high-speed memory requirement is exceeded. A program to implement state increment dynamic programming requires 2,500 locations. Therefore, in this case up to 162 layers, which corresponds to 651 quantization levels in each state variable could be handled.

The program was run several times with varying increment sizes in x_1, x_2, and u. The interpolation and extrapolation formulas in all cases were the simplest linear formulas. In this problem an exact solution can be obtained [Ref. 8]. The solutions from the program for optimal control and minimum cost agreed with the exact solution within the errors due to the inexact integration routine, the interpolation and extrapolation procedures and round-off.

One run was made with increment sizes $\Delta x_1 = 0.2$, $\Delta x_2 = 0.2$ and with the set of admissible controls $U = \{-1, +1\}$. The initial conditions for the run were obtained by using an iterative procedure to obtain minimum cost at the 9 quantized states in the block centered at the origin. The optimal control is shown in Fig. 9.7, and the minimum cost is shown in Fig. 9.8. These figures can be compared with the exact solution; optimal control is shown in Fig. 9.9, and minimum cost is shown in Fig. 9.10.

The control is seen to be a good approximation to the true solution. In both cases the plane can be divided into two regions separated by a switching line; the control is $+1$ below the line and -1 above it. The shape of the two switching lines is basically the same, namely parabolic, but the line for the computer run is flatter in x_1 than is the exact solution. This can be largely accounted for by a bias which the inexact integration formulas introduce.

In addition to this discrepency, there are isolated areas on both sides of the switching line where the computer run calculates the wrong value of optimal control. This is due mostly to the accumulation of errors in the minimum cost function which the simple interpolation and extrapolation formulas are not capable of smoothing out.

A comparison of the computed minimum cost with the exact minimum cost shows that, as a whole, the two agree fairly well. The percentage error in the last layer, which is the least accurate of all the layers, is an average of 16% error. The average error over the entire space is about 13%. This error is well within that expected on the basis of the errors due to round-off, the inexact integration routine, the simple interpolation and extrapolation formulas, the limited set of initial values, and the instability of the system being controlled.

The various sources of error were examined experimentally in some detail to see if their effects corresponded to those expected.

The first experiments were related to the initial conditions. It is clear that as the area in which values of minimum cost are determined accurately by an

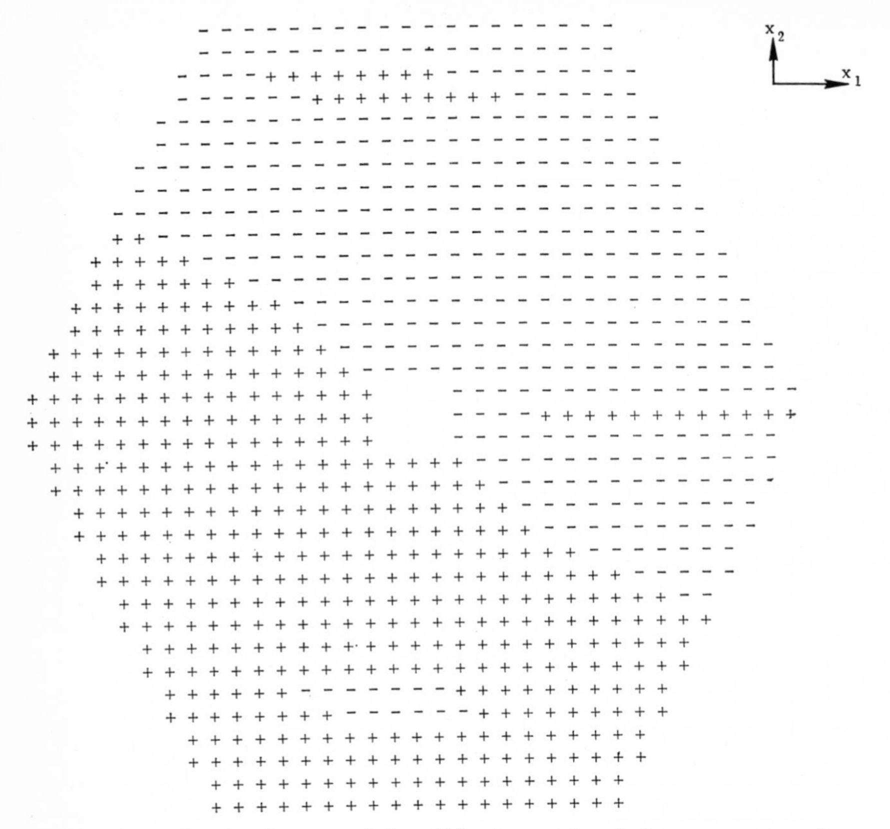

Fig. 9.7. Optimal control for $1/s^2$ plant, $|u| \leq 1$, $J = t_f - t_0$, as obtained from computer run with $\Delta x_1 = 0.2$, $\Delta x_2 = 0.2$, $U = \{-1, +1\}$, initial area = 1 layer. (+ = +1, − = −1).

initial calculation increases, the accuracy of the total result increases, but at the cost of additional computing time. In order to see how much accuracy improves as this region is increased, a calculation was made where the exact minimum costs were provided for both the first two layers of hexagons, a total of 41 values as compared to 9 in the original run. The optimal control and minimum cost for this run are shown in Figs. 9.11 and 9.12. The optimal control improves somewhat; the switching curve is slightly closer to the true switching curve and there are less isolated points where the wrong control is computed. The minimum cost also improves significantly; the average error computed over the same region as that covered by the original run* was 9.5% as compared with 13% for the original run.

* Although the points shown in Fig. 9.12 only cover 7 layers, the computations were actually made over 9 layers; the figure 9.5% refers to all 9 layers.

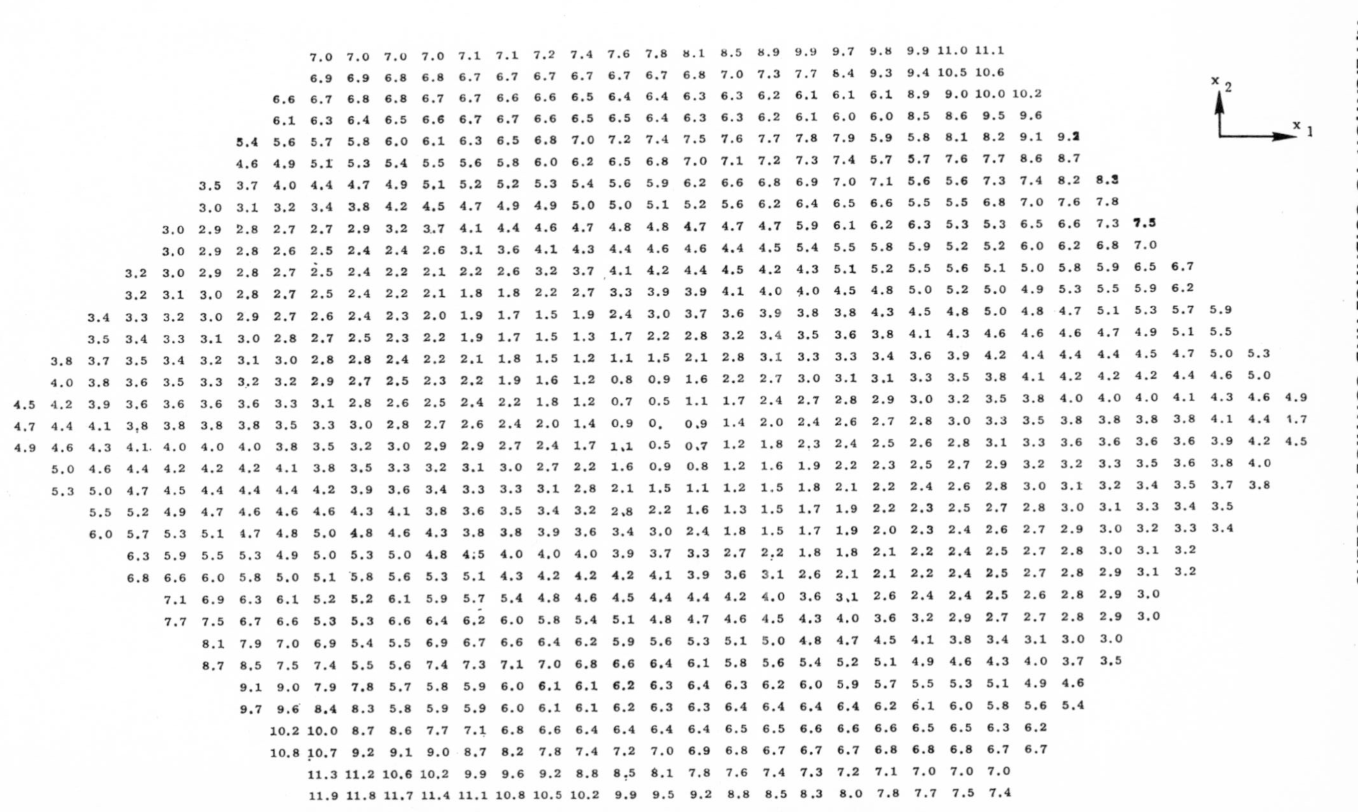

Fig. 9.8. Minimum cost in seconds corresponding to optimal control in Fig. 9.7.

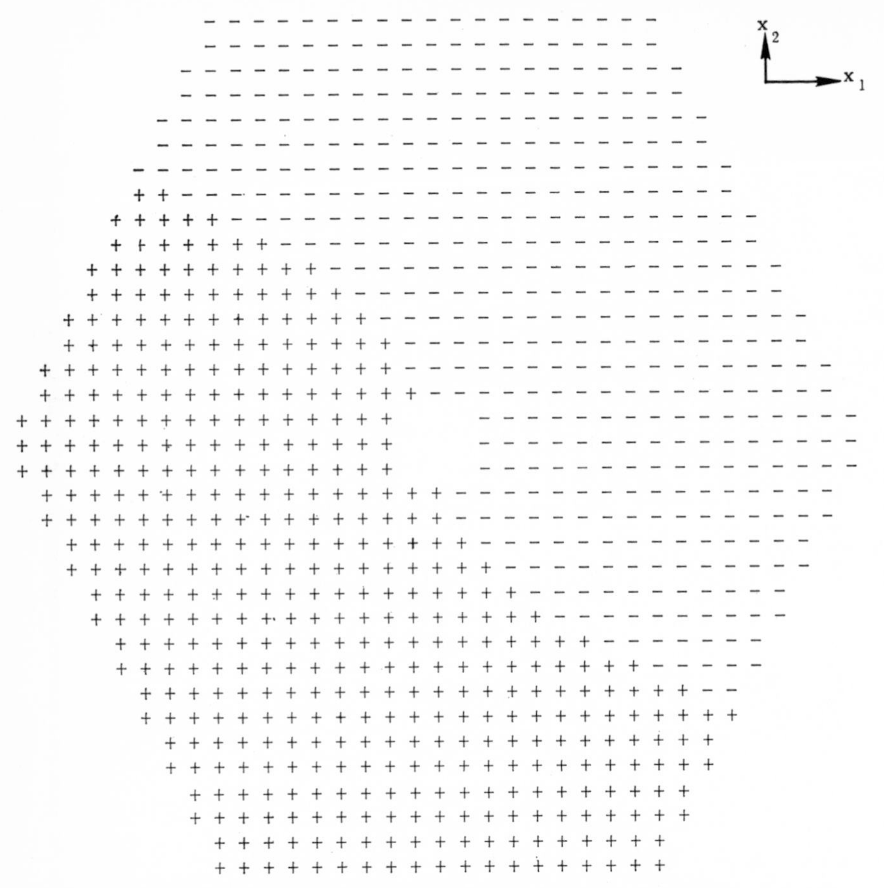

Fig. 9.9. Exact solution for optimal control of $1/s^2$ plant, $|u| \leq 1$, $J = t_f - t_0$, $(+ = +1, - = -1)$.

Another experiment which was performed along these lines is shown in Figs. 9.13, 9.14, and 9.15. Here the increment sizes are $\Delta x_1 = .001$ and $\Delta x_2 = .006$, while U is still $\{-1, +1\}$. The initialization is performed by using the correct values over the first seven layers. In Fig. 9.13 the values obtained along one side of the eighth layer are shown. In Fig. 9.14 the true values are presented, while in Fig. 9.15 the errors are illustrated. The average percentage error is a very low 1.5%.

Still another experiment was performed to determine the sensitivity to incorrect initial conditions. In this run a minimum cost of 0 was assigned to the 9 points in the block centered at the origin; with this as the initial set of points, the results shown in Figs. 9.16 and 9.17 were obtained. Clearly, the incorrect initial conditions lead to results vastly different from the true values.

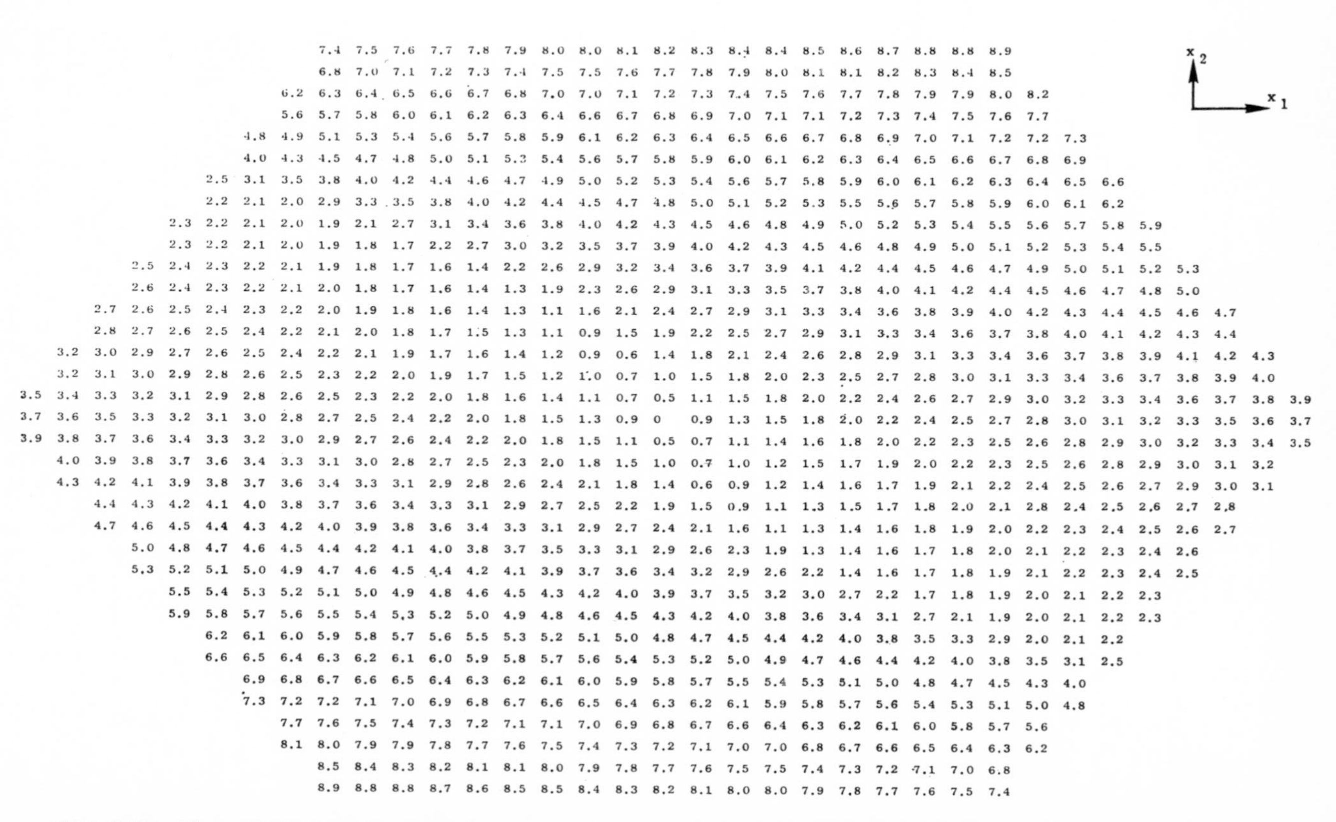

Fig. 9.10. Exact solution for minimum cost in seconds corresponding to optimal control in Fig. 9.9.

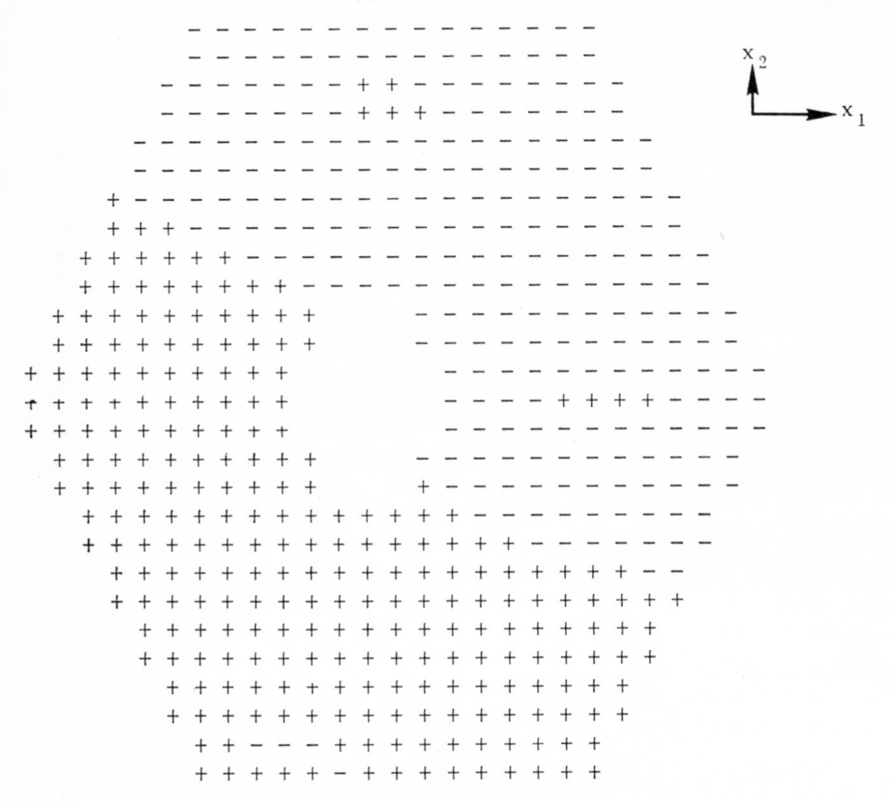

Fig. 9.11. Optimal control for $1/s^2$ plant, $|u| \leq 1$, $J = t_f - t_0$, as obtained from computer run with expanded initial conditions (initial area = 2 layers), $\Delta x_1 = 0.2$, $\Delta x_2 = 0.2$, $U = \{-1, +1\}$. (+ = +1, − = −1).

Some tests were then performed to see if the results obtained in the original program could be improved by iteration, using the values of minimum cost determined from the program as an initial guess. After just one iteration, the number of incorrect controls in isolated areas was reduced from 42 (out of 937 points) to 8. In most cases a significantly better estimate of the minimum cost was also obtained. This indicated that with very few iterations results very close to the true values can be obtained.

The error introduced by the inexact integration routines was isolated and found to be greater than the error due all other sources, even the interpolation and extrapolation errors. This error is a bias error that occurs because the next state is determined as

$$\begin{aligned} x_1(t + \delta t) &= x_1(t) + x_2(t)\,\delta t \\ x_2(t + \delta t) &= x_2(t) + u(t)\,\delta t \end{aligned} \tag{9.54}$$

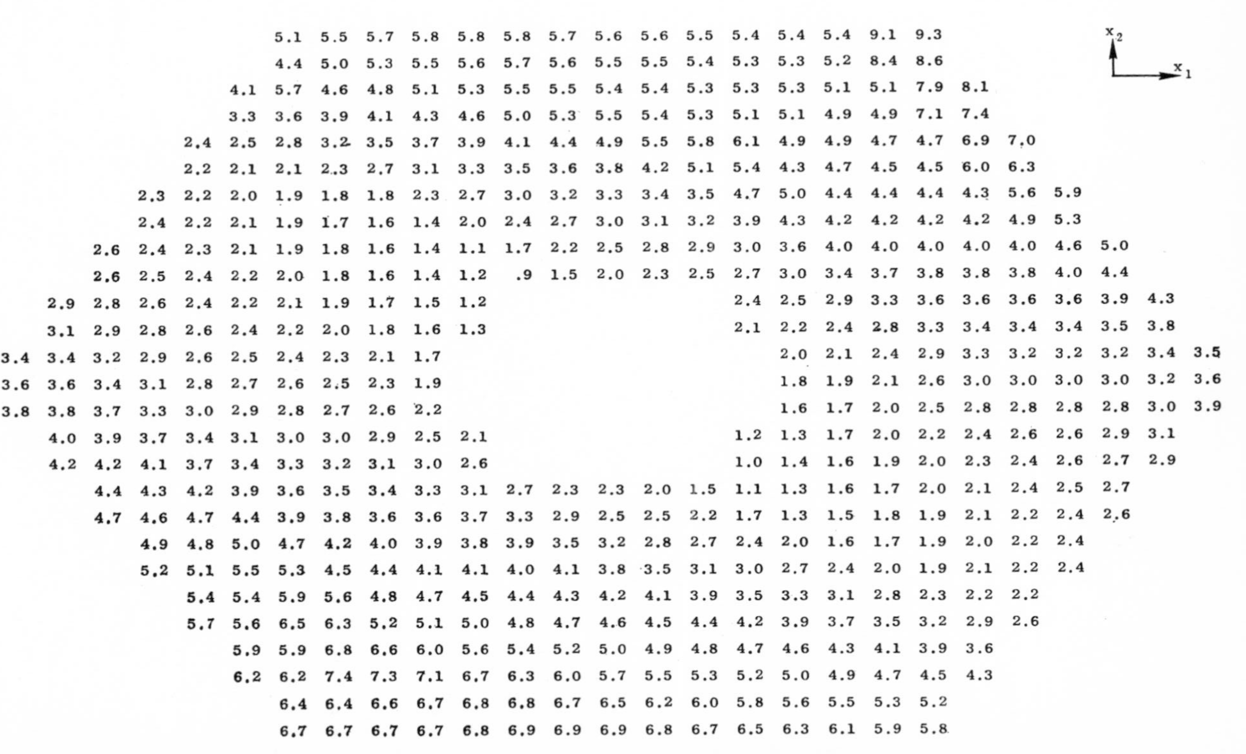

Fig. 9.12. Minimum cost in seconds corresponding to optimal control in Fig. 9.11.

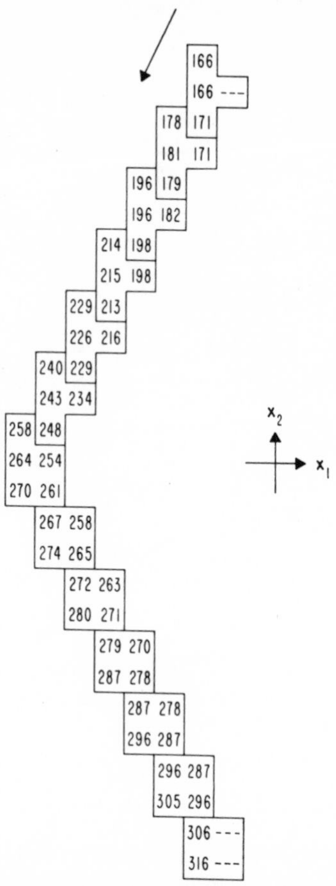

Fig. 9.13. Minimum cost in thousandths of a second for $1/s^2$ plant, $|u| \leq 1$, $J = t_f - t_0$ at layer 8 as obtained from computer run with initial area covering layers 1–7, where $\Delta x_1 = .001$, $\Delta x_2 = .006$, $U = \{-1, +1\}$.

rather than using a more accurate formula that accounts for the fact that $x_2(t)$ actually varies over the interval δt. This error causes the next state to be in error, and hence the minimum cost, even if interpolated exactly, takes on the wrong value. The error is consistent in the sense that it always has the same sign as the sign of $u(t)$. This causes the switching line to be uniformly flatter as a function of x, than is the true switching line, an effect that is readily noticeable in Figs. 9.7–9.14.

Because the $1/s^2$ system is not stable, the stability problems of the integration routine are compounded. As a result errors made early in the procedure tend to grow rather than damp out. This at least partially accounts for the high

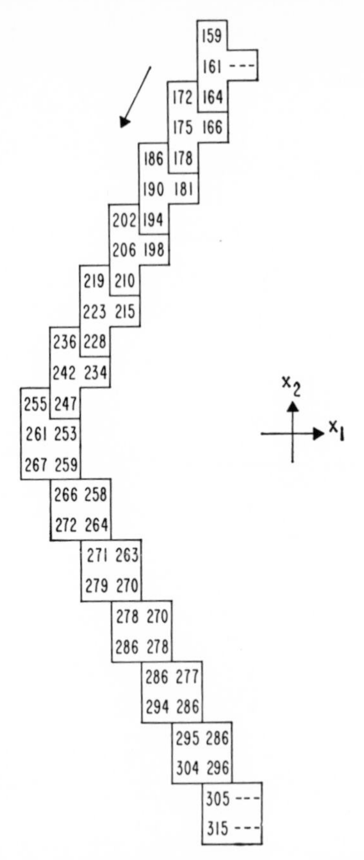

Fig. 9.14. Exact solution for minimum cost in thousandths of a second for problem shown in Fig. 9.13.

sensitivity to incorrect initial conditions and for the fact that the results get worse farther away from the origin.

As a check on the influence of the instability of the system, the same program (same performance criterion, same constraints, same increment sizes, same block processing procedure, and same initialization region) was run for the $1/(s^2 + 4s + 1)$ plant, a stable system described by

$$\ddot{x} + 4\dot{x} + x = u \tag{9.55}$$

or in the standard form,

$$\begin{aligned} \dot{x}_1 &= x_2 \\ \dot{x}_2 &= -x_1 - 4x_2 + u. \end{aligned} \tag{9.56}$$

The optimal control obtained is shown in Fig. 9.18 and the true solution for optimal control shown in Fig. 9.19. Note that there are only two isolated

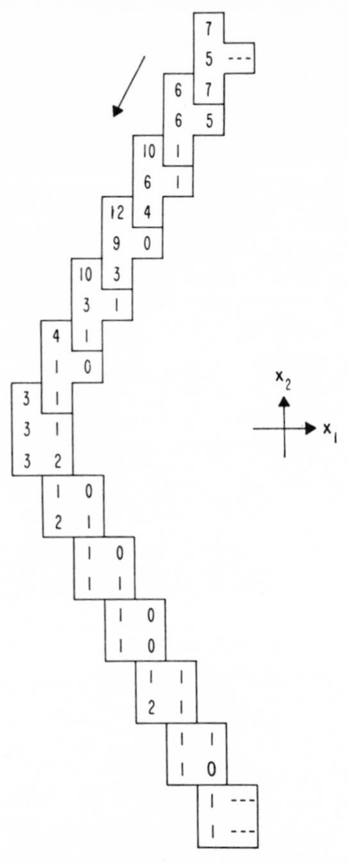

Fig. 9.15. Error between exact solution of Fig. 9.14 and computed solution of Fig. 9.13 of thousandths of a second.

points at which the wrong optimal control is obtained, as contrasted to 42 for the unstable $1/s^2$ plant. Also, note that the switching line obtained from the computer run has almost exactly the same shape as the true switching line, except that it is displaced from it by one increment in x_1. This displacement can be accounted for by the bias error due to the inexact integration routine. Thus, the agreement between computed optimal control and true optimal control [obtained in Ref. 8] for this stable system is significantly better than for the unstable $1/s^2$ system. A check of the minimum cost also shows better agreement with the true value than in the $1/s^2$ plant case.

The program has also been applied to a number of other problems including nonlinear problems with other types of performance criteria and with different constraints. Good results have been obtained in all cases. As a final

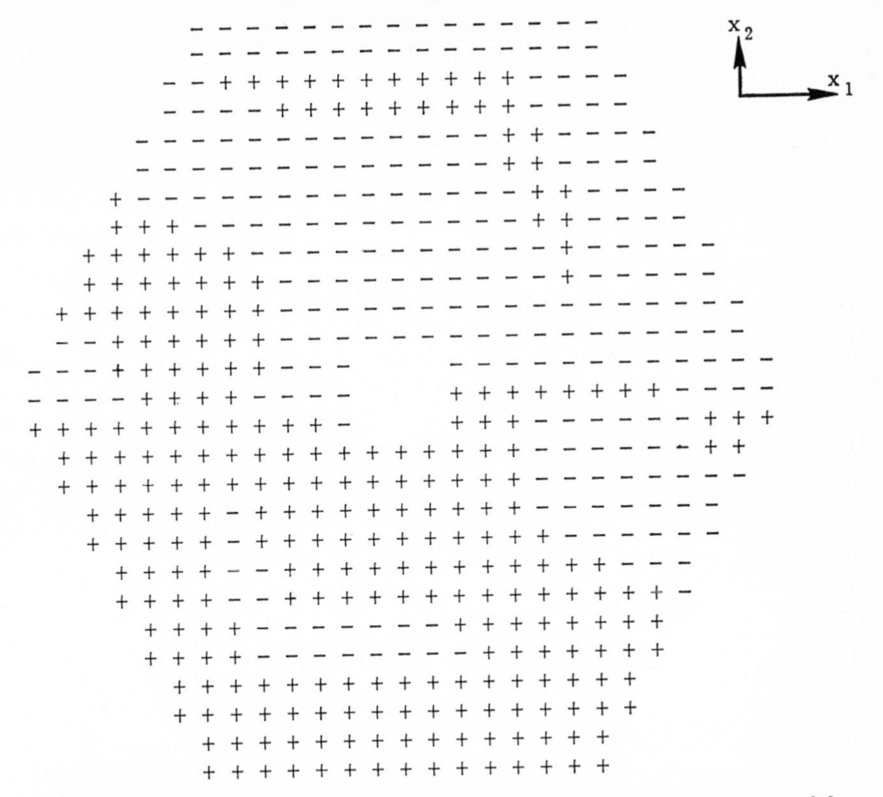

Fig. 9.16. Optimal control for $1/s^2$ plant, $|u| \leq 1$, $J = t_f - t_0$, with incorrect (all zero) initial costs, $\Delta x_1 = 0.2$, $\Delta x_2 = 0.2$, $U = \{-1, +1\}$, initial area = 1 layer. ($+ = +1$, $- = -1$).

illustrative example for this section, optimal control was computed for the $1/(s^2 + 3s + 2)$ plant with performance criterion

$$J = \int_0^\infty (5x^2 + 5\dot{x}^2 + u^2)\, dt \tag{9.57}$$

and with the constraint $|u| \leq 1$. Writing $x_1 = x$ and $x_2 = \dot{x}$ as before, the system equation is

$$\begin{aligned} \dot{x}_1 &= x_2 \\ \dot{x}_2 &= -2x_1 - 3x_2 + u. \end{aligned} \tag{9.58}$$

The performance criterion is

$$J = \int_0^\infty (5x_1^2 + 5x_2^2 + u^2)\, dt \tag{9.59}$$

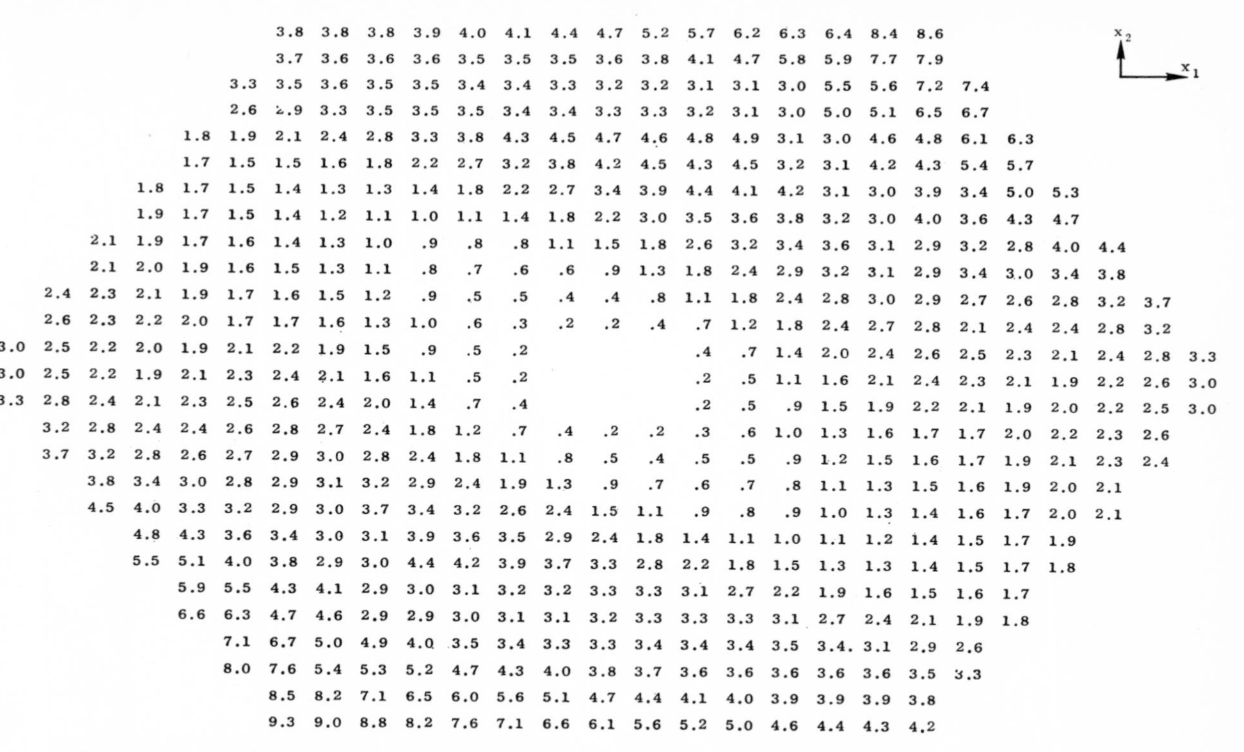

Fig. 9.17. Minimum cost in seconds corresponding to optimal control in Fig. 9.16.

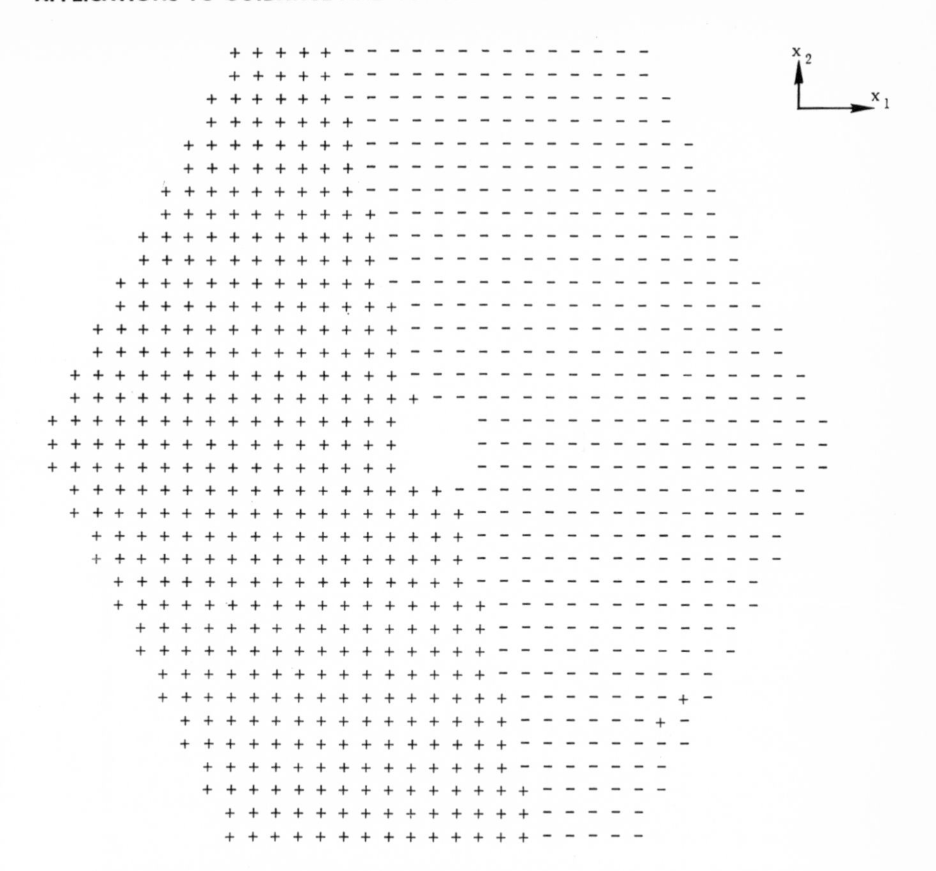

Fig. 9.18. Optimal control for $1/s^2 + 4s + 1$ plant, $|u| \leq 1$, $J = t_f - t_0$, as obtained from computer run with $\Delta x_1 = 0.2$, $\Delta x_2 = 0.2$, $U = \{-1, +1\}$, initial area = 1 layer. ($+ = +1$, $- = -1$).

The computer solutions for optimal control and minimum cost are shown in Figs. 9.20 and 9.21, and the true optimal control [Ref. 9] is shown in Fig. 9.22. The agreement between the two is excellent. A check of the true minimum cost [Ref. 9] at isolated points also showed excellent agreement with the values in Fig. 9.21.

9.7 CONCLUSIONS

From the results of experiments with actual computer programs, such as the one described in this chapter and in Chapter 8, it is concluded that state increment dynamic programming can be successfully applied to optimum guidance and control problems. Even when very simple interpolation and

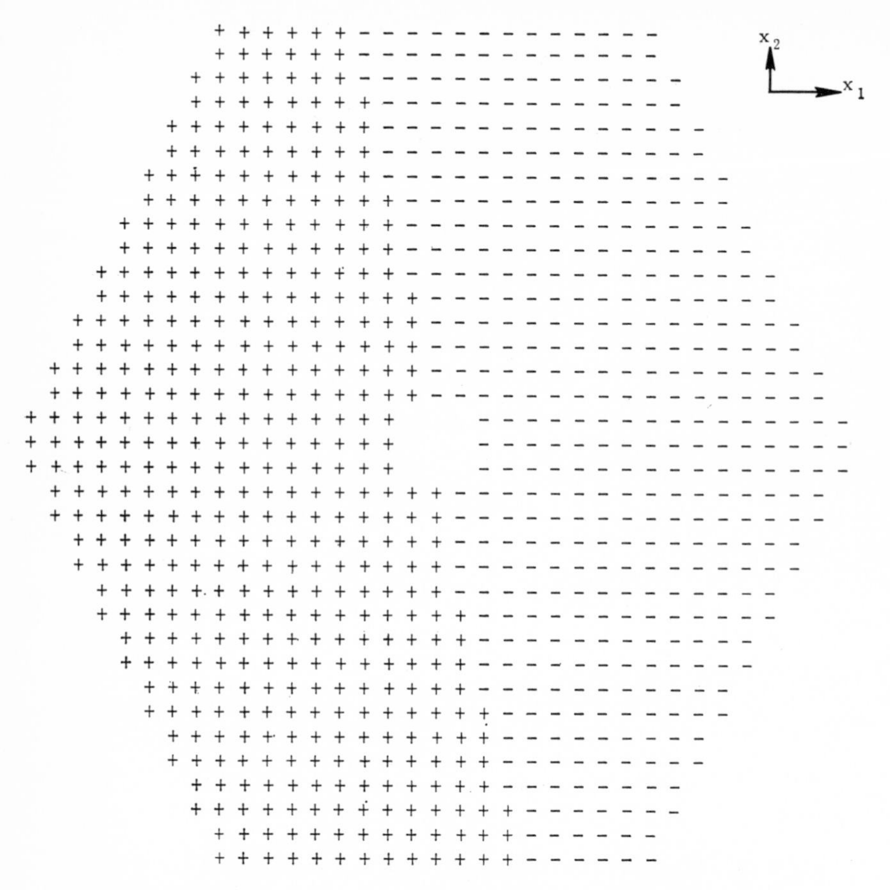

Fig. 9.19. Exact solution for optimal control shown in Fig. 9.18.

extrapolation routines are used and when the preferred direction of motion is known only approximately, excellent results are obtained. In most of the problems considered, errors introduced by inexact integration routines, by the propagation of previously made errors through unstable system equations, and by other error sources encountered with conventional dynamic programming, proved to be larger than the errors which are peculiar to state increment dynamic programming; in fact, for the reasons discussed in Chapter 7, the errors with state increment dynamic programming were in some cases actually less than for conventional dynamic programming. Thus, in optimal trajectory problems of the kind considered in this chapter, the computational advantages of state increment dynamic programming are obtained with little or no sacrifice in the accuracy of the results.

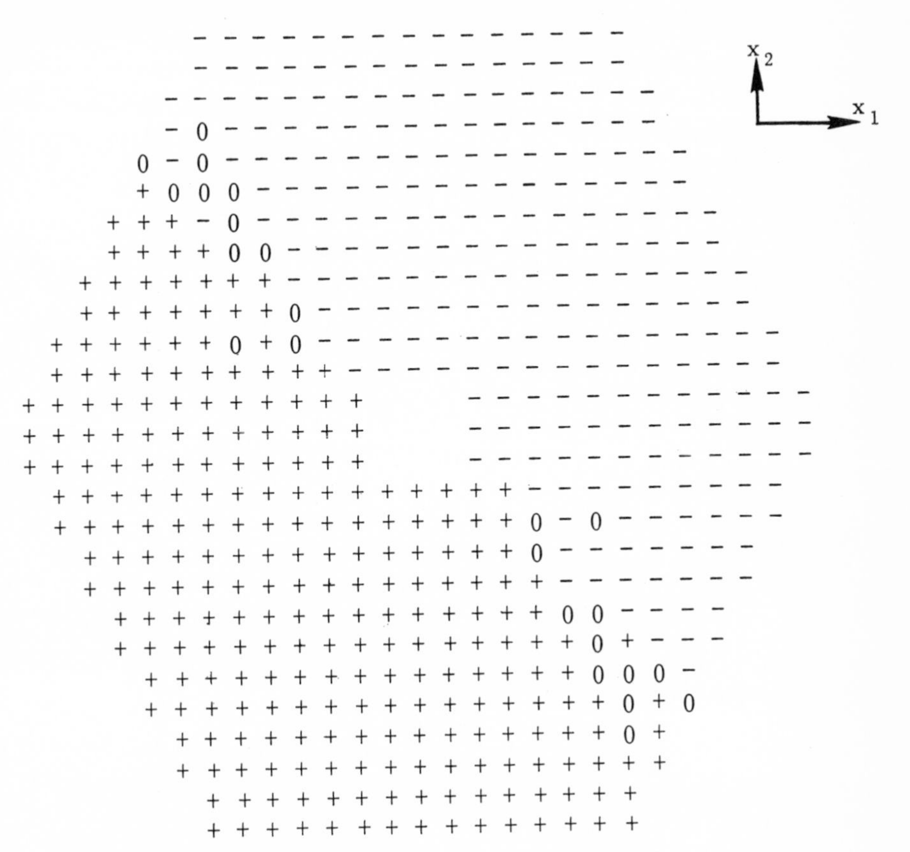

Fig. 9.20. Optimal control for $1/s^2 + 3s + 2$ plant, $|u| \leq 1$,

$$J = \int_{t_0}^{\infty} (5x^2 + 5\dot{x}^2 + u^2)\, dt,$$

as obtained from computer run with $\Delta x_1 = 0.5$, $\Delta x_2 = 0.5$, $U = \{-1, 0, +1\}$, initial area = 1 layer. ($+ = +1, 0 = 0, - = -1$).

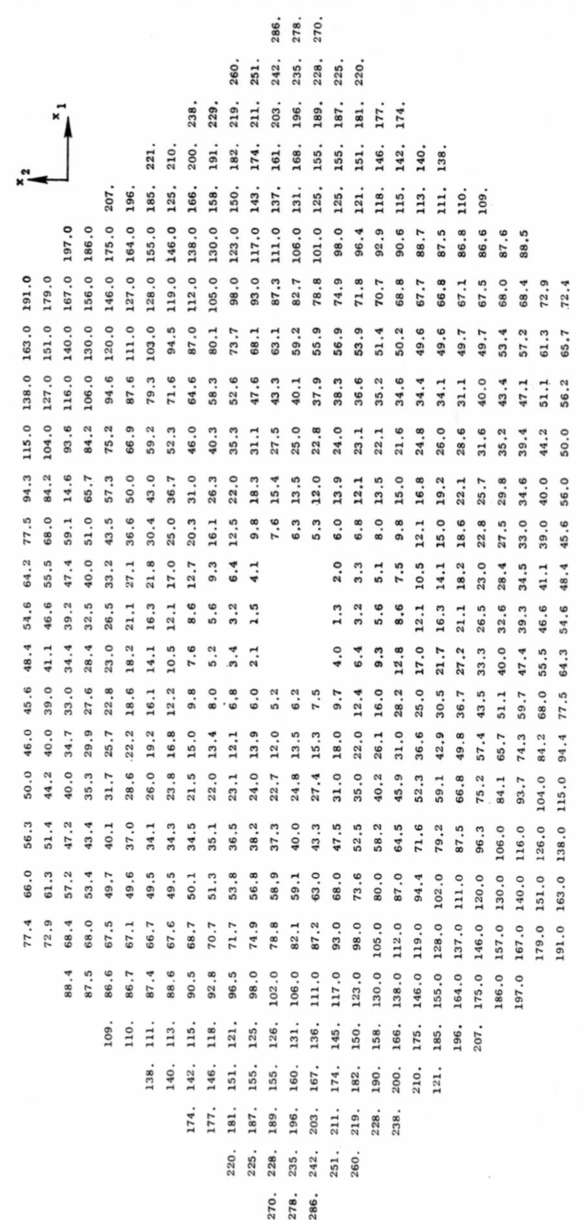

Fig. 9.21. Minimum cost corresponding to optimal control shown in Fig. 9.20.

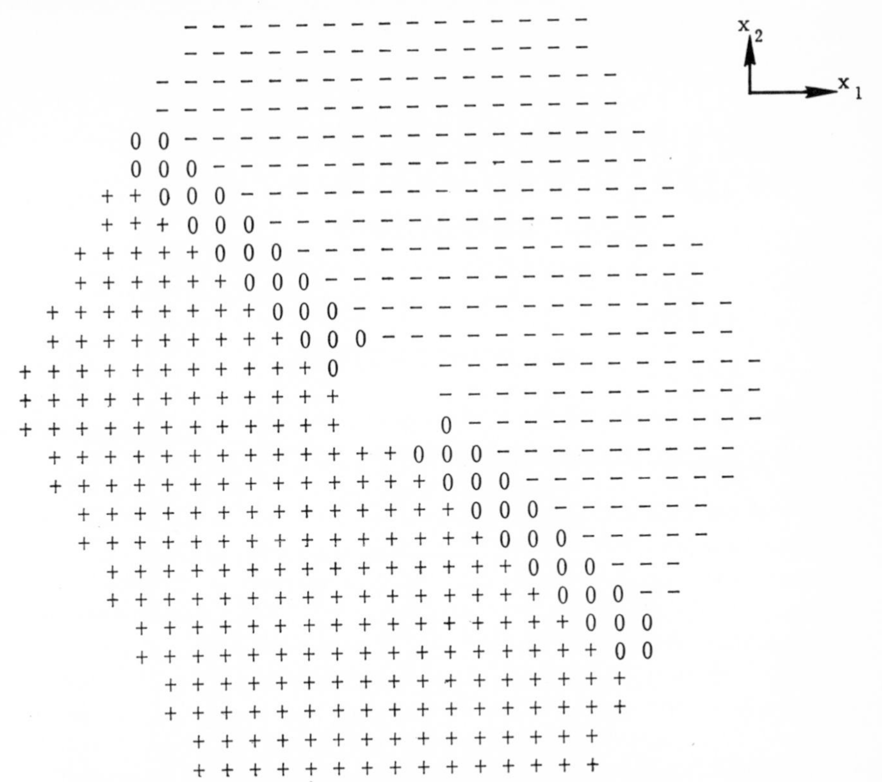

Fig. 9.22. Exact solution for optimal control shown in Fig. 9.20.

REFERENCES

1. Richardson, D. W., Hechtman, E. P., Larson, R. E., and Rosen, M. H., "Control Data Investigation for Optimization of Fuel on Supersonic Transport Vehicles," Phase I Report, Contract AF 33(657)-8822, Project 9056, Hughes Aircraft Company, Culver City, California (January 1963).
2. Richardson, D. W., Hechtman, E. P., and Larson, R. E., "Control Data Investigation for Optimization of Fuel on Supersonic Transport Vehicles," Phase II Report, Contract AF 33(657)-8822, Project 9056, Hughes Aircraft Company, Culver City, California (June 1963).
3. Larson, R. E., "Dynamic Programming with Continuous Independent Variable," TR6302-6, Stanford Electronics Laboratories, Stanford, California (April 1964).
4. Bellman, R., and Dreyfus, S., *Applied Dynamic Programming* (Princeton University Press, Princeton, New Jersey, 1962).
5. Cartaino, S. and Dreyfus, S., "Application of Dynamic Programming to the Minimum Time-to-Climb Problem," *Aero. Engr. Rev.*, Vol. 16, No. 6 (June 1957), pp. 74–77.
6. Keckler, W. G., "Optimization About a Nominal Trajectory via Dynamic Programming," Engineer's Thesis, Department of Electrical Engineering, Stanford University, Stanford, California (September 1967).
7. Keckler, W. G., and Larson, R. E., "Dynamic Programming for the Pre-Launch Calculation," Preliminary Report, Contract DA-01-021-AMC-90006(y), SRI Project 5188, Stanford Research Institute, Menlo Park, California (July 1967).
8. Bellman, R., Glicksberg, I, and Gross, O., "On the Bang-Bang Control Problem," *Q. Appl. Math.*, vol. 14, 1956, p. 11–18.
9. Stubberud, A. R., and Swiger, J. M., "Minimum Energy Control of a Linear Plant with Magnitude Constraint on the Control Input Signals," 1965 JACC (June 1965), pp. 398–406.

Chapter Ten

APPLICATIONS TO OPERATIONS RESEARCH AND PROCESS CONTROL PROBLEMS

10.1 INTRODUCTION

In addition to guidance and control problems associated with aerospace systems, state increment dynamic programming has been applied to a number of other areas. In this section problems from two other areas, namely operations research and process control, are discussed. Some related computational techniques of particular interest in these areas are also discussed.

In the next section forward dynamic programming is described. This technique has already received brief mention in Sec. 9.4 in connection with the minimum-time-to-intercept problem for an anti-missile missile. Particular attention is given to defining the class of problems where this technique is particularly advantageous. In the section after that a class of scheduling problems, typified by the aircraft routing problem, is discussed. Then, the optimization of distributed-parameter systems is considered. First, a methodology for formulating these problems in terms such that state increment dynamic programming can be applied is given. Then, a transformation of the resulting system equations that results in reducing the dimensionality of the high-speed memory requirement from that of the state vector to that of the control vector is described; the generalization of this result to other classes of problems is also indicated. Finally, the application to optimum dispatching in natural gas pipeline networks is discussed.

10.2 FORWARD DYNAMIC PROGRAMMING

In all the discussions of dynamic programming computational procedures presented so far, the computations begin at the final time $t = t_f$ and work backwards toward the initial time, $t = t_0$. It is possible to start instead with the initial time and work towards the final time using a procedure called *forward dynamic programming*. By interpreting the computations in a suitable manner, it can be shown that this procedure is more useful than the normal method, referred to here as *backward dynamic programming*, for a number of applications.

The iterative equation based on the principle of optimality for forward dynamic programming is derived in a manner similar to that for backward dynamic programming. The minimum cost function is defined somewhat differently; instead of being the minimum cost that can be achieved starting at the present state and going to the end of the process, it is the minimum cost that can be achieved starting in an admissible initial state and arriving at the present state at the present time. Writing this minimum cost as $I'(\mathbf{x}, t)$ to distinguish it from the minimum cost function for backward dynamic programming, the definition in the discrete time case becomes

$$I'(\mathbf{x}, k) = \underset{\mathbf{u}(0),\ldots,\mathbf{u}(k-1)}{\text{Min}} \left\{ \sum_{j=0}^{k-1} l[\mathbf{x}(j), \mathbf{u}(j), j] \right\} \tag{10.1}$$

where

$$g[\mathbf{x}(k-1), \mathbf{u}(k-1), k-1] = \mathbf{x}. \tag{10.2}$$

Note that in this definition $\mathbf{x}$ is the last state along a trajectory, rather than the initial state, as in backward dynamic programming.

The iterative equation for this case is derived exactly as in Chapter 2. First the quantity $\mathbf{h}[\mathbf{x}, \mathbf{u}(k-1), k]$ is defined by

$$\mathbf{g}[\mathbf{h}[\mathbf{x}, \mathbf{u}(k-1), k], \mathbf{u}(k-1), (k-1)] = \mathbf{x}. \tag{10.3}$$

Then, the equation can be derived as

$$I'(\mathbf{x}, k) = \underset{\mathbf{u}(k-1)}{\text{Min}} \{ l[\mathbf{h}[\mathbf{x}, \mathbf{u}(k-1), k], \mathbf{u}(k-1), k-1] + I'[\mathbf{h}[\mathbf{x}, \mathbf{u}(k-1), k], k-1] \}. \tag{10.4}$$

Note that this iterative procedure determines $I(\mathbf{x}, k)$ in terms of $I(\mathbf{x}, k-1)$; thus, the calculations do indeed go forward in time rather than backward. The initial condition to start the iterations is

$$I'(\mathbf{x}, 0) = 0 \tag{10.5}$$

i.e., no cost is incurred before the system begins to operate. Note also that the optimal control which minimizes Eq. (10.4) tells what control should have been applied at the last time increment to result in state $\mathbf{x}(k) = \mathbf{x}$, rather than which control to apply at $\mathbf{x}$. This control is denoted as $\hat{\mathbf{u}}'(\mathbf{x}, k)$, where the prime is again used to distinguish this quantity from the corresponding quantity in the backward procedure.

One way of carrying out the computations is to use a procedure exactly analogous to that for backward dynamic programming: at a given quantized $\mathbf{x}$ and k try all possible quantized controls $\mathbf{u}(k-1)$; find the corresponding previous state $\mathbf{x}(k-1)$; evaluate the quantity inside brackets in Eq. (10.4), using interpolation procedures if necessary; and pick the optimal control and minimum cost. This procedure has some deficiencies. In the first place, there is no a priori way of insuring that a given control $\mathbf{u}(k-1)$ is in fact an admissible control at $\mathbf{x}(k-1) = \mathbf{h}(\mathbf{x}, \mathbf{u}(k-1), k)$. Second, it may be difficult to compute the functional $\mathbf{h}$ if $\mathbf{g}$ is a nonlinear time-varying functional.

A procedure which overcomes these difficulties and offers other advantages as well is the following: at each quantized state $\mathbf{x}(k-1)$ where $I'[\mathbf{x}(k-1), k-1]$ has been computed apply all the quantized admissible controls; for each corresponding next state, $\mathbf{x}(k) = \mathbf{g}[\mathbf{x}(k-1), \mathbf{u}(k-1), k-1]$, check to see if it has been the next state for any control applied at previous values of $\mathbf{x}(k-1)$; if it has not previously been a next state, then store the quantity in braces in Eq. (10.4) as the tentative minimum cost at that point; if it has, then compare the quantity in braces in Eq. (10.4) with the tentative minimum cost already computed at that point, and if it is less than this minimum cost and optimal control, replace the values stored there. This procedure continues until the quantized admissible controls have been applied at every quantized state $\mathbf{x}(k-1)$. The tentative minimum costs and optimal controls at each $\mathbf{x}(k) = \mathbf{g}[\mathbf{x}(k-1), \mathbf{u}(k-1), k-1]$ are then the true minimum costs and optimal controls at these points.

This procedure is seen more clearly in a simple one-dimensional example. The system equation for this problem is

$$x(k+1) = x(k) + u(k). \tag{10.6}$$

The performance criterion is

$$J = \sum_{j=0}^{4} [x^2(j) + u^2(j)]. \tag{10.7}$$

The constraints are

$$-1 \le u(k) \le 1 \tag{10.8}$$

and

$$0 \le x(k) \le 2. \tag{10.9}$$

The initial state is known to be

$$x(0) = 2. \tag{10.10}$$

The quantization increments are taken to be $\Delta x = 1$ and $\Delta u = 1$, as in previous examples. The first step in the procedure is to apply the three quantized admissible controls, $u = -1$, $u = 0$, and $u = +1$, at the initial state $x(0) = 2$. The minimum cost and optimal control for admissible states at $x(1)$ are found without need for comparison. These values are shown in Fig. 10.1, where the minimum cost appears above and to the right and the optimal control appears below and to the right of the quantized state and stage to which these quantities correspond. Note that the optimal control determines what state to have come from, not what state to go to.

In going from $I'(x, 1)$ to $I'(x, 2)$ a comparison is required in some of the calculations. If the controls are first applied at $x(1) = 2$, the tentative minimum costs and optimal controls are as shown in Fig. 10.2.

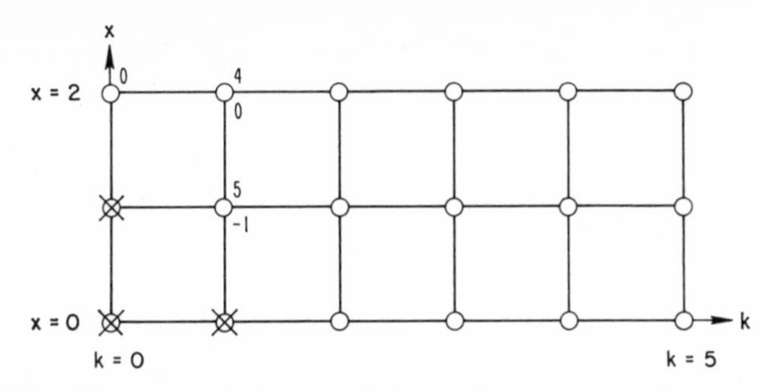

Fig. 10.1. First step in forward dynamic programming procedure.

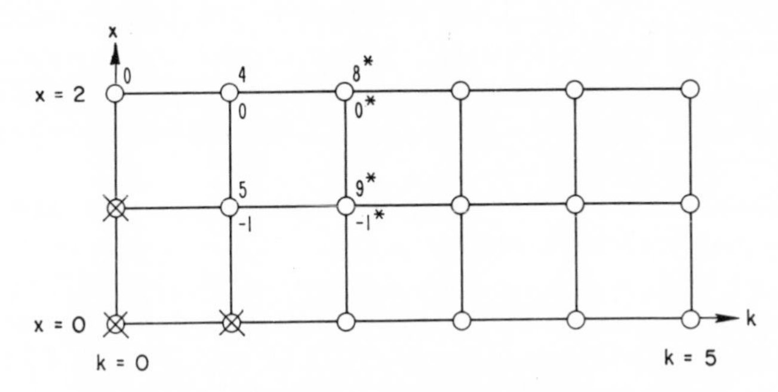

Fig. 10.2. Tentative minimum costs at $k = 2$.

Asterisks are placed beside these values to show that they are tentative.

When the controls at $x(1)$ are applied, the states $x(2) = 2$ and $x(2) = 1$ are possible next states. The minimum costs coming from $x(1) = 1$ are compared with the values already there in Fig. 10.2; in both cases, less cost is obtained when $x(1) = 1$. The complete results at $k = 2$ are thus as shown in Fig. 10.3.

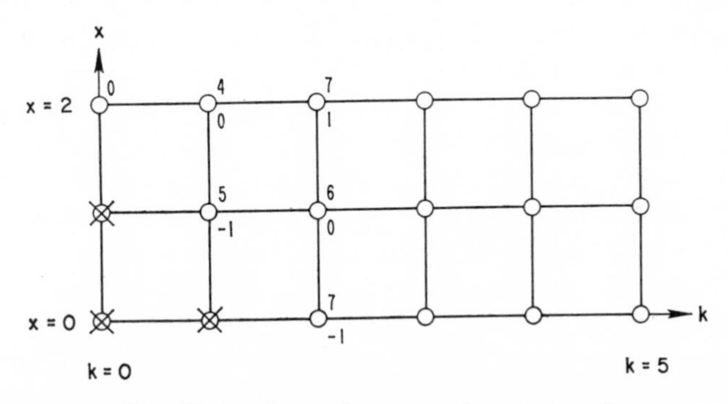

Fig. 10.3. Complete results at $k = 2$.

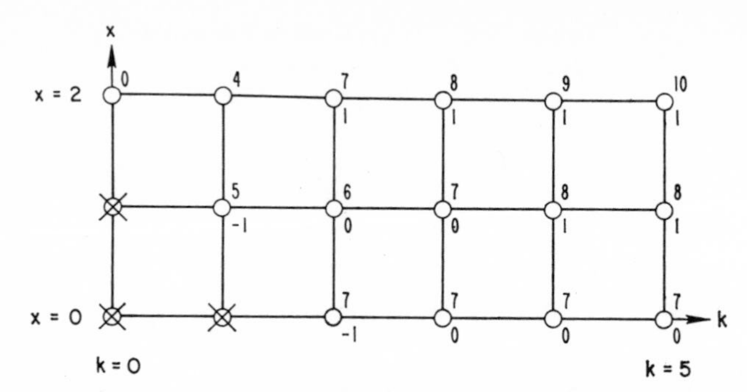

Fig. 10.4. Complete results for forward dynamic programming example.

This procedure continues until minimum cost and optimal control have been computed at all quantized values of x and k. The results are shown in Fig. 10.4

These results can be used in a number of different ways. First, suppose that the final state can be anywhere in the region of $0 \leq x \leq 2$. In this case a second search is made over the minimum costs at the quantized final states, and the final state is taken to be the one for which the minimum cost is least. In this example, the minimum cost is $I' = 7$ at $x(5) = 0$. The optimal trajectory is then as shown as in Fig. 10.5.

Next, suppose that a terminal cost function is added at $t = 5$. Let this terminal cost function be

$$\psi[x(5), 5] = 2.5[x(5) - 2]^2 \tag{10.11}$$

This function can then be added to the minimum costs $I'[x(5), 5]$ shown in Fig. 10.4. The resulting total cost is obtained for $x = 0, 1, 2$ in Table 10.1. The minimum *total* cost is thus seen to be 10, corresponding to $x(5) = 2$. The optimal trajectory corresponding to this final state is as shown in Fig. 10.6.

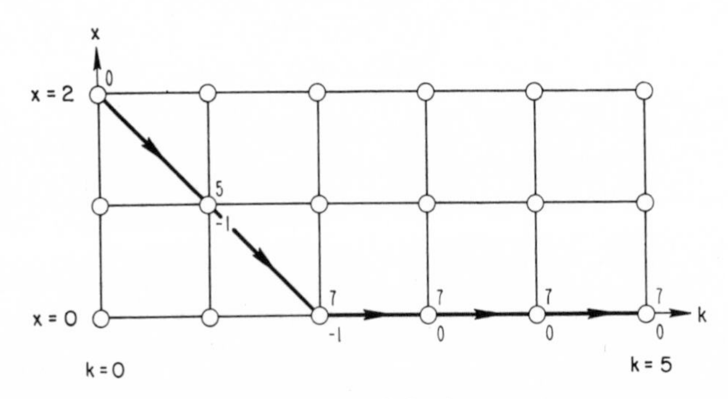

Fig. 10.5. Optimal trajectory if final state is not constrained.

Table 10.1. Total Cost in Forward Dynamic Programming Example

x	$I'(x, 5)$	$\psi(x, 5)$	*Total cost*
2	10	0	10
1	8	2.5	10.5
0	7	10	17

The same problem with the same terminal cost function was solved by backward dynamic programming in Sec. 2.9. From the results presented there, it can be seen that for the initial state $x = 2$, the minimum cost and optimal trajectory are exactly the same as shown in Fig. 10.6. It is true in general that for the same problem with the same initial and final states, forward dynamic programming and backward dynamic programming obtain the same minimum cost and optimal trajectory. Of course, the costs and controls at intermediate states are completely different; this is due to the difference in definition between $I(\mathbf{x}, k)$ and $I'(\mathbf{x}, k)$ and between $\hat{\mathbf{u}}(\mathbf{x}, k)$ and $\hat{\mathbf{u}}'(\mathbf{x}, k)$.

This example illustrates a number of the useful properties of the forward dynamic programming solution. In the first place, there is great flexibility in the terminal cost function and/or terminal constraints that can be applied. In particular, a terminal cost function can be added *after* all the computations have been made, and the terminal state that minimizes the total cost can then be selected. It is then possible to assess the effects of using a number of different terminal cost functions without repeating the forward dynamic programming calculations. On the other hand, in backward dynamic programming it is necessary to repeat the entire computation for each different terminal cost function.

Using the forward dynamic programming solution it is easy to examine trajectories reaching many different terminal states: thus, if an explicit

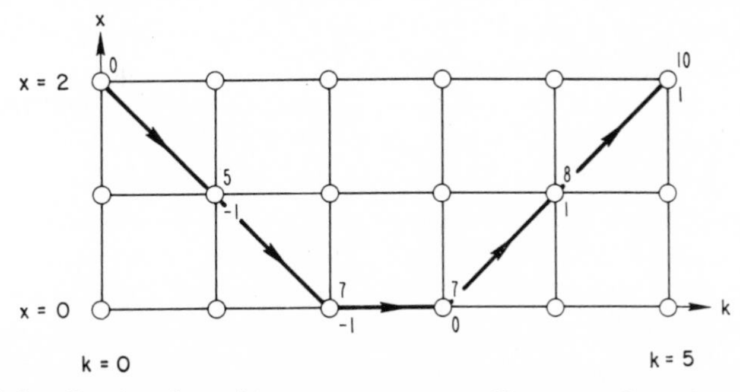

Fig. 10.6. Optimal trajectory corresponding to $x(5) = 2$ (optimal trajectory for the given terminal cost function).

terminal cost function is not known, the trajectories corresponding to several of the lowest values of minimum cost can be examined, and the optimal trajectory selected on the basis of more subjective grounds, such as simplicity of realization, reliability, convenience to humans associated with the system, etc. If a trajectory other than the one corresponding to the lowest cost is chosen, then the added cost in terms of the quantities in the performance criterion is given explicitly.

The application of terminal constraints can be done in exactly the same way with similar results.

Another useful property of forward dynamic programming is that the initial state can easily be constrained to one state, while with backward dynamic programming optimal trajectories for all admissible states at the initial time are found. This is true even when in backward dynamic programming the computations at $t = t_0$ are made for only one initial state; in this case computations have still been done at a number of points through which no trajectory from the initial state will pass. Thus, when it is desired to have an optimal trajectory from a given initial state, a saving in computing time can be achieved by using forward dynamic programming.

The extension of this forward dynamic programming procedure to problems for which the next states $\mathbf{x}(k) = \mathbf{g}[\mathbf{x}(k-1), \mathbf{u}(k-1), k-1]$ do not occur exactly at quantized values is straightforward. This can be done in a particularly useful way by associating the next state with the nearest quantized state and interpolating in the n-dimensional state space to compare the cost in brackets in Eq. (10.4) with the tentative minimum cost at the quantized state. Thus, the interpolations can be used for comparison purposes only, and the minimum cost and optimal control can be evaluated exactly along a true trajectory. This eliminates the need for interpolations in reconstructing optimal trajectories after the computations have been completed. The only purpose which quantization in the state variables serves is to divide the state space into areas over which the minimum costs are compared; the minimum cost and optimal control associated with a given quantized state are not necessarily evaluated exactly at the quantized state, but can be at any state within this area.

One disadvantage of forward dynamic programming is that the feedback control property of backward dynamic programming is not retained. The optimal controls determine what the previous state should have been, rather than the next state. Thus, if a deviation occurs from the selected optimal trajectory, a new optimal trajectory cannot be easily found.

The type of problem to which forward dynamic programming is best suited is one in which the initial state is specified, flexibility in choosing the terminal state is desirable, and computation of a new optimal trajectory if deviations occur from the original trajectory is feasible. One class of problems where these conditions are met is on-line control and dispatching applications.

The aircraft routing problem discussed in the next section is one such example. Another class of problem occurs when it is desired to pre-compute the response of a system which always starts from a known initial state, to a large number of possible conditions. The problem of determining minimum-time-to-intercept trajectories for an anti-missile missile, which was discussed in Sec. 9.4, falls into this category. Still another class of problems is optimum system planning; in these cases the present system configuration is known, and flexibility in assessing the results of different terminal cost functions is often very valuable.

State increment dynamic programming can be combined with forward dynamic programming in a straightforward manner. All the advantages of both techniques can be preserved in a combined procedure. The only non-trivial step in a combined procedure is developing an interpolation procedure which preserves the useful properties of the forward dynamic programming interpolation procedure. This is accomplished by associating *both* next state and next time with the nearest quantized state and quantized value of t. The interpolations, which are again for comparison purposes only, are then in $(n - 1)$ state variables and t. Thus, in this procedure it is again true that the optimal control and minimum cost can be specified along an exact trajectory rather than including interpolated portions.

10.3 THE AIRCRAFT ROUTING PROBLEM

In this section a computational procedure for the optimal short-term allocation of aircraft to meet a fixed schedule is discussed. This problem is commonly referred to as the aircraft routing problem. The same basic procedure can be applied to similar problems, such as crew scheduling, optimal planning of a time table, and optimal scheduling of maintenance. It also has applications to a wide range of other scheduling problems.

In aircraft routing, a limited resource, namely the fleet of airplanes, is to be utilized in such a way that the best system performance is obtained while meeting certain constraints. System performance can be measured by the net profit of the airline, the percentage of scheduled flights that are completed on time, or some other relevant measure. The constraints can include crew availability, regular maintenance of the aircraft at specified locations, the unavailability of a given airplane due to mechanical failure, the inaccessibility of a given airport due to weather conditions, and any other desired conditions.

The formulation of this problem in terms suitable for the application of dynamic programming proceeds as follows. First, the time interval over which the optimization is to be performed is quantized into increments; typically, this period is 24 hours and the time increment is one hour. The state variables for the system are then taken to be the positions of each of the airplanes. A complete description of position specifies the origin and destination of the flight as well as how far along the flight the airplane is. This information is conveniently summarized in a state diagram such as the one shown in Fig. 10.7. Here it is assumed that there are three airports, labeled as nodes 1, 2, and 3. The flight time from 1 to 2 is taken to be 4 hours; from 2 to 3 it is 2 hours; and from 1 to 3 it is 6 hours. If the flight time between two airports is k hours, then $(k - 1)$ intermediate states are defined between them; in this manner, if the airplane takes off at a quantized time, then at each of the next k hours it passes from one state to the next until it arrives after k hours at the destination. Two sets of intermediate states are defined between each pair of airports, corresponding to flights in both directions.

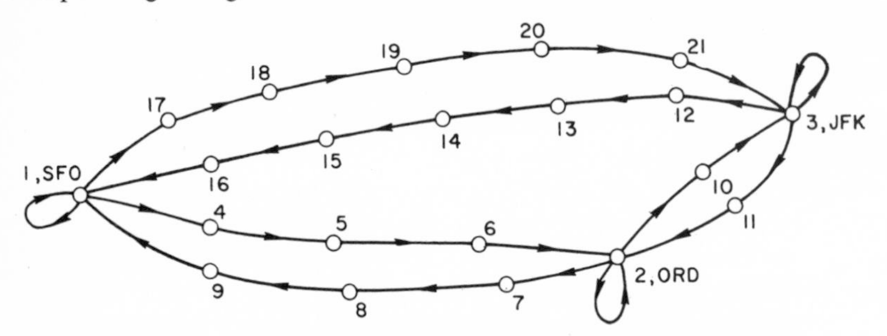

Fig. 10.7. State diagram for aircraft routing problem.

For the moment, it is assumed that once an airplane takes off for a given destination, it flies on to that particular airport. Consequently, the only time that a decision can be made concerning a specific airplane is when it is on the ground at an airport. Then, the airplane has three choices: it can stay where it is or it can take off for either of the other two airports. Except for these decisions, the next state for each airplane is completely determined by knowing the present state.

The n-dimensional state vector, $\mathbf{x}$, consists of the number of the state for each of the n airplanes. The n-dimensional control vector, $\mathbf{u}$, specifies which airport each airplane heads for if it is presently on the ground. The system equations then take the form

$$\mathbf{x}(k + 1) = \mathbf{g}[\mathbf{x}(k), \mathbf{u}(k), k] \tag{10.12}$$

If the ith airplane obeys the state diagram in Fig. 10.7, then the ith component of $\mathbf{g}$ has the form shown in Table 10.2.

A general cost function of the following form can be used:

$$J = \sum_{k=0}^{K} l[\mathbf{x}(k), \mathbf{u}(k), k] \tag{10.13}$$

As already mentioned, this function can be maximum profit, minimum disruption of service, etc. Constraints, such as those mentioned previously,

Table 10.2. State Transition Function $x_i(k+1) = g_i[x_i(k), u_i(k), k]$ for System Shown in Fig. 10.7

$x_i(k)$ *Present state*	$u_i(k)$* *Control*	$x_i(k+1) = g_i[x_i(k), u_i(k), k]$ *Next state*
1	1	1
	2	4
	3	17
2	1	7
	2	2
	3	10
3	1	12
	2	11
	3	3
6	—	2
9	—	1
10	—	3
11	—	2
16	—	1
21	—	3
All other states, j	—	$(j+1)$

* A blank in the control column indicates that the next state is independent of the control; in this problem this means that the airplane is in the air and has no choice of control.

can easily be implemented. These constraints simply cut down the number of alternatives that an airplane has; for example, if it is due for maintenance at airport 1 after seven more flying hours and it is presently at airport 2, it has no choice but to fly to 1.

With minor modifications, the program can handle nonintegral times of flight between airports, nonintegral departure times, the possibility of diversion to other airports, varying flight times due to wind, and other constraints and conditions that arise in the actual operation of an airline.

A computer program using forward dynamic programming to optimize the system of Fig. 10.7 has been written for the SDS-910. One input to the program is the schedule of flights over the time period, including estimates of the number of passengers on each flight. The present state of each airplane in the fleet is also given. Finally, the constraints are specified. The program then uses forward dynamic programming to determine the order of flights that each aircraft should make in order to optimize the performance criterion within the given constraints.

In Tables 10.3–10.7, the results for an actual problem are summarized. This particular run took about five seconds on the SDS-910. The performance criterion is taken to be total net profit, defined as total revenue minus total operating cost. The fleet for this problem consists of two airplanes; one of

Table 10.3. Fares and Flight Times Between Airports

Origin/Destination pair	*Fare*	*Flight time*
1(SFO)–2(ORD)	$r_{12} = r_{21} = \$100$	$t_{12} = t_{21} = 4$ hours
1(SFO)–3(JFK)	$r_{13} = r_{31} = \$150$	$t_{13} = t_{31} = 6$ hours
2(ORD)–3(JFK)	$r_{23} = r_{32} = \$\ 50$	$t_{23} = t_{32} = 2$ hours

which has a capacity of 100 passengers and an operating cost of \$1,250/hour, and the other a capacity of 60 passengers and an operating cost of \$750/hour. Both airplanes are assumed to fly at the speed on which the state transition diagram in Fig. 10.7 is based. The fare schedule is shown in Table 10.3 along with the time of flights between airports. The net profit obtained in a flight from Airport j to Airport m using airplane i at time k can be expressed as

$$l(j, m, i, k) = p_{jm}(k)r_{jm} - c_i t_{jm} \tag{10.14}$$

where

$$i = 1, 2$$

$$j = 1, 2, 3$$

$$m = 1, 2, 3$$

$$k = 0, 1, 2, \ldots, 24$$

$p_{jm}(k)$ = number of passengers going from j to m at time k: constrained to be less than P_i, the capacity of airplane i ($P_1 = 100, P_2 = 60$)

r_{jm} = fare from j to m

c_i = operating cost per hour of airplane i

($c_1 = \$1250$/hr, $c_2 = \$750$/hr)

t_{jm} = flight time from j to m

$l(j, m, i, k)$ = net profit obtained if airplane i is flown from j to m at time k.

The profit is evaluated once when the flight takes off, but not again during the flight; also, the profit is 0 if the airplane stays at the same airport. These conditions are expressed by

$$\begin{aligned} l(j, m, i, k) &= 0 \qquad (j \neq 1, 2, 3, \text{ or } m \neq 1, 2, 3, \text{ all } i, k) \\ l(j, j, i, k) &= 0 \qquad (\text{all } i, j, k.) \end{aligned} \tag{10.15}$$

The performance criterion can then be expressed in the usual form

$$J = \sum_{k=0}^{K} l(j, m, i, k). \tag{10.16}$$

Table 10.4. Schedule for the Day, Including Expected Passenger Loads

Destination \ Origin	1(SFO)	2(ORD)	3(JFK)
1(SFO)	—	8:00–90 20:00–85	6:00–80
2(ORD)	0:00–40 16:00–90	—	6:00–65 22:00–60
3(JFK)	12:00–110	2:00–45 20:00–90	—

A typical schedule for this particular routine is shown in Table 10.4. The initial state is specified as $x_1(0) = 2$, $x_2(0) = 1$, i.e., airplane 1, the large airplane, is at airport 2 and airplane 2 is at airport 1. The program then determines the maximum profit for every possible state at the end of 24 hours. In this particular program the final state was constrained to have both airplanes on the ground at $k = 24$. The maximum profit for each of these nine states is shown in Table 10.5. If there is no additional constraint on the final state, then the final state having the largest maximum profit among these nine values is chosen. This maximum profit is \$23,000, corresponding to the state $x_1(24) = 1$, $x_2(24) = 2$; the schedule for each airplane which achieves this profit is shown in Table 10.6. Note that in this schedule the airplane flies empty or with so few passengers that it loses money in some flights, but that it then arrives at a state where it is able to make a net profit later on. Note also that in this case an airplane is provided for all 10 flights.

If instead airplane 2 is constrained to be in state 1 at $k = 24$ in order to have some maintenance performed, then the maximum profit that can be obtained is \$22,500, corresponding to $x_1(24) = 2$, $x_2(24) = 1$. The schedule that results in this profit is shown in Table 10.7.

Table 10.5. Maximum Profits at $k = 24$

	$x_1(24)$		
$x_2(24)$	1(SFO)	2(ORD)	3(JFK)
1(SFO)	\$17,000	\$22,500	\$22,000
2(ORD)	\$23,000	\$19,500	\$22,000
3(JFK)	\$21,500	\$21,000	\$20,500

Table 10.6. Optimum Schedule for Airplanes with $x_1(24) = 1$, $x_2(24) = 2$ (Maximum Profit With No Final Constraints)

Airplane 1					
Departs	*Time*	*Arrives*	*Time*	*Passengers*	*Profit*
2	2:00	3	4:00	45	−$ 250
3	6:00	2	8:00	65	750
2	8:00	1	12:00	90	4,000
1	12:00	3	18:00	100	7,500
3	18:00	2	20:00	0	− 2,500
2	20:00	1	24:00	85	3,500
			Total Profit, Airplane 1		$13,000
Airplane 2					
Departs	*Time*	*Arrives*	*Time*	*Passengers*	*Profit*
1	0:00	2	4:00	40	$ 1,000
2	4:00	3	6:00	0	− 1,500
3	6:00	1	12:00	60	4,500
1	16:00	2	20:00	60	3,000
2	20:00	3	22:00	90	1,500
3	22:00	2	24:00	60	1,500
			Total Profit, Airplane 2		$10,000
			TOTAL PROFIT		$23,000

It is possible to print out the best N trajectories in order to compare their performance on the basis of quantities not directly measured by the profit, such as percentage of flights actually made. Also, if sometime during the day different constraints arise, such as the temporary mechanical failure of one of the airplanes, these can be entered into the program and a new optimization performed over the rest of the day, starting at the present positions of the airplanes. As already mentioned, the program can be expanded to include other constraints, other performance criteria, diversions to other airports, more airplanes, more airports, nonintegral flight times and departure times, etc. This program can be used for planning the aircraft utilization for a given day; for real-time dispatching when a new situation arises; or for off-line determination of an optimum time table (flight schedule) based on the distribution of passenger demand.

This technique is related to state increment dynamic programming in that the state always changes by one increment over one time interval. However, even when this is exploited to reduce the memory requirements, the computer time is such that extension of this technique to more than five or six airplanes

Table 10.7. Optimum Schedule for Airplanes with $x_1(24) = 2$, $x_2(24) = 1$ (Maximum Profit When Airplane 2 Must Go to Airport 1 for Maintenance at $k = 24$)

Airplane 1					
Departs	*Time*	*Arrives*	*Time*	*Passengers*	*Profit*
2	2:00	3	4:00	45	−$ 250
3	6:00	2	8:00	65	750
2	8:00	1	12:00	90	4,000
1	12:00	3	18:00	100	7,500
3	22:00	2	24:00	60	500
			Total Profit, Airplane 1		$12,500
Airplane 2					
Departs	*Time*	*Arrives*	*Time*	*Passengers*	*Profit*
1	0:00	2	4:00	40	$ 1,000
2	4:00	3	6:00	0	− 1,500
3	6:00	1	12:00	60	4,500
1	16:00	2	20:00	60	3,000
2	20:00	1	24:00	60	3,000
			Total Profit, Airplane 2		$10,000
			TOTAL PROFIT		$22,500

appears quite difficult. In Chapter 12 another approach, which is capable of expansion to problems involving a few hundred aircraft, is given. This latter approach is also of considerable interest to scheduling problems in other fields.

10.4 OPTIMIZATION OF DISTRIBUTED PARAMETER SYSTEMS

Distributed parameter systems have always presented a difficult optimization problem to the control engineer. Although a rather complete problem formulation and a large number of optimization methods exist for systems described by ordinary differential equations, little in the way of concrete results has been obtained for systems described by partial differential equations.

The most often used method for dealing with these systems is to convert the partial differential equations to a set of ordinary differential equations and then to apply the methods developed for these equations. In theory, one partial differential equation would require an infinite number of ordinary differential equations to represent it exactly. However, in any practical application it is necessary to use an approximation based on a finite number of equations.

A general procedure for making this approximation is to use a lumped model for the distributed parameter system. In this approach all the independent variables except time are quantized, and the space containing these variables is sub-divided into small units. The dependent variables are assumed to be fixed within a unit except for a variation with time. It is then possible to write a set of coupled ordinary differential equations, where the values of the dependent variables over a unit are the state variables for the system. Unfortunately, computational difficulties often arise when a large number of units are required to represent the distributed parameter system with sufficient accuracy. Nonetheless, this procedure can conceptually be applied to almost any distributed parameter system.

In order to illustrate the general method, attention will be focused on problems with a single dependent variable, y, a single control variable, u, and two independent variables, z and t. It will be assumed for the moment that the highest partial derivative with respect to t is $\partial y(z, t)/\partial t$. This partial differential equation can then be written as

$$\frac{\partial y}{\partial t} = f\left[u, z, t, y, \frac{\partial y}{\partial z}, \frac{\partial^2 y}{\partial z^2} \cdots\right] \tag{10.17}$$

where f is a nonlinear scalar functional. The first step in the procedure is to quantize z into n increments of size Δz. These quantized values of z can then be indexed as

$$\begin{gathered} z_i = z^- + (i - 1)\,\Delta z \qquad (i = 1, 2, \ldots, n) \\ (n - 1)\,\Delta z = z^+ - z^- \end{gathered} \tag{10.18}$$

where z^- = minimum value of z

z^+ = maximum value of z.

The values of y and u at $z = z_i$ and t can then be denoted as $y_i(t)$ and $u_i(t)$ respectively, i.e.,

$$y_i(t) = y(z_i, t) \tag{10.19}$$

$$u_i(t) = u(z_i, t). \tag{10.20}$$

The partial derivatives on the right side of Eq. (10.17) can then be approximated by finite difference equations. If central differences are used (Ref. 1), then the approximations become

$$\begin{aligned} \frac{\partial y(z_i, t)}{\partial z} &= \frac{y(z_{i+1}, t) - y(z_{i-1}, t)}{2\,\Delta z} \\ \frac{\partial^2 y(z_i, t)}{\partial z^2} &= \frac{y(z_{i+1}, t) - 2y(z_i, t) + y(z_{i-1}, t)}{(\Delta z)^2}. \\ &\cdots \end{aligned} \tag{10.21}$$

Using these expressions and the definitions in Eqs. (10.18)–(10.20), the single partial differential equation, Eq. (10.17), is transformed to the n ordinary differential equations

$$\dot{y}_i = f_i(y_1, y_2, \ldots, y_n, u_1, u_2, \ldots, u_n, t) \qquad (i = 1, 2, \ldots, n). \quad (10.22)$$

If an n-dimensional state vector $\mathbf{x}$ is defined as

$$x_i = y_i \qquad (i = 1, 2, \ldots, n) \quad (10.23)$$

and if the n-dimensional vector $\mathbf{u}$ is defined to have components u_i, $i = 1, 2, \ldots, n$, the set of equations in Eq. (10.22) can be written more compactly as

$$\dot{\mathbf{x}} = \mathbf{f}(\mathbf{x}, \mathbf{u}, t) \quad (10.24)$$

These equations are now in the standard form to apply dynamic programming and other optimization techniques. This procedure can easily be extended to the case where higher partial derivatives are present.

The same approach can be applied to the performance criterion. If the original performance criterion is

$$J = \int_{t_0}^{t_f} l\left(u, z, t, y, \frac{\partial y}{\partial z}, \frac{\partial^2 y}{\partial z^2}, \ldots\right) dt \quad (10.25)$$

where l is a scalar functional, then use of Eqs. (10.18)–(10.20) and (10.23) converts the function to

$$J = \int_{t_0}^{t_f} l(\mathbf{x}, \mathbf{u}, t)\, dt. \quad (10.26)$$

Similarly, constraints can be put in the form

$$\mathbf{x} \in X(t) \quad (10.27)$$

$$\mathbf{u} \in U(\mathbf{x}, t). \quad (10.28)$$

10.5 REDUCTION OF THE HIGH-SPEED MEMORY REQUIREMENT IN THE OPTIMIZATION OF A LUMPED APPROXIMATION TO A DISTRIBUTED-PARAMETER SYSTEM

In many applications involving lumped approximation to partial differential equations, the change in the variables over a given unit depends on only the values of variables in neighboring units; in fact, the partial differential equation is often derived from such relations in the first place. If such a relation cannot be easily deduced on physical grounds, then it can be obtained by using central differencing techniques involving only values at neighboring units.

Another condition that is frequently met in practice is that the control, u, is applied not everywhere in the system, but only on a boundary.

If both these conditions are satisfied for the problem described by the partial differential equation in Eq. (10.17), then the corresponding system differential equation, Eq. (10.24), can be written in the special form

$$\begin{aligned}
\dot{x}_1 &= f_1(x_1, x_2, u, t) \\
\dot{x}_i &= f_i(x_{i-1}, x_i, x_{i+1}, t) \qquad (i = 2, 3, \ldots, n-1) \\
\dot{x}_n &= f_n(x_{n-1}, x_n, t)
\end{aligned} \tag{10.29}$$

where the control u is applied at $z = z^-$.

This structure, which occurs in a great many problems, can be exploited to obtain significant reductions in the high-speed memory requirement. First, the differential equations in Eq. (10.29) are approximated by the equations

$$\begin{aligned}
x_1(t + \delta t) &= x_1(t) + f_1[x_1(t), x_2(t), u(t), t]\, \delta t \\
x_i(t + \delta t) &= x_i(t) + f_i[x_{i-1}(t), x_i(t), x_{i+1}(t), t]\, \delta t \\
&\qquad (i = 2, 3, \ldots, n-1) \\
x_n(t + \delta t) &= x_n(t) + f_n[x_{n-1}(t), x_n(t)]\, \delta t.
\end{aligned} \tag{10.30}$$

If δt is a fixed value, Δt, as in the conventional procedure then the following difference equations can be written

$$\begin{aligned}
x_1(t + \Delta t) &= g_1[x_1(t), x_2(t), u(t), t] \\
x_i(t + \Delta t) &= g_i[x_{i-1}(t), x_i(t), x_{i+1}(t), t] \qquad (i = 2, 3, \ldots, n-1) \\
x_n(t + \Delta t) &= g_n[x_{n-1}(t), x_n(t), t]
\end{aligned} \tag{10.31}$$

where the functionals g_i are easily determined from Eq. (10.30) with $\delta t = \Delta t$. This approximation is referred to in the literature as an explicit integration formula [Refs. 2, 3].

An alternative state description can be given for this problem: the variables $x_1, x_2, \ldots, x_n$ are replaced by the values of x_n at t, $t + \Delta t$, $t + 2\,\Delta t, \ldots,$ $t + (n-1)\,\Delta t$. The new state vector, $\tilde{\mathbf{x}}$, can then be defined to have components

$$\tilde{x}_i(t) = x_n[t + (i-1)\,\Delta t]. \qquad (i = 1, 2, \ldots, n). \tag{10.32}$$

The system equation is thus

$$\begin{aligned}
\tilde{x}_i(t + \Delta t) &= \tilde{x}_{i+1}(t) \qquad (i = 1, 2, \ldots, n) \\
\tilde{x}_n(t + \Delta t) &= \tilde{g}[\tilde{x}(t), u(t), t]
\end{aligned} \tag{10.33}$$

The functional $\tilde{g}$ is computed in the following steps. First, it is noted that for a given initial condition, $\mathbf{x}(0)$, it is possible to compute directly $\tilde{\mathbf{x}}(0)$. This is

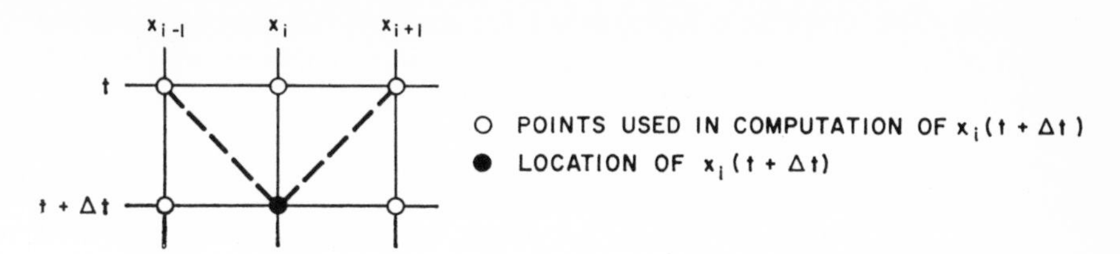

Fig. 10.8. Representation of explicit integration formula.

due to the nature of the explicit integration formula. In Fig. 10.8 the quantized values of z and t for which values of $\mathbf{x}$ are used in computing $x_i(t + \Delta t)$ are indicated. These values are $x_{i-1}(t)$, $x_i(t)$, and $x_{i+1}(t)$. The set of all four points forms a triangle, as shown in the figure. If the values $x_i(0)$, $i = 1, 2, \ldots, N$, are known, then values of $x_i(\Delta t)$ for $i = 2, 3, \ldots, N$ can immediately be obtained by use of Eq. (10.31).

Next, $x_i(2\,\Delta t)$, $i = 3, 4, \ldots, n$, can be obtained from $x_i(\Delta t)$, $i = 2, 3, \ldots, n$. In general, then, $x_i(k\,\Delta t)$, $i = k + 1, \ldots, n$, can be determined from the original set $x_i(0)$, $i = 1, 2, \ldots, n$. The values of z and t where $x_i(t)$ is known thus form the large triangle in the (z, t) plane shown in Fig. 10.9. The desired values of $x_n(t)$, $t = 0, \Delta t, \ldots, (n - 1)\,\Delta t$ are on the right-hand side of this initial triangle.

The second step in the computation of $\tilde{g}$ is related to computing additional triangles as t increases. In general, if $\mathbf{x}(t)$ is known, a triangle of values, $x_i[t + (i - 1)\,\Delta t]$, $i = 1, 2, \ldots, n$, is known. The next triangle is computed

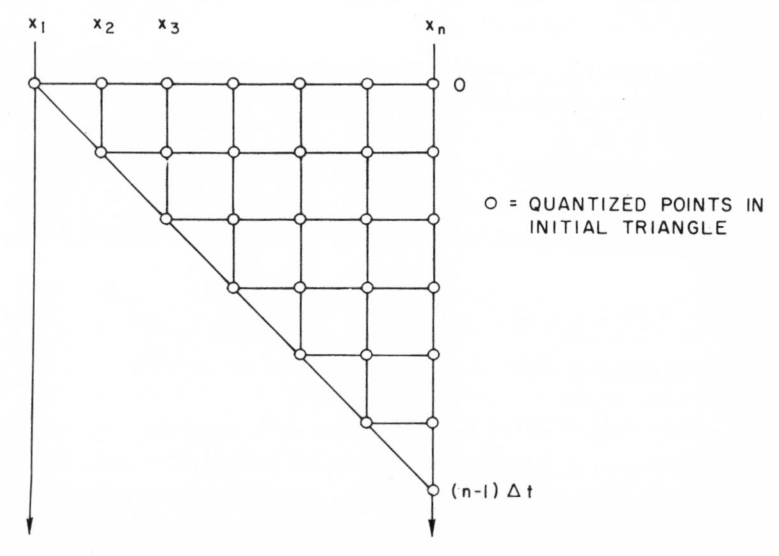

Fig. 10.9. Initial triangle $x_i(t)$ is determined from $\mathbf{x}(0)$.

for a given value of u by applying the equations

$$x_1(t + \Delta t) = g_1[x_1(t), x_2(t), u(t), t]$$
$$x_i(t + i\,\Delta t)$$
$$= g_i[x_{i-1}[t + (i-1)\,\Delta t], x_i[t + (i-1)\,\Delta t], x_{i+1}[t + (i-1)\,\Delta t], t + (i-1)\,\Delta t]$$
$$(i = 2, 3, \ldots, n-1).$$
$$x_n(t + n\,\Delta t) = g_n[x_{n-1}[t + (n-1)\,\Delta t], x_n[t + (n-1)\,\Delta t], t + (n-1)\,\Delta t]. \tag{10.34}$$

The quantities on the right-hand side of Eq. (10.34) are obtained either from points within the present triangle or else from previously computed points

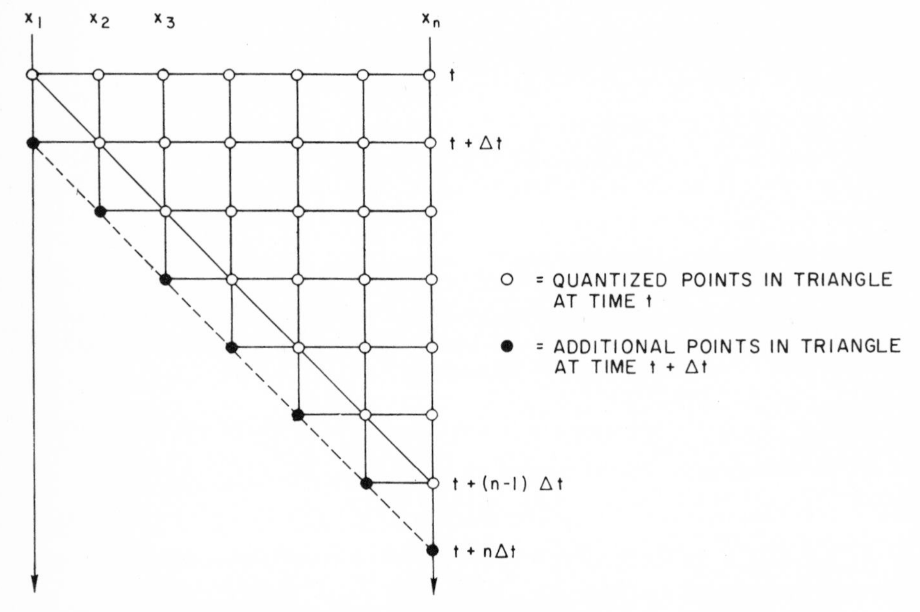

Fig. 10.10. Computation of a new triangle at $t + \Delta t$ based on the triangle at t.

along the new triangle. The computations are pictured in Fig. 10.10, where the quantized points in the triangle at t are connected by solid lines, and the quantized points in the new triangle are connected by a dotted line.

Finally, when the new triangle has been computed, the desired value $\tilde{x}_n(t + \Delta t) = x_n(t + n\,\Delta t)$, has been computed as the last of Eqs. (10.34).

The entire set of operations in Eq. (10.33), once the initial triangle has been computed, involves about the same amount of computations as Eq. (10.31). However, the form of Eq. (10.33) has some special advantages. In particular, for a given set of values of $\tilde{x}_2(t), \tilde{x}_3(t), \ldots, \tilde{x}_n(t)$, the values of $\tilde{x}_1(t + \Delta t)$, $\tilde{x}_2(t + \Delta t), \ldots, \tilde{x}_{n-1}(t + \Delta t)$ are immediately determined. The single control variable thus affects only the one variable $\tilde{x}_n(t + \Delta t)$. As a result, in searching for the minimum cost at the next state, $I[\tilde{\mathbf{x}}(t + \Delta t), t + \Delta t]$, it is necessary only to bring in a one-dimensional array of values, namely the set with values $\tilde{x}_i(t + \Delta t)$ fixed at $x_{i+1}(t)$, $i = 1, 2, \ldots, (n - 1)$, but with $\tilde{x}_n(t + \Delta t)$ allowed to vary over its entire range. It is then possible to compute a one-dimensional array of values of $I(\tilde{\mathbf{x}}, t)$ based on this one-dimensional array of values of $I(\tilde{\mathbf{x}}, t + \Delta t)$. Consequently the high-speed memory

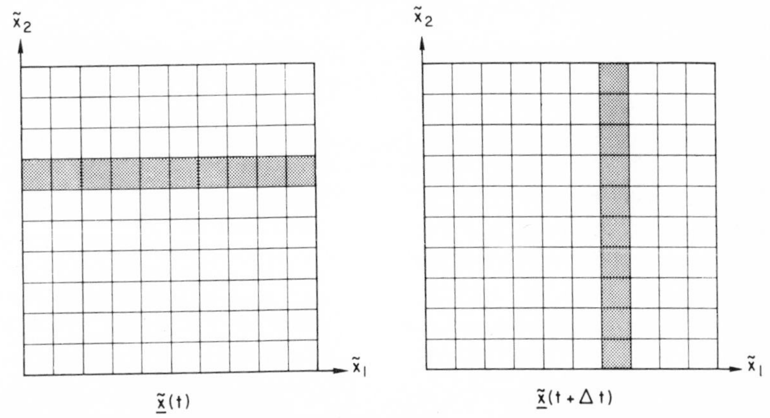

Fig. 10.11. Reduction of the high-speed memory requirement in a two-dimensional problem to the value for a one-dimensional problem.

requirement is N_n, the number of quantization level in the nth coordinate, rather than

$$\prod_{i=1}^{n} N_i = (N_n)^n.$$

As a useful by-product, interpolations of $I(\tilde{\mathbf{x}}, t + \Delta t)$ are in one dimension rather than in n dimensions.

This result is illustrated in Fig. 10.11 for a two-dimensional example. For any state $\tilde{\mathbf{x}}(t + \Delta t)$ in the shaded seventh *column* on the right, the previous state $\tilde{\mathbf{x}}(t)$ must lie in the corresponding shaded seventh *row* on the left; this is true because the fixed value $\tilde{x}_1(t + \Delta t)$ must be the *same* as $\tilde{x}_2(t)$. This figure also illustrates that the next state lies in the *same* one-dimensional set for an entire one-dimensional set of present states; in two dimensions, for every next state $\tilde{\mathbf{x}}(t + 1)$ in a given column, the present state must lie in a corresponding row. Thus, by putting values of the minimum cost at $(t + \Delta t)$ for a whole one-dimensional set into the high-speed memory, it is possible to determine optimal control and minimum cost at t for a corresponding one-dimensional set.

The generalization of this result to q-dimensional controls has been made by Wong [Ref. 4]. In this reference it is shown that the high-speed memory requirement can always be reduced from the order of $(N_n)^n$ to $(N_n)^q$. If q is smaller than n, this saving can be very significant. In Ref. 4 the extension of this result to other forms of the differential equation, Eq. (10.24), is treated as well. Finally the use of this method in combination with state increment dynamic programming is also considered in Ref. 4.

10.6 SHORT-TERM OPTIMIZATION OF NATURAL GAS PIPELINE NETWORKS

In Refs. 5 and 6 the application of these methods to the short-term optimization of natural gas pipeline networks is discussed. In this problem it is desired to find the operating policy for the compressors in the pipelines such that over a 24-hour period, the cost of compressor operation is minimized while the demanded flow of gas is delivered at a pressure higher than contract pressure.

The partial differential equations for the flow of gas in a pipeline are

$$\frac{\partial P}{\partial t} = -c_1 \frac{\partial Q}{\partial z} \tag{10.35}$$

$$\frac{\partial P^2}{\partial z} = -c_2 Q^2 \tag{10.36}$$

where $P(z, t)$ = pressure inside the pipeline
$Q(z, t)$ = instantaneous mass flow rate in the pipeline
z = spatial coordinate along the pipeline
t = time
c_1, c_2 = constants that depend on the pipeline characteristics.

These equations can be written as the single equation

$$\frac{\partial x}{\partial t} = \varphi(x) \frac{\partial^2 x}{\partial z^2} \tag{10.37}$$

where $x(z, t) = P^2(z, t)$

$$\varphi(x) = \frac{c_1}{\sqrt{c_2}} \left(\frac{x}{\left| \dfrac{\partial x}{\partial z} \right|} \right)^{1/2}$$

For a single pipeline of length L, with a compressor located at $z = 0$ and a demand point at $z = L$, the boundary conditions are

$$x(0, t) = u^2(t) \tag{10.38}$$

$$\frac{\partial x(L, t)}{\partial z} = -c_2\, d^2(t) \tag{10.39}$$

where $u(t)$ = control variable
= pressure generated by the compressor at the input to the pipeline
$d(t)$ = demanded flow at the end of the pipeline.

If the pipeline is divided into n sections of length Δz, as in Fig. 10.12, the

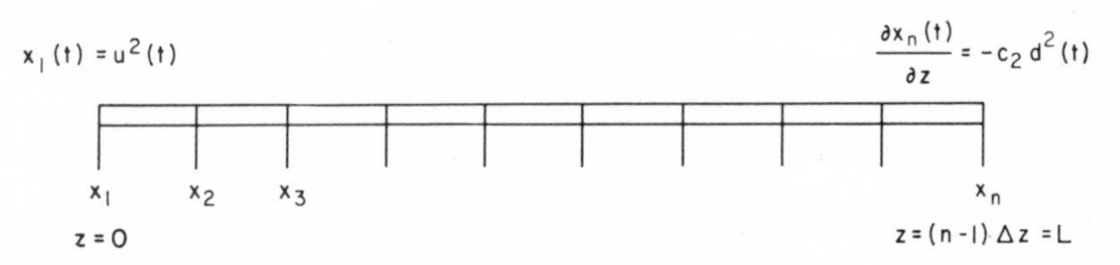

Fig. 10.12. A single pipeline with compressor at $z = 0$ and demand point at $z = L$.

approximations of Eqs. (10.21) can be applied. After suitable manipulation the system equations for the explicit integration formula become

$$x_1(t + \Delta t) = u^2(t + \Delta t)$$

$$x_i(t + \Delta t) = x_i(t) + \frac{\Delta t}{(\Delta z)^2}\frac{c_1}{\sqrt{c_2}}\left\{\frac{(2\,\Delta z)x_i(t)}{|x_{i+1}(t) - x_{i-1}(t)|}\right\}^{1/2}\{x_{i+1}(t) - 2x_i(t) + x_{i-1}(t)\}$$

$$(i = 2, 3, \ldots, n - 1)$$

$$x_n(t + \Delta t) = x_{n-1}(t + \Delta t) - c_2(\Delta z)\, d^2(t + \Delta t). \tag{10.40}$$

The performance criterion is generally taken to be compressor fuel cost, which is proportional to compressor energy. Under suitable assumptions this criterion can be written as (Ref. 5)

$$J = \sum_{t=0}^{K\Delta t}\left(\frac{x_1(t) - x_2(t)}{c_2(\Delta z)}\right)^{1/2}\left(c_3\left(\frac{x_1}{P_0^2}\right)^{c_4} - 1\right) \tag{10.41}$$

where $K\,\Delta t$ = time interval over which optimization is performed
P_0 = suction pressure at input to compressor
c_3, c_4, c_5 = constants for the specific pipeline

The constraints are that demand flow is to be supplied at a pressure greater than or equal to contract pressure, P_{con}, and that pressure inside the pipeline can never exceed a value P_M, determined from safety considerations. The constraint on demanded flow is already incorporated into the last of Eqs. (10.40). The constraint on contract pressure can be written as

$$x_n(t) \geq P_{\text{con}}^2. \tag{10.42}$$

The constraint on maximum pressure in the pipeline is

$$x_i(t) \leq P_M^2 \qquad (i = 1, 2, \ldots, n). \tag{10.43}$$

A computer program based on the theory of the preceding section was run for a particular pipeline (Ref. 5). The number of sections was determined by setting $n = 4$.* The constants for the program were

$$\begin{aligned} \Delta t &= 10 \text{ minutes} \\ K &= 18 \\ L &= 15.95 \text{ miles} \\ P_{\text{con}} &= 450 \text{ psi} \\ P_M &= 500 \text{ psi} \\ c_1 &= 113.414 \\ c_2 &= 22.631 \times 10^{-9} \\ c_3 &= 1.0088 \\ c_4 &= 0.1085. \end{aligned} \tag{10.44}$$

A representative solution for a particular demand curve is shown in Fig. 10.13. The demanded flow rate, $d(t)$, is shown as the lower solid line in the figure. The corresponding input flow rate $Q_{\text{in}}(t)$ is shown as the lower dotted line. This latter quantity is plotted rather than input pressure, $\sqrt{x_1(t)}$, because in this program it was preferable for numerical reasons to use input flow rate as the control variable, instead of input pressure.

A very interesting result of this program is that output pressure $P_{\text{out}}(t) = \sqrt{x_n(t)}$, was very close to contract pressure, P_{con}, over the entire run. This same phenomenon was observed in many other cases. Thus, a near-optimal policy is clearly to adjust the compressor operation so as to maintain P_{out} as near P_{con} as possible.

Further work on the short-term optimization problem is being directed towards determining an algorithm that maintains P_{out} as close to P_{con} as possible while still meeting the demanded flow and satisfying the constraint on maximum pressure in the pipeline. Some extremely fast and accurate algorithms that have been developed are described in Refs. 5 and 6. Thus in this application dynamic programing was used as a guide for developing and evaluating near-optimal policies, and an on-line control algorithm is now being designed on the basis of a policy that was shown to be both simple to implement and very close to optimum in performance.

* Because of the way the boundary conditions were handled, the number of state variables in the program was not $4 = n$, but 2.

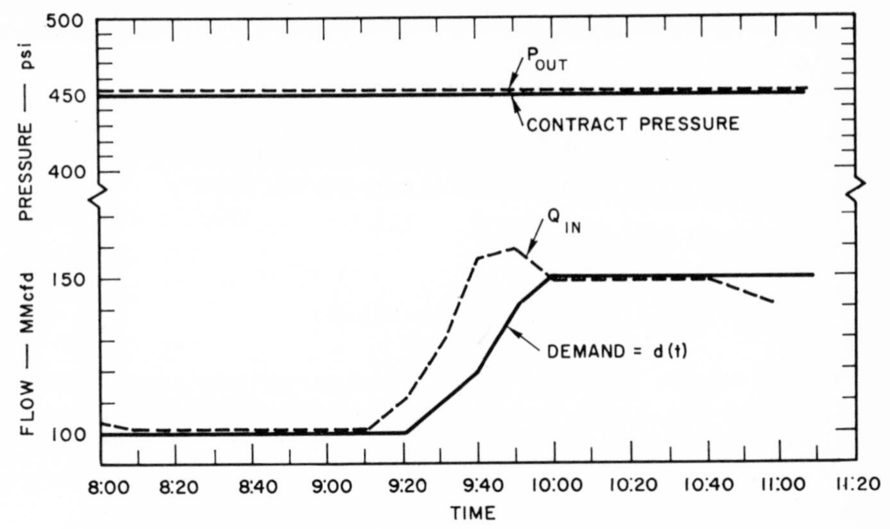

Fig. 10.13. Solution to the pipeline optimization problem.

10.7 CONCLUSIONS

In this section it was shown that state increment dynamic programming can be applied to problems quite different from the guidance and control problems of the previous chapter. It was also shown that in many applications other dynamic programming methods can be combined with state increment dynamic programming to produce an overall computational algorithm of great power and generality. Finally, it was indicated that dynamic programming in general, and state increment dynamic programming in particular, has the flexibility to be a valuable tool in applications ranging from fast-response on-line control to long-term off-line evaluation and planning studies.

REFERENCES

1. F. B. Hildebrand, "Introduction to Numerical Analysis," pp. 91–99 (McGraw-Hill Book Company, Inc., New York, N.Y., 1956).
2. V. K. Savlèv, "Integration of Equations of Parabolic Type by the Method of Nets," Transl. by G. S. Tee (Macmillan Co., New York, N.Y., 1964).
3. R. D. Richtmyer, "Difference Methods for Initial Value Problems," pp. 89–120 (Interscience Publishers, Inc., New York, N.Y., 1957).
4. P. J. Wong, "Dynamic Programming Using Shift Vectors," Ph.D. Thesis, Department of Electrical Engineering, Stanford University, Stanford, California (September, 1967).
5. R. E. Larson, T. L. Humphrey and P. J. Wong, "Short-Term Optimization of a Single Pipeline," Interim Report, SRI Project 5975, Stanford Research Institute, Menlo Park, California (February 1967).
6. R. E. Larson, T. L. Humphrey and P. J. Wong, "Short-Term Optimization of Gas Pipeline Networks," Final Report, SRI Project 5975, Stanford Research Institute, Menlo Park, California (April 1967).

Chapter Eleven

APPLICATIONS TO SYSTEMS CONTAINING UNCERTAINTY

11.1 INTRODUCTION

In all of the discussions so far it has been assumed that there are no random forces present in the system. In this chapter the extension of the basic dynamic programming formulation to problems where these forces are present is discussed. Also, the modifications of the state increment dynamic programming computational procedure for these cases are presented.

In the next section Bellman's stochastic control problem [Refs. 1, 2] is discussed. In this problem the present values of the state variables are known, but random forces are present which affect the state variables at future times. The problem formulation is given, and an iterative functional equation is derived. The state increment dynamic programming computational procedure for this case is presented, and some examples are worked out.

In the section after that the optimum estimation problem is discussed. In this problem it is desired to find the best estimate of the present values of the state variables based on past noisy measurements of some or all of these variables. Again, the problem formulation is given, and an iterative functional equation is derived [Ref. 3]. The applicability of state increment dynamic programming is indicated.

Finally, the combined optimum control and estimation problem is discussed [Refs. 4–6]. In this problem both of the effects mentioned above are present—i.e., future values of the state variables are affected by random forces, and the only information given about the present values is a set of noisy measurements. This problem formulation includes many cases of great theoretical and practical interest, such as optimum adaptive control. It is shown that an iterative functional equation can again be derived. The use of state increment dynamic programming in these problems is mentioned briefly.

11.2 STOCHASTIC CONTROL PROBLEMS

One of the most important areas to which dynamic programming has been applied is stochastic control problems. Bellman [Refs. 1, 2] has developed a general problem formulation which leads to a stochastic principle of optimality.

In this section this formulation is briefly reviewed, and the computational procedure of state increment dynamic programming is extended to cover it. Some examples are presented.

The general stochastic control problem is concerned with systems for which the present values of the state variables are known, but future values are affected by random forces. The system equations, instead of depending only on the state, control, and time, depend as well on a set of random variables, $w_j, j = 1, 2, \ldots, b$. These variables are written as a b-dimensional random forcing vector, $\mathbf{w}$. In the discrete-time case the system difference equation is

$$\mathbf{x}(k+1) = \mathbf{g}[\mathbf{x}(k), \mathbf{u}(k), \mathbf{w}(k), k] \tag{11.1}$$

The probability density function of $\mathbf{w}$ is assumed to be known. This function may or may not vary with k; in either case, it will be written as $p[\mathbf{w}(k)]$.

It can be assumed without loss of generality that the random variables $\mathbf{w}(k)$ are independent from one time instant to the next. If there is correlation, it can be represented by defining a new set of random variables that *are* independent between samples and defining additional state variables, as in Ref. 4. Under this assumption the joint probability density function, $p[\mathbf{w}(0), \mathbf{w}(1), \ldots, \mathbf{w}(K)]$, can be rewritten as the product of the individual probability densities, $p[\mathbf{w}(k)]$,

$$p[\mathbf{w}(0), \mathbf{w}(1), \ldots, \mathbf{w}(K)] = p[\mathbf{w}(0)]p[\mathbf{w}(1)], \ldots, p[\mathbf{w}(K)] \tag{11.2}$$

It is no longer possible to optimize a given function of $\mathbf{x}$, $\mathbf{u}$, and k, because the present state $\mathbf{x}(k)$ and the present and future controls, $\mathbf{u}(j), k \le j \le K$, do *not* completely specify the future states, $\mathbf{x}(j)$, $k+1 \le j \le K$, but instead determine only the probability distribution of these states. The quantity that is instead optimized is the *expected value* of such a function. The performance criterion is thus written as

$$J = \underset{\mathbf{w}(0),\mathbf{w}(1),\ldots,\mathbf{w}(K)}{E}\left\{\sum_{k=0}^{K} l[\mathbf{x}(k), \mathbf{u}(k), \mathbf{w}(k), k]\right\}. \tag{11.3}$$

Constraints are exactly as in the deterministic case; they are written

$$\mathbf{x} \in X(k) \tag{11.4}$$

$$\mathbf{u} \in U(\mathbf{x}, k). \tag{11.5}$$

The stochastic principle of optimality can now be derived. First, $I(\mathbf{x}, k)$ is defined to be the expected minimum cost function

$$I(\mathbf{x}, k) = \underset{\mathbf{u}(k),\ldots,\mathbf{u}(K)}{\text{Min}}\left\{\underset{\mathbf{w}(k),\ldots,\mathbf{w}(K)}{E}\left[\sum_{j=k}^{K} l[\mathbf{x}(j), \mathbf{u}(j), \mathbf{w}(j), j]\right]\right\} \tag{11.6}$$

where $\mathbf{x}(k) = \mathbf{x}$. Using this definition and substituting Eq. (11.6) into the expected value operator, the derivation proceeds very much as in the deterministic case. The functional equation that is obtained is

$$I(\mathbf{x}, k) = \operatorname*{Min}_{\mathbf{u}} \left\{ \operatorname*{E}_{\mathbf{w}}[l(\mathbf{x}, \mathbf{u}, \mathbf{w}, k) + I[\mathbf{g}(\mathbf{x}, \mathbf{u}, \mathbf{w}, k), k + 1]] \right\} \tag{11.7}$$

For details of the derivation the reader is referred to Refs. 1 and 2.

The computational procedure based on this equation is very similar to that for the deterministic case. The random forcing vector $\mathbf{w}$ is quantized, and a discrete probability distribution is obtained. In order to evaluate the quantity inside the brackets in Eq. (11.7) for a given state $\mathbf{x}$ and a given control $\mathbf{u}$, each quantized value of $\mathbf{w}$ is substituted, the resulting quantity is multiplied by the probability of that value of $\mathbf{w}$, and the expected value is found by adding these quantities. If the quantized values of $\mathbf{w}$ are denoted as $\mathbf{w}^{(a)}$, $a = 1, 2, \ldots, A$, then the expected value can be written as

$$\begin{aligned} &\operatorname*{E}_{\mathbf{w}}[l(\mathbf{x}, \mathbf{u}, \mathbf{w}, k) + I[\mathbf{g}(\mathbf{x}, \mathbf{u}, \mathbf{w}, k), k + 1]] \\ &\qquad = \sum_{a=1}^{A} [l(\mathbf{x}, \mathbf{u}, \mathbf{w}^{(a)}, k) + I[\mathbf{g}(\mathbf{x}, \mathbf{u}, \mathbf{w}^{(a)}, k), k + 1]] p(\mathbf{w}^{(a)}). \end{aligned} \tag{11.8}$$

Once this quantity has been computed for each given value of $\mathbf{x}$ and $\mathbf{u}$, the procedure is exactly the same as in the deterministic case.

It can be seen from this simplified explanation that the stochastic nature of the problem increases the computing time by a factor of A, where A is the number of quantized values of $\mathbf{w}$. On the other hand, the high-speed memory requirement is not increased over the value of deterministic control. Thus, in most stochastic control problems of a reasonable size both high-speed memory requirement and computing time cause difficulties, while in many deterministic problems the high-speed memory requirement is the more severe problem.

State increment dynamic programming can be applied to stochastic control problems. The application is most clearly explained in terms of the continuous-time case, where the equations of motion are the differential equations

$$\dot{\mathbf{x}} = \mathbf{f}(\mathbf{x}, \mathbf{u}, \mathbf{w}, t) \tag{11.9}$$

where $\mathbf{w}$ is now a continuous-time random function. The performance criterion is

$$J = E\left\{ \int_{t_0}^{t_f} l(\mathbf{x}, \mathbf{u}, \mathbf{w}, \sigma)\, d\sigma + \psi[\mathbf{x}(t_f), t_f] \right\}. \tag{11.10}$$

If the differential equation is converted to a difference equation in the usual manner, then the stochastic principle of optimality becomes

$$I(\mathbf{x}, t) = \operatorname*{Min}_{\mathbf{u}} \left\{ \operatorname*{E}_{\mathbf{w}}[l(\mathbf{x}, \mathbf{u}, \mathbf{w}, t)\, \delta t + I[\mathbf{x} + \mathbf{f}(\mathbf{x}, \mathbf{u}, \mathbf{w}, t)\, \delta t, t + \delta t] \right\}. \tag{11.11}$$

Because the above equation is based on finite difference approximations to Eqs. (11.9) and (11.10), the question of the physical meaning of continuous-time random functions which are uncorrelated from time instant to time instant is avoided. However, it is assumed that samples of $\mathbf{w}$ at t and $t + \delta t$ are independent for all admissible values of δt over all t, $t_0 \leq t \leq t_f$.

In evaluating the term inside brackets in Eq. (11.9) at a given value of $\mathbf{x}$ and $\mathbf{u}$, the vector $\mathbf{w}$ is quantized into A values, $\mathbf{w}^{(a)}$, $a = 1, 2, \ldots, A$. The corresponding time interval over which the control $\mathbf{u}$ is applied, $\delta t^{(a)}$, is determined as

$$\delta t^{(a)} = \operatorname*{Min}_{i=1,2,\ldots,n} \left\{ \left| \frac{\Delta x_i}{f_i(\mathbf{x}, \mathbf{u}, \mathbf{w}^{(a)}, t)} \right| \right\}. \tag{11.12}$$

The expected value in Eq. (11.11) is then evaluated as

$$\begin{aligned} &\operatorname*{E}_{\mathbf{w}}[l(\mathbf{x}, \mathbf{u}, \mathbf{w}, t)\, \delta t + I[\mathbf{x} + \mathbf{f}(\mathbf{x}, \mathbf{u}, \mathbf{w}, t)\, \delta t, t + \delta t] \\ &\quad = \sum_{a=1}^{A} [l(\mathbf{x}, \mathbf{u}, \mathbf{w}^{(a)}, t)\, \delta t^{(a)} + I[\mathbf{x} + \mathbf{f}(\mathbf{x}, \mathbf{u}, \mathbf{w}^{(a)}, t)\, \delta t^{(a)}, t + \delta t^{(a)}]]p(\mathbf{w}^{(a)}). \end{aligned} \tag{11.13}$$

The interpolation for I is again in $(n - 1)$ state variables and t.

With this definition of the expected value term inside the brackets, the search procedure at a given value of $\mathbf{x}$ is exactly as in the deterministic case. As in the conventional procedure, the number of computations is increased by a factor of A. The high-speed memory requirement is thus the same in both the deterministic and the stochastic case, so that state increment dynamic programming results in the same reduction in both cases.

State increment dynamic programming has been applied to a number of stochastic control problems. The results for one illustrative example are shown in Fig. 11.1. Here the system equations are

$$\dot{x} = u + w \tag{11.14}$$

where x = scalar state variable
u = scalar control variable
w = scalar random forcing function.

The probability density function of w is taken to be gaussian with zero mean and variance $= \frac{1}{2}$.

$$p(w) = \frac{1}{\sqrt{\pi}} e^{-w^2}. \tag{11.15}$$

The performance criterion is

$$J = E\left\{ \int_0^{10} (x^2 + u^2)\, d\sigma + 2.5(x(10))^2 \right\}. \tag{11.16}$$

The constraints are

$$-3 \leq x \leq 3 \tag{11.17}$$

$$-1 \leq u \leq 1. \tag{11.18}$$

The state variable is quantized uniformly in increments of $\Delta x = 1$, the control is quantized in increments of $\Delta u = 1$, and Δt is taken to be 1. The probability density of the random forcing function is quantized so that

$$\begin{aligned} p(w = 1) &= \tfrac{1}{4} \\ p(w = 0) &= \tfrac{1}{2} \\ p(w = -1) &= \tfrac{1}{4}. \end{aligned} \tag{11.19}$$

This distribution has the same mean and variance as the original density function.

The system equation in discrete form becomes

$$x(t + \delta t) = x(t) + [u(t) + w(t)]\, \delta t \tag{11.20}$$

where δt is determined by Eq. (11.12). The iterative equation is

$$I(x, t) = \underset{u=-1,0,1}{\mathrm{Min}} \left\{ (x^2 + u^2)\, \delta t + \underset{w}{E}[I(x + (u + w)\, \delta t, t + \delta t)] \right\}. \tag{11.21}$$

The block size is taken to be 2 Δx by 5 Δt. The preferred direction of motion is taken to be towards the origin. The results of the computations are shown in Fig. 11.1.

The same problem with the same increment sizes was also solved by the conventional procedure. This solution appears in Fig. 11.2. As in the deterministic problem of Chapter 7, the minimum costs are lower in the state increment dynamic programming solution; again, this can be accounted for because this procedure effectively applies more controls when the state is changing rapidly. The optimal controls determined by both procedures are identical.

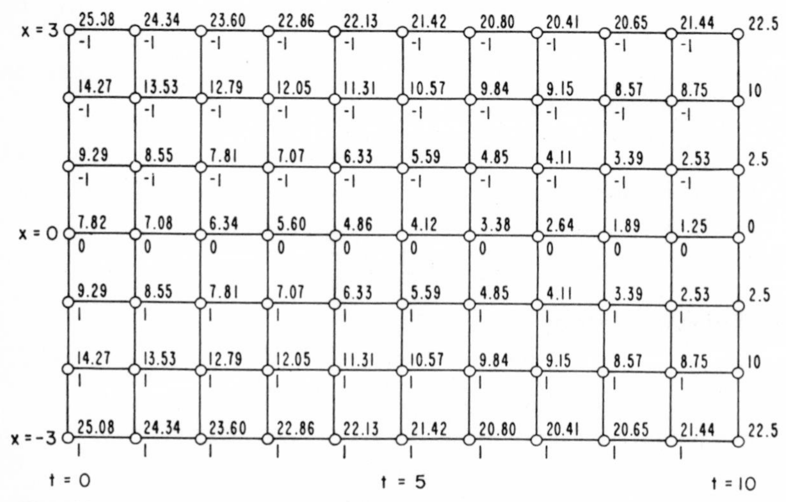

Fig. 11.1. Solution to stochastic control problem as obtained by state increment dynamic programming.

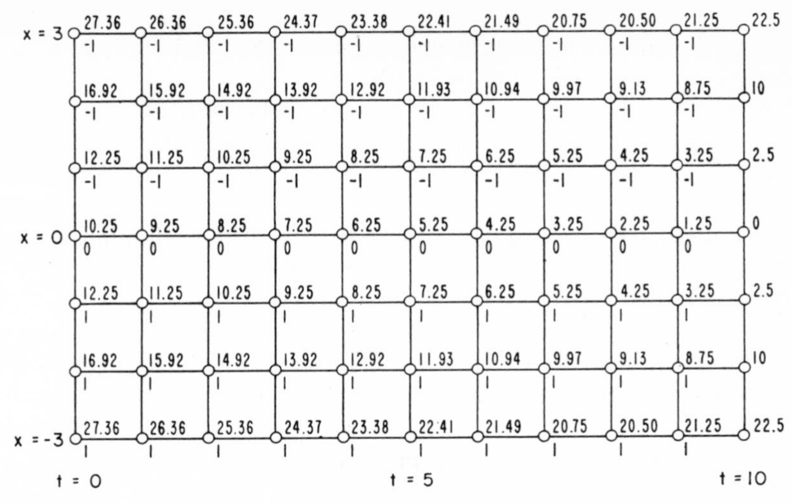

Fig. 11.2. Solution to stochastic control problem as obtained by the conventional procedure.

11.3 OPTIMUM ESTIMATION

A problem closely related to optimal control is that of optimal estimation. In this problem, it is desired to find the best estimate of the state of a system on the basis of a number of noisy observations of it. In this section a new computational procedure based on dynamic programming for solving a general class of optimum estimation problems is presented [Ref. 3]. The use of the concepts of state increment dynamic programming to reduce the computational requirements of the procedure is also mentioned.

The optimum estimation problem is formulated as follows. The system equation for the discrete case is

$$\mathbf{x}(k+1) = \mathbf{g}[\mathbf{x}(k), \mathbf{w}(k), k] \tag{11.22}$$

where $\mathbf{x}$ = n-dimensional state vector
$\mathbf{w}$ = b-dimensional random forcing function
$\mathbf{g}$ = n-dimensional vector functional

The equations for the measurement system are

$$\mathbf{z}(k) = \mathbf{h}[\mathbf{x}(k), \mathbf{v}(k), k] \tag{11.23}$$

where $\mathbf{z}$ = c-dimensional measurement vector, where in general $c \leq n$.
$\mathbf{h}$ = c-dimensional vector functional.
$\mathbf{v}$ = d-dimensional vector of random noise.

The probability density functions $p[\mathbf{w}(k)]$ and $p[\mathbf{v}(k)]$ are assumed to be known and independent from sample to sample. The probability density function of the initial state before any measurements are received, denoted as $p[\mathbf{x}(0/-1)]$, is also assumed to be known.

The problem which is solved in this section can be stated as follows: given the system described by Eq. (11.22), the measurement system described by Eq. (11.23), the probability density functions $p[\mathbf{w}(j-1)]$ and $p[\mathbf{v}(j)]$, $j = 1, 2, \ldots, k$, the a priori probability density $p[\mathbf{x}(0/-1)]$, and the noisy measurements, $\mathbf{z}(1), \ldots, \mathbf{z}(k)$, find the maximum likelihood estimate of the entire trajectory, $\mathbf{x}(0), \mathbf{x}(1), \ldots, \mathbf{x}(k)$. This estimate of the trajectory is denoted as $\hat{\mathbf{X}}(k/k)$.

This problem is somewhat different from the one customarily solved by using Bayes' theorem [Refs. 7, 8]. In the latter problem the maximum likelihood estimate of $\mathbf{x}(k)$, the present state, is found, rather than the maximum likelihood estimate for the entire trajectory. The difference between these estimates is discussed in more detail later in this section.

An iterative relation for the maximum likelihood trajectory estimate is derived by applying Bayes' rule. The notation will be such that it is easy to see the similarity between this iterative relation and the dynamic programming functional equation. First, in order to simplify the algebra, the following definitions are made:

$$\begin{aligned} \mathbf{X}(k) &= \mathbf{x}(k), \mathbf{x}(k-1), \ldots, \mathbf{x}(0) \\ \mathbf{Z}(k) &= \mathbf{z}(k), \mathbf{z}(k-1), \ldots, \mathbf{z}(1). \end{aligned} \tag{11.24}$$

The function $I[\mathbf{x}(k), k]$ is then defined as

$$\begin{aligned} I[\mathbf{x}(k), k] &= \max_{\mathbf{x}(0),\mathbf{x}(1),\ldots,\mathbf{x}(k-1)} \{p[\mathbf{x}(k), \mathbf{x}(k-1), \ldots, \mathbf{x}(0)/\mathbf{z}(k), \mathbf{z}(k-1) \cdots \mathbf{z}(1)]\} \\ &= \max_{\mathbf{x}(0),\mathbf{x}(1),\ldots,\mathbf{x}(k-1)} \{p[\mathbf{X}(k)/\mathbf{Z}(k)]\}. \end{aligned} \tag{11.25}$$

The desired relation is obtained by deriving $I[\mathbf{x}(k+1), k+1]$ in terms of $I[\mathbf{x}(k), k]$.

Bayes' rule can be written as

$$p[\mathbf{X}(k+1)/\mathbf{Z}(k+1)] = \frac{p[\mathbf{z}(k+1)/\mathbf{X}(k+1), \mathbf{Z}(k)]p[\mathbf{X}(k+1)/\mathbf{Z}(k)]}{p[\mathbf{z}(k+1)/\mathbf{Z}(k)]}. \tag{11.26}$$

Since, by Eq. (11.23), $\mathbf{z}(k+1)$ depends only on $\mathbf{x}(k+1)$, Eq. (11.26) can be simplified as

$$p[\mathbf{X}(k+1)/\mathbf{Z}(k+1)] = \frac{p[\mathbf{z}(k+1)/\mathbf{x}(k+1)]p[\mathbf{X}(k+1)/\mathbf{Z}(k)]}{p[\mathbf{z}(k+1)/\mathbf{Z}(k)]}. \tag{11.27}$$

The term $p[\mathbf{X}(k+1)/\mathbf{Z}(k)]$ can be re-written as

$$p[\mathbf{X}(k+1)/\mathbf{Z}(k)] = p[\mathbf{x}(k+1)/\mathbf{x}(k)]p[\mathbf{X}(k)/\mathbf{Z}(k)]. \tag{11.28}$$

Substituting this relation into Eq. (11.27),

$$I[\mathbf{x}(k+1), k+1] = \underset{\mathbf{x}(0),\mathbf{x}(1),\ldots,\mathbf{x}(k)}{\text{Max}} \{p[\mathbf{X}(k+1)/\mathbf{Z}(k+1)]\}$$

$$= \underset{\mathbf{x}(k)}{\text{Max}} \left\{ \frac{p[\mathbf{z}(k+1)/\mathbf{x}(k+1)]p[\mathbf{x}(k+1)/\mathbf{x}(k)]}{p[\mathbf{z}(k+1)/\mathbf{Z}(k)]} I[\mathbf{x}(k), k] \right\}. \quad (11.29)$$

This equation yields the desired iterative relation; the maximization operation is over the single value $\mathbf{x}(k)$, rather than over the entire set of past states, $\mathbf{x}(0)$, $\mathbf{x}(1)$, . . . , $\mathbf{x}(k)$.

The quantity $p[\mathbf{z}(k+1)/\mathbf{x}(k+1)]$ is determined by Eq. (11.23) and knowledge of $p[\mathbf{v}(k+1)]$, while $p[\mathbf{x}(k+1)/\mathbf{x}(k)]$ is determined by Eq. (11.22) and $p[\mathbf{w}(k)]$. The function $p[\mathbf{z}(k+1)/\mathbf{Z}(k)]$ serves only as a normalization factor, and the maximum likelihood estimate can be determined without explicitly computing it. The iterative relation can then be rewritten as

$$I^*[\mathbf{x}(k+1), k+1] = \max_{\mathbf{x}(k)} \{p[\mathbf{z}(k+1)/\mathbf{x}(k+1)]p[\mathbf{X}(k+1)/\mathbf{x}(k)]I^*[\mathbf{x}(k), k]\} \quad (11.30)$$

where the function $I^*[\mathbf{x}(k), k]$ is proportional to, but not equal to $I[\mathbf{x}(k), k]$.

If the system equation, Eq. (11.22), is linear in $\mathbf{x}(k)$ and $\mathbf{w}(k)$, if the measurement system, Eq. (11.23), is linear in $\mathbf{x}(k)$ and $\mathbf{v}(k)$, and if the probability density functions $p[\mathbf{x}(0/-1)]$, $p[\mathbf{w}(k)]$, and $p[\mathbf{v}(k)]$ are all gaussian; then the Kalman-Bucy estimator [Ref. 9] can be obtained directly from Eq. (11.25) [Ref. 3].

For the general case a computational procedure analogous to forward dynamic programming (see Sec. 10.2) can be implemented. The state vector $\mathbf{x}(k)$ is quantized. The a priori probability density function is used as an initial condition

$$I^*[\mathbf{x}(0), 0] = p[\mathbf{x}(0/-1)]. \quad (11.31)$$

At each quantized value of $\mathbf{x}(1)$ the quantity inside brackets in Eq. (11.30) is evaluated for every quantized value of $\mathbf{x}(0)$. The maximum value is selected as $I^*[\mathbf{x}(1), 1]$, and the corresponding value $\mathbf{x}(0)$ is stored along with it. This procedure, in which $I^*[\mathbf{x}(j), j]$ is computed on the basis of $I[\mathbf{x}(j-1), j-1]$ and the corresponding value of $\mathbf{x}(j-1)$ is carried along, is repeated until $I^*[\mathbf{x}(k), k]$ is reached. The maximum likelihood trajectory is then determined by computing the value of $\mathbf{x}(k)$ which maximizes $I^*[\mathbf{x}(k), k]$, retrieving the $\mathbf{x}(k-1)$ that corresponds to this value, finding the $\mathbf{x}(k-2)$ corresponding to this value of $I[\mathbf{x}(k-1), k-1]$, and repeating until all values of $\mathbf{x}(j)$ along the trajectory have been found.

The relation between the estimates obtained from this procedure and from the Bayesian estimation procedure of Refs. 7 and 8 is illustrated for a two-stage problem in Fig. 11.3. The state of the system is assumed to be one

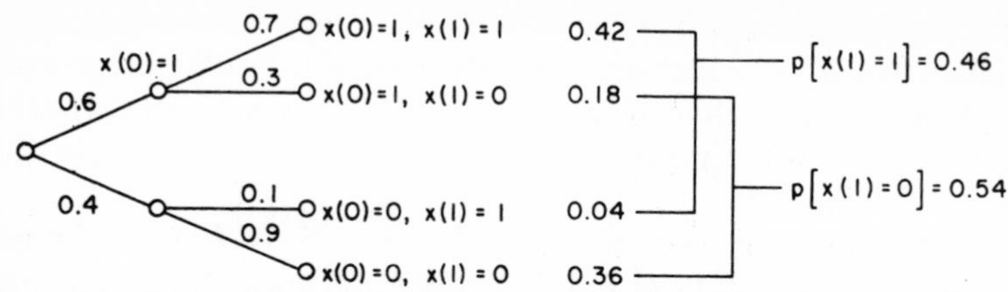

Fig. 11.3. The difference between maximum likelihood trajectory and maximum likelihood state.

of two values, 0 and 1. The a priori probabilities of the two states are $p[x(0) = 0] = 0.4$ and $p[x(0) = 1] = 0.6$.

The system equations are such that $p[x(1)/x(0)]$ is given in Table 11.1. The

Table 11.1. System Equation for First Optimum Estimation Problem

$x(0)$	$x(1)$	$p[x(1)/x(0)]$
0	0	0.9
0	1	0.1
1	0	0.3
1	1	0.7

observation at $k = 1$ is also assumed to have two values, 0 and 1. The observation equations specify $p[z(1)/x(1)]$ as in Table 11.2.

If the observation $z(1) = 1$ is received, the conditional probabilities of the four possible trajectories are as shown in Fig. 11.3 (the same results are achieved if $z(1) = 0$). The maximum likelihood trajectory is found to be $x(0) = 1$, $x(1) = 1$, which has probability 0.42. On the other hand, the probability that $x(1) = 1$, which is obtained by adding the probabilities of trajectories $x(0) = 1$, $x(1) = 1$, and $x(0) = 0$, $x(1) = 1$, is 0.46, while the probability that $x(1) = 0$ is 0.54. Thus, the maximum likelihood present state is $x(1) = 0$. In this simple example, the final state along the maximum likelihood trajectory differs from the maximum likelihood state.

Table 11.2 Measurement Equation for First Optimum Estimation Problem

$x(1)$	$z(1)$	$p[z(1)/x(1)]$
0	0	0.5
0	1	0.5
1	0	0.5
1	1	0.5

This example also clarifies the relation between these two estimates. The maximum likelihood trajectory is the trajectory with the largest probability, while the maximum likelihood present state is found by adding the probabilities of all trajectories with the same present state and choosing the state with the largest probability.

A second example is specified in Tables 11.3–11.5. The state of the system, the random forcing function, and the observation noise all take on only two values, 0 and 1. The system difference equation can be summarized in Table 11.3.

Table 11.3. System Equation for Second Optimum Estimation Example

$x(k)$	$w(k)$	$x(k+1)$
0	0	0
0	1	1
1	0	1
1	1	0

The probability density function for $w(k)$ is specified as

$$p[w(k) = 0] = 0.875$$
$$p[w(k) = 1] = 0.125. \tag{11.32}$$

The measurement equation is shown in Table 11.4.

The probability density function for $v(k)$ is given by

$$p[v(k) = 0] = 0.75$$
$$p[v(k) = 1] = 0.25. \tag{11.33}$$

The a priori probability density for x is given by

$$p[x(0/-1) = 0] = I^*(0, 0) = 0.20$$
$$p[x(0/-1) = 1] = I^*(1, 0) = 0.80. \tag{11.34}$$

Table 11.4. Measurement Equation for Second Optimum Estimation Example

$x(k)$	$v(k)$	$z(k)$
0	0	0
0	1	1
1	0	1
1	1	1

Table 11.5. Maximum Likelihood Trajectory and the Sequence of Maximum Likelihood States in Second Optimum Estimation Example

					Max. likelihood traj. calculation					Bayesian calculation		
k	$x(k)$	$w(k)$	$v(k)$	$z(k)$	$l^*(0, k)$	Value of $\hat{x}(k-1)$ for $l^*(0, k)$	$l^*(1, k)$	Value of $\hat{x}(k-1)$ for $l^*(1, k)$	Max. likelihood traj. for $l^*(x, 25)$	$p[x(k) = 0)]$	$p[x(k) = 1)]$	Most likely value $x(k)$
0	1	0			1	—	4	—	1	.200	.800	1
1	1	1	1	1	1	0	16	1	1	.087	.913	1
2	0	0	1	1	1	1	28	1	1	.055	.945	1
3	0	0	0	0	1	1	0	—	0	1.000	.000	0
4	0	0	1	1	1	0	0.57	0	0	.636	.364	0
5	0	0	1	1	1	0	2.29	1	0	.274	.726	1
6	0	0	0	0	1	0	0	—	0	1.000	.000	0
7	0	0	0	0	1	0	0	—	0	1.000	.000	0
8	0	1	0	0	1	0	0	—	0	1.000	.000	0
9	1	0	0	1	1	0	0.57	0	1	.636	.364	1
10	1	0	0	1	1	0	2.29	1	1	.274	.726	1
11	1	0	0	1	1	0	9.14	1	1	.111	.889	1
12	1	0	1	1	1	1	28	1	1	.062	.938	1
13	1	0	0	1	1	1	28	1	1	.048	.952	1
14	1	0	1	1	1	1	28	1	1	.046	.954	1
15	1	0	0	1	1	1	28	1	1	.045	.955	1
16	1	1	0	1	1	1	28	1	1	.045	.955	1
17	0	0	0	0	1	1	0	—	0	1.000	.000	0
18	0	0	0	0	1	0	0	—	0	1.000	.000	0
19	0	0	0	0	1	0	0	—	0	1.000	.000	0
20	0	0	1	1	1	0	0.57	0	0	.636	.364	0
21	0	1	0	0	1	0	0	—	1	1.000	.000	0
22	1	0	0	1	1	0	0.57	0	1	.636	.364	1
23	1	0	0	1	1	0	2.29	1	1	.274	.726	1
24	1	0	0	1	1	0	9.14	1	1	.111	.889	1
25	1		1	1	1	1	28	1	1	.062	.938	1

The maximum likelihood trajectory was actually computed for a sequence of 25 measurements, where the measurements were obtained by using a random number generator to specify $x(0/-1)$, $w(k)$, and $v(k)$. The results for one particular run are shown in Table 11.5. The first four columns list the values of $x(k)$, $w(k)$, $v(k)$, and $z(k)$. The next four columns list $I^*(0, k)$, the corresponding $\hat{x}(k-1)$, $I^*(1, k)$, and the corresponding $\hat{x}(k-1)$. The values $I^*(x, k)$ are normalized so that $I^*(0, k)$ always equals 1. The next column lists the maximum likelihood trajectory corresponding to $I^*(x, 25)$. The final three columns list the results of the Bayesian estimation procedure [Refs. 7, 8]. By comparing columns 2, 10, and 13, it is seen that in this example the maximum likelihood trajectory is in error at one point, namely $k = 2$, while the Bayesian procedure makes four errors.

The result that the maximum likelihood trajectory is more accurate than the trajectory obtained from the Bayesian estimation procedure was found for many runs with this example and for many other examples. The explanation for this is that the maximum likelihood trajectory uses *all* received measurements to estimate $x(k)$, while the Bayesian approach customarily uses only measurements up to time k. In other words, the maximum likelihood trajectory is a smoothing solution, while the sequence of maximum likelihood states is not. By using this additional information the maximum likelihood trajectory approach is able to make a better estimate of past values of the state. It is possible to modify the Bayesian approach to obtain smoothed estimates, i.e., the best estimates of past states using information up to the present time [Ref. 10]; however, this requires a substantial increase in the amount of computation. The computational requirements for the maximum likelihood trajectory estimation are about the same as for the Bayesian estimation procedure without smoothing. Thus, the increased accuracy for estimates of past states is obtained with little increase in computing time or high-speed memory requirement.

An application where the system equations and measurement equations are nonlinear, but where the random signals are approximately gaussian, is the estimation of ballistic missile trajectories. In this problem it is desired to estimate three position coordinates, three velocity coordinates, and one ballistic parameter. The use of a modified version of the Kalman-Bucy filter [Ref. 9] for this problem is discussed in Refs. 13–14.

Because of the similarity of the recursive equation, Eq. (11.29) or (11.30), and the dynamic programming functional equation, it is possible to use the concepts of state increment dynamic programming in a computational procedure which obtains the maximum likelihood trajectory with reduced computational requirements. As in the previous section, the system equation is treated as the discrete approximation to a continuous system. The only difficulty with the procedure is that measurements between sampling instants are required in the calculations. If the measurements are continuously available, this presents no problem. However, some sort of interpolation must be done in the case where the measurements are received only at discrete times.

11.4 COMBINED OPTIMUM CONTROL AND ESTIMATION

The general problem of finding optimal control for a system on the basis of noisy measurements has been solved by dynamic programming [Refs. 4–6]. This problem covers a great many cases of both practical and theoretical interest, including the general problem of optimal synthesis of adaptive and learning systems. It is called the "combined" optimum control and estimation

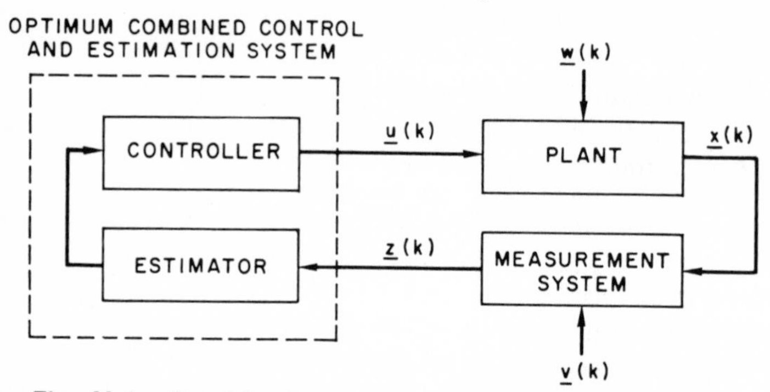

Fig. 11.4. Combined optimum control and estimation.

problem because the solution for certain special cases [Refs. 11–12] is first to find a best estimate of the state of the system from the measurements and then to apply the control that is optimal if the state of the system is equal to this estimate. In this section the general solution of the combined optimization problems as obtained by Meier [Ref. 4] is presented. The use of state increment dynamic programming to obtain a feasible computational procedure is also discussed.

The problem in the discrete-time case can be formulated in terms of Fig. 11.4. The system that is to be controlled, denoted as the plant, is described by the difference equation

$$\mathbf{x}(k+1) = \mathbf{g}[\mathbf{x}(k), \mathbf{u}(k), \mathbf{w}(k), k] \tag{11.35}$$

where $\mathbf{x}(k)$ = state vector of the plant
k = index of the time instants
$\mathbf{u}(k)$ = control vector
$\mathbf{w}(k)$ = random forcing vector, which is statistically independent from time instant to time instant*
$\mathbf{g}$ = vector functional.

* The case where $\mathbf{w}(k)$ and $\mathbf{w}(j)$, $k \neq j$ are correlated can be handled as in Sec. 11.2.

The measurement system is described by

$$\mathbf{z}(k) = \mathbf{h}[\mathbf{x}(k), \mathbf{v}(k), k] \tag{11.36}$$

where $\mathbf{z}(k)$ = measurement vector, which generally has dimension less than that of $\mathbf{x}(k)$
$\mathbf{v}(k)$ = measurement noise vector, which is statistically independent from time instant to time instant*
$\mathbf{h}$ = vector functional.

The performance criterion, which is to be minimized, is the expected value of a variational performance measure,

$$J = E\sum_{k=0}^{K} l[\mathbf{x}(k). \mathbf{u}(k), \mathbf{w}(k), k] \tag{11.37}$$

where the expectation is over the a priori probability density for $\mathbf{x}(0)$, $p[\mathbf{x}(0/-1)]$, and the probability densities $p[\mathbf{w}(k-1)]$ and $p[\mathbf{v}(k)]$, $k = 1, 2, \ldots, K$. Constraints are present in the form

$$\mathbf{x} \in X(k) \tag{11.38}$$

$$\mathbf{u} \in U(\mathbf{x}, k). \tag{11.39}$$

The combined optimization problem can be stated as follows: given the plant described by Eq. (11.35), the measurement system described by Eq. (11.36), and the probability density functions $p[\mathbf{x}(0/-1)]$, $p[\mathbf{w}(k-1)]$, and $p[\mathbf{v}(k)]$, $k = 1, 2, \ldots, K$, find the control sequence $\mathbf{u}(k)$, $k = 0, 1, 2, \ldots, K$ which minimizes J in Eq. (11.37) subject to the constraints in Eqs. (11.38) and (11.39).

As an indication of the generality of this problem, the deterministic control problem, the stochastic control problem, and the optimum estimation problem are all special cases. The stochastic control problem is obtained if $\mathbf{z}(k) \equiv \mathbf{x}(k)$; the deterministic control problem is obtained if in addition $\mathbf{w}(k) \equiv 0$; while the optimum estimation problem is obtained if $\mathbf{g}$ is independent of $\mathbf{u}(k)$, and $\mathbf{u}(k)$ then represents the best estimate. Furthermore, the control of adaptive and learning systems, in which system parameters are not known, can be treated by including the unknown parameters as additional state variables; the results of this section thus provide a general solution to the optimal control problem for these systems.

The algebra in the derivation of the solution to this problem is simplified if the following notation is used.

$$\begin{aligned} \mathbf{Z}(k) &= \mathbf{z}(k), \mathbf{z}(k-1), \ldots, \mathbf{z}(1) \\ \mathbf{U}(k) &= \mathbf{u}(k), \mathbf{u}(k-1), \ldots, \mathbf{u}(1), \mathbf{u}(0). \end{aligned} \tag{11.40}$$

* The case where $\mathbf{v}(k)$ and $\mathbf{v}(j)$, $k \neq j$, are correlated can be handled as in Sec. 11.2.

Because the state of the system is not known, but instead is a probabilistic function, it is no longer appropriate to write an iterative functional equation directly in terms of the state. Instead, it becomes appropriate to speak of an "information state," which summarizes all the information known about the state. In the general case, the information state is the set of all measurements up to the present, $z(1), \ldots, \mathbf{z}(k)$, and all past controls, $\mathbf{u}(0), \mathbf{u}(1), \ldots, u(k-1)$. The minimum expected cost function can then be defined as the minimum value of the expected cost based on the past and present measurements and past controls. Using Eq. (11.40), this function can be written as

$$I^*[\mathbf{Z}(k), \mathbf{U}(k-1), k] = \min_{\mathbf{u}(k),\ldots,\mathbf{u}(K)} \left\{ \underset{\mathbf{x}(k),\mathbf{w}(k),\ldots,\mathbf{w}(K),\mathbf{v}(k+1),\ldots,\mathbf{v}(K)}{E} \left[\sum_{j=k}^{K} l[\mathbf{x}(j), \mathbf{u}(j), \mathbf{w}(j), j] / \mathbf{Z}(k), \mathbf{U}(k-1) \right] \right\} \quad (11.41)$$

where the expectation is conditioned on $\mathbf{Z}(k)$ and $\mathbf{U}(k-1)$.

The iterative relation is found by a process similar to that used in the stochastic and deterministic control cases. The details are contained in Ref. 4. The iterative relation is

$$\begin{aligned} I^*[\mathbf{Z}(k), \mathbf{U}(k-1), k] = \operatorname*{Min}_{\mathbf{u}(k)} \Big\{ \underset{\mathbf{x}(k),\mathbf{w}(k),\mathbf{v}(k+1)}{E} & [l[\mathbf{x}(k), \mathbf{u}(k), \mathbf{w}(k), k] \\ & + I^*(\mathbf{h}[\mathbf{g}[\mathbf{x}(k), \mathbf{u}(k), \mathbf{w}(k)], \mathbf{v}(k+1), k+1], \\ & \mathbf{Z}(k), \mathbf{u}(k), \mathbf{U}(k-1))/\mathbf{Z}(k), \mathbf{U}(k-1)] \Big\} \end{aligned} \quad (11.42)$$

The initial condition for the iterative procedure is obtained as

$$I^*[\mathbf{Z}(K), \mathbf{U}(K-1), K] = \operatorname*{Min}_{\mathbf{u}(K)} \left\{ \underset{\mathbf{x}(K),\mathbf{w}(K)}{E} [l[\mathbf{x}(K), \mathbf{u}(K), \mathbf{w}(K), k]/\mathbf{Z}(K), \mathbf{U}(K-1)] \right\}. \quad (11.43)$$

In order to carry out the computations, it is necessary to have the conditional probability distribution $p[\mathbf{x}(k)/\mathbf{Z}(k), \mathbf{U}(k-1)]$. This may be calculated iteratively using Bayes' rule. The iterative relation is

$$p[\mathbf{x}(k+1)/\mathbf{Z}(k+1), \mathbf{U}(k)] = \frac{p[\mathbf{z}(k+1)/\mathbf{x}(k+1)] \int_{x(k)} p[\mathbf{x}(k+1)/\mathbf{x}(k), \mathbf{u}(k)] \times p[\mathbf{x}(k)/\mathbf{Z}(k), \mathbf{U}(k-1)]\, d\mathbf{x}(k)}{\int_{x(k+1)} p[\mathbf{z}(k+1)/\mathbf{x}(k+1)] \int_{x(k)} p[\mathbf{x}(k+1)/\mathbf{x}(k), \mathbf{u}(k)] \times p[\mathbf{x}(k)/\mathbf{Z}(k), \mathbf{U}(k-1)]\, d\mathbf{x}(k)\, d\mathbf{x}(k+1)} \quad (11.44)$$

where

$$\int_{x(k+1)} \cdots d\mathbf{x}(k+1) \quad \text{and} \quad \int_{x(k)} \cdots d\mathbf{x}(k)$$

denote integration over all possible values of $\mathbf{x}(k+1)$ and $\mathbf{x}(k)$ respectively. The probability density $p[\mathbf{z}(k+1)/\mathbf{x}(k+1)]$ is determined by Eq. (11.36) and $p[\mathbf{v}(k+1)]$, while $p[\mathbf{x}(k+1)/\mathbf{x}(k), \mathbf{u}(k)]$ is known from Eq. (11.37) and $p[\mathbf{w}(k)]$. The iterations begin with

$$p[\mathbf{x}(1)/\mathbf{z}(1), \mathbf{u}(0)] = \frac{p[\mathbf{z}(1)/\mathbf{x}(1)] \int_{x(0/-1)} p[\mathbf{x}(1)/\mathbf{x}(0/-1), \mathbf{u}(0)] p[\mathbf{x}(0/-1)] \, d\mathbf{x}(0/-1)}{\int_{x(1)} p[\mathbf{z}(1)/\mathbf{x}(1)] \int_{x(0/-1)} p[\mathbf{x}(1)/\mathbf{x}(0/-1), \mathbf{u}(0)] p[\mathbf{x}(0/-1)] \, d\mathbf{x}(0/-1) \, d\mathbf{x}(1)} \tag{11.45}$$

where $p[\mathbf{x}(0/-1)]$ is the a priori probability density of $\mathbf{x}(0)$.

The difficulty with applying the iterative relation in Eq. (11.42) is that the information state, $[\mathbf{Z}(k), \mathbf{U}(k-1)]$ grows with time. As more past observations and controls accumulate, the dimension of this state grows linearly; if the dimension of $\mathbf{z}(k)$ is d and the dimension of $\mathbf{u}(k)$ is q, then the dimension of the information state at time k is $k(d+q)$. Clearly, even if d and q are reasonably small, after a few time instants the dimension of the information state will become excessive.

One method of keeping the dimensionality of the system within bounds is to use an alternate description of the information state. One alternative is to use the moments of the conditional probability density function $p[\mathbf{x}(k)/\mathbf{Z}(k), \mathbf{U}(k-1)]$. It has been shown [Ref. 11–12] that if the system equations, Eq. (11.35), are linear in $\mathbf{x}(k)$, $\mathbf{u}(k)$, and $\mathbf{w}(k)$; if the measurement equations, Eq. (11.36), are linear in $\mathbf{x}(k)$ and $\mathbf{v}(k)$; if the performance criterion, Eq. (11.37), is a quadratic function of $\mathbf{x}(k)$, $\mathbf{u}(k)$, and $\mathbf{w}(k)$; and if the probability density functions $p[\mathbf{x}(0/-1)]$, $p[\mathbf{w}(k)]$, and $p[\mathbf{v}(k)]$ are all gaussian, then this conditional probability density function is also gaussian. Consequently, it is completely specified by the mean and covariance of this distribution. Thus, the information state has dimension $(n + n(n+1)/2)$, where n is the dimension of $\mathbf{x}(k)$; n of the quantities are the components of mean vectors while the other $n(n+1)/2$ are terms in the covariance matrix. In this special case the optimum combined system consists of finding the maximum likelihood estimate of $\mathbf{x}(k)$ and applying the control that would be optimal if this estimate were actually the state [Refs. 11–12].

Even if the mean and covariance are not a set of sufficient statistics, it is often convenient to approximate the optimal system by assuming that they are. Another approximation technique is to assume that I^* depends on only the last few measurements and controls, rather than on all past values.

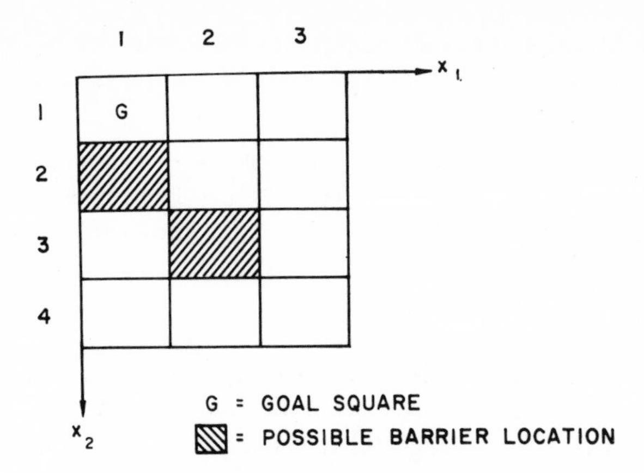

Fig. 11.5. Automaton environment.

A simple problem that illustrates some of the features of this formulation is shown pictorially in Figs. 11.5–6. In this example an automaton is attempting to find its way through the environment shown in Fig. 11.5 to a goal located at the upper left-hand square, which has coordinates ($x_1 = 1, x_2 = 1$). As shown in Fig. 11.6, the automaton can move one square either up, down, left, or right. There are two squares on which it is possible that barriers are present; namely ($x_1 = 1, x_2 = 2$) and ($x_1 = 2, x_2 = 3$). The automaton is not allowed to pass through these barriers. Initially, the automaton does not know if these barriers are present; instead, it knows that there is a barrier in ($x_1 = 1, x_2 = 2$) with probability 0.4 and a barrier in ($x_1 = 2, x_2 = 3$) with probability 0.5. The robot can always "see" one move ahead—i.e., if it is within one move of a barrier location, it can find out if the barrier is there or not. For a certain price, which is expressed as a specified fraction of a move, the automaton can make an observation of all squares that are two moves away (see Fig. 11.6). The objective is to find the policy for the automaton that reaches the square ($x_1 = 1, x_2 = 1$) from any initial square while minimizing the sum of moves and penalties for making observations.

	2	
2	1	2
1	0	1
2	1	2

0 = PRESENT SQUARE
1 = SQUARES THAT CAN BE REACHED IN ONE MOVE
2 = SQUARES THAT CAN BE REACHED IN TWO MOVES

Fig. 11.6. Automaton moves.

This problem can most easily be put into the framework of this section by defining the *information state* of the system to be the four-dimensional vector $\mathbf{x}$,

$$\mathbf{x} = \begin{bmatrix} x_1 \\ x_2 \\ x_3 \\ x_4 \end{bmatrix} \tag{11.46}$$

where x_1 = horizontal coordinate in Fig. 11.5
x_2 = vertical coordinate in Fig. 11.5
x_3 = state of knowledge about barrier at $(x_1 = 1, x_2 = 2)$
x_4 = state of knowledge about barrier at $(x_1 = 2, x_2 = 3)$.

The first two variables are quantized to the values

$$\begin{aligned} x_1 &= 1, 2, 3 \\ x_2 &= 1, 2, 3, 4. \end{aligned} \tag{11.47}$$

The latter two variables are quite different from the usual state variables associated with dynamic systems. Each variable can take on three different values as follows

$$\begin{aligned} x_3 &= P, A, Q \\ x_4 &= P, A, Q \end{aligned} \tag{11.48}$$

where P = barrier is known to be present
A = barrier is known to be absent
Q = absence or presence of barrier is not known.

The control vector, $\mathbf{u}$, has three components. They are

$$\mathbf{u} = \begin{bmatrix} u_1 \\ u_2 \\ u_3 \end{bmatrix} \tag{11.49}$$

where u_1 = change in x_1
u_2 = negative change in x_2*
u_3 = decision to make an observation.

The variable u_3 takes on two values

$$u_3 = L, N \tag{11.50}$$

where L = an observation is made
N = no observation is made.

* This definition of u_2 means that a positive move is up and to the right.

The set of admissible controls is

$$U = \begin{bmatrix}1\\0\\N\end{bmatrix}, \begin{bmatrix}-1\\0\\N\end{bmatrix}, \begin{bmatrix}0\\1\\N\end{bmatrix}, \begin{bmatrix}0\\-1\\N\end{bmatrix}, \begin{bmatrix}0\\0\\L\end{bmatrix} \tag{11.51}$$

corresponding to move right, move left, move up, move down, and make an observation. The first two system equations can be written as

$$\begin{aligned} x_1(k+1) &= x_1(k) + u_1(k) \\ x_2(k+1) &= x_2(k) - u_2(k). \end{aligned} \tag{11.52}$$

The uncertainty about the barriers is taken into account by defining a random forcing function vector $\mathbf{w}$ with two components,

$$\mathbf{w} = \begin{bmatrix}w_1\\w_2\end{bmatrix} \tag{11.53}$$

where w_1 = presence or absence of barrier at ($x_1 = 1, x_2 = 2$)
w_2 = presence or absence of barrier at ($x_1 = 2, x_2 = 3$).

These variables can take on the values

$$\begin{aligned} w_1 &= B, R \\ w_2 &= B, R \end{aligned} \tag{11.54}$$

where B = barrier is present
R = barrier is absent.

This vector affects only the state variables x_3 and x_4. In writing the system equations for these two variables, it is useful to define two auxiliary variables, m_1 and m_2. The variable m_1 takes on the value 1 if the control is such that the presence or absence of the barrier at ($x_1 = 1, x_2 = 2$) will be determined; and $m_1 = 0$ otherwise. The variable $m_2 = 1$ if the control will determine the presence or absence of the barrier at ($x_1 = 2, x_2 = 3$), and $m_2 = 0$ otherwise.

For $u_3 = N, m_1 = 1$ if the move chosen causes the next square to be within one move of the barrier at ($x_1 = 1, x_2 = 2$). For $u_3 = L, m_1 = 1$ if the barrier at ($x_1 = 1, x_2 = 2$) is within two moves of the present square. Otherwise $m_1 = 0$. Similar conditions can be written for m_2.

The system equation for x_3 can thus be written in the form

$$x_3(k+1) = f_3[x_3(k), u_3(k), m_1(k), w_1(k)] \tag{11.55}$$

Table 11.6. Next Value of x_3 for Automaton

$x_3(k)$	$u_3(k)$	$m_1(k)$	$w_1(k)$	$x_3(k+1)$
P	—	—	—	P
A	—	—	—	A
Q	N	0	—	Q
Q	N	1	B	P
Q	N	1	R	A
Q	L	0	—	Q
Q	L	1	B	P
Q	L	1	R	A

— = value of $x_3(k+1)$ is the same for all values of this variable

where f_3 is defined by Table 11.6, and where

$$\begin{aligned} p(w_1 = B) &= 0.4 \\ p(w_1 = R) &= 0.6 \\ p(w_2 = B) &= 0.5 \\ p(w_2 = R) &= 0.5. \end{aligned} \tag{11.56}$$

The equation for x_4 is

$$x_4(k+1) = f_4[x_4(k), u_4(k), m_2(k), w_2(k)] \tag{11.57}$$

where f_4 is specified by a table similar to that in Table 11.6.

The performance criterion, which is to be minimized, is the sum of moves and penalties for observations. This criterion can be written as

$$J = \sum_{k=0}^{\infty} l[\mathbf{x}(k), \mathbf{u}(k)] \tag{11.58}$$

where $l[\mathbf{x}(k), \mathbf{u}(k)]$ is specified as in Table 11.7. In this table it is assumed that the penalty for an observation is 0.3 moves.

Table 11.7. Value of $l[\mathbf{x}(k), \mathbf{u}(k)]$ for Automaton

$x_1(k)$	$x_2(k)$	$u_3(k)$	$l[\mathbf{x}(k), \mathbf{u}(k)]$
1	1	—	0
1	$\neq 1$	N	1.0
1	$\neq 1$	L	0.3
$\neq 1$	1	N	1.0
$\neq 1$	1	L	0.3
$\neq 1$	$\neq 1$	N	1.0
$\neq 1$	$\neq 1$	L	0.3

The constraints are that the set of admissible controls and the set of admissible states are as defined above. Also, if a barrier is present, a move to that square is forbidden.

Because the system equations and the constraints are time-invariant, and because the performance criterion is the sum of time-invariant terms over an

Table 11.8. Solution to Automaton Problem

present state				optimal control			minimum cost
x_1	x_2	x_3	x_4	u_1	u_2	u_3	I
1	1	—	—	—	—	—	0
2	1	—	—	−1	0	N	1
3	1	—	—	−1	0	N	2
1	2	—	—	0	1	N	1
2	2	—	—	0	1	N	2
3	2	—	—	0	1	N	3
1	3	A	—	0	1	N	2
1	3	P	A	1	0	N	4
1	3	P	P	0	−1	N	8
2	3	—	—	0	1	N	3
3	3	—	—	0	1	N	4
1	4	A	—	0	1	N	3
1	4	P	A	0	1	N	5
1	4	P	P	1	0	N	7
1	4	P	Q	1	0	N	6
1	4	Q	A	0	1	N	3.8
1	4	Q	P	0	0	L	4.9
1	4	Q	Q	0	0	L	4.5
2	4	—	A	0	1	N	4
2	4	A	P	−1	0	N	4
2	4	P	P	1	0	N	6
2	4	Q	P	−1	0	N	5.9
3	4	—	—	0	1	N	5

— = Optimal control and minimum cost are the same for all values of this variable.

infinite number of stages, a modified version of approximation in policy space* [see Sec. 9.5 and Refs. 1, 2] can be used. The details of this procedure are discussed in Ref. 15. The solution is shown in Table 11.8.

A number of interesting effects can be observed in Table 11.8. The first is Feldbaum's dual control effect [Ref. 17]; this effect is said to occur whenever the optimal control is used to gain more information about the system, rather than to optimize performance directly. This effect can be observed in this

* Because the cost at state $x_1 = 1$, $x_2 = 1$, is identically zero, independent of the random forcing vector, it is not necessary to use Howard's iteration in policy space [Ref. 16].

example whenever the optimal control is to make an observation. By imposing a penalty for the observation, the system is able to make a decision whether to gather more information or to continue moving on the basis of the information gathered so far. This occurs at $(x_1, x_2, x_3, x_4) = (1, 4, Q, P)$ and $(x_1, x_2, x_3, x_4) = (1, 3, Q, Q)$.

Another effect that can be observed is Howard's value of information. [Ref. 18]. This effect is related to the cost that should be paid in order to gain information. Again, this is illustrated here in the decision of whether or not to make an observation; if the cost of 0.3 moves does not pay for the increase in performance, then the observation should not be made. At the two points

Table 11.9. Comparison of Optimal Control and Minimum Cost With and Without the Observation Control

				minimum cost with observation	minimum cost without observation	optimal control with no observation		
x_1	x_2	x_3	x_4			u_1	u_2	u_3
1	4	Q	P	4.9	5.4	0	1	N
1	4	Q	Q	4.5	4.6	0	1	N

where the observation is made, it is interesting to note the optimal control and minimum cost for the case where the decision to observe is *not* allowed. It is seen from Table 11.9 that at $(x_1, x_2, x_3, x_4) = (1, 4, Q, P)$ the net profit of making the observation is 0.5 moves, while at $(x_1, x_2, x_3, x_4) = (1, 4, Q, Q)$ the net profit is 0.1 moves.

In addition to these effects, many other elements of adaptive and learning control problems can be observed either in this example directly or by a slight modification.

The only difference between this approach and the direct use of Eqs. (11.42)–(11.45) is in the computational details. All the properties of the solution via the latter approach are retained. Often the former approach is more computationally feasible. A larger version of the automaton problem is worked out in Ref. 15, while the problem of optimum design and operation of reliable and maintainable systems is treated by this approach in Ref. 19.

The state increment dynamic programming computational procedures indicated in the previous two sections can be used in either approach to the combined problem as a method of reducing computational requirements.

11.5 CONCLUSIONS

In this chapter three problem formulations involving uncertainty were examined. These were the optimal stochastic control problem, the optimum estimation problem, and the combined optimum control and estimation problem. In each case an iterative functional equation was derived. It was shown that the computational requirements of the conventional procedure are even greater in these cases than they were in the deterministic case. Thus, the importance of state increment dynamic programming in obtaining a feasible solution by reducing computational requirements is even more apparent.

REFERENCES

1. R. Bellman, *Dynamic Programming* (Princeton University Press, Princeton, New Jersey, 1957).
2. R. Bellman, *Adaptive Control Processes* (Princeton University Press, Princeton, New Jersey, 1962).
3. R. E. Larson and J. Peschon, "A Dynamic Programming Approach to Trajectory Estimation," *IEEE Trans. Auto. Cont.*, Vol. AC-10, No. 3, pp. 537–540 (July 1966).
4. L. Meier, "Combined Optimum Control and Estimation," paper presented at 1965 Allerton Conference, University of Illinois (October 1965), pp. 109–120.
5. M. Aoki, "Optimal Bayesian and Min-Max Controls of a Class of Stochastic and Adaptive Dynamic Systems," Preprints, pp. II-21-II-31, paper presented at IFAC Tokyo Symposium on Systems Engineering for Control System Design, Tokyo, Japan (September 1965).
6. R. Sussman, "Optimal Control of Systems with Stochastic Disturbances," Technical Report Series 63, No. 20, Electronics Research Laboratory, University of California, Berkeley, California (November 1963).
7. Yu-Chi Ho and R. C. K. Lee, "A Bayesian Approach to Problems in Stochastic Estimation and Control," 1964 JACC, (June 1964), pp. 382–387.
8. R. C. K. Lee, *Optimal Estimation, Identification and Control*, MIT Press, p. 59–72 (1964).
9. R. E. Kalman and R. S. Bucy, "New Results in Linear Filtering and Prediction Theory," *Trans. ASME J. Basic Engr.*, Vol. 83, Series D, No. 1, pp. 95–108 (March 1961).
10. H. E. Rauch, F. Tung, and C. B. Streibel, "Maximum Likelihood Estimates of Linear Dynamic Systems," *AIAA J.*, Vol. 3, No. 8, pp. 1445–1450 (August 1965).
11. T. L. Gunckel, "Optimum Design of Sampled-Data Systems with Random Parameters," S.E.L. T.R. No. 2101-2, Stanford Electronics Laboratory, Stanford, California.
12. P. D. Joseph and J. T. Tou, "On Linear Control Theory," *Trans. AIEE*, Pt. II, Vol. 80, No. 56, pp. 193–6 (September 1961).
13. J. Peschon and R. E. Larson, "Analysis of an Intercept System," Stanford Research Institute, Menlo Park, Calif. Final Report to NIKE-X Project office, Redstone Arsenal, Alabama, Contract DA-01-021-AMC-90006(Y), SRI Project 5188-7 (December 1965).
14. R. E. Larson, R. M. Dressler, and R. S. Ratner, "Application of the Extended Kalman Filter to Ballistic Trajectory Estimation" Stanford Research Institute, Menlo Park, Calif. Final Report to NIKE-X Project Office, Redstone Arsenal, Alabama, Contract DA-01-021-AMC-90006(Y), SRI Project 5188-103 (January 1967).
15. R. E. Larson and W. G. Keckler, "Optimum Control of a Robot in a Partially Unknown Environment," submitted to *Automatica*.
16. R. A. Howard, *Dynamic Programming and Markov Processes*, John Wiley & Sons, Inc., New York, New York (1960).
17. A. A. Feldbaum, "Dual Control Theory—I," *Automation and Remote Control*, Vol. 21, No. 9 (April 1961), pp. 874–880.
18. R. A. Howard, "Information Value Theory," *IEEE Trans. Sys. Sci. Cyber.*, Vol. SSC-2, No. 1 (August 1966), pp. 22–26.
19. R. S. Ratner, "Performance-Adaptive Renewal Policies for the Operation of Linear Systems Subject to Failure," Ph.D. Thesis, Department of Electrical Engineering, Stanford University, Stanford, California (June 1968).

Chapter Twelve

SUCCESSIVE APPROXIMATIONS

12.1 INTRODUCTION

In preceding chapters it has continually been emphasized that state increment dynamic programming reduces computational requirements over those of the conventional procedure and thus makes possible the solution of problems of higher dimensionality. In particular, a computer program applicable to a large class of problems having four or less state variables was described in Chapter 8. However, there are still many other problems of even higher dimensionality which state increment dynamic programming cannot solve directly. The question then arises as to what benefit the ideas in this book are in these larger problems.

The answer to this question lies in the area of approximation techniques. The basic approach is to attack a high-dimensional problem by solving a number of appropriately chosen lower-dimensional problems. The utility of state increment dynamic programming is that the size of lower-dimension problem for which a complete solution can be obtained is increased.

In this chapter one of the most promising approximation techniques, Bellman's successive approximations [Refs. 1, 2], will be discussed. First, the general method will be described. Next, a number of illustrative examples will be solved. Then, a four-dimensional problem, namely the optimum operation of a system of four multipurpose water reservoirs, will be treated. Finally, the solution to an airline scheduling problem with 50 state variables will be presented.

12.2 THE GENERAL METHOD

The philosophy behind the method is to break up a problem with several control variables into sub-problems in which there is only one control variable. Thus, for a given number of state variables, the size of sub-problem, and hence the computational requirements, becomes smaller as the number of control variables is increased. The method is most clearly explained when there are as many control variables as state variables, i.e., the case of complete state-transition control discussed in Chapter 5. For clarity, therefore, the application of the method to this case will be discussed first, and the extension to other problems will be indicated later in this section.

For the discrete-time case the system equations are

$$\mathbf{x}(k+1) = \mathbf{g}[\mathbf{x}(k), \mathbf{u}(k), k] \tag{12.1}$$

the performance criterion is

$$J = \sum_{k=0}^{K} l[\mathbf{x}(k), \mathbf{u}(k), k] \tag{12.2}$$

the constraints are

$$\mathbf{x} \in X(k)$$

$$\mathbf{u} \in U(\mathbf{x}, k) \tag{12.3}$$

and the initial state is given by

$$\mathbf{x}(0) = \mathbf{c} \tag{12.4}$$

where all quantities are as defined in Chapter 2, except that the dimensionality of both $\mathbf{x}$ and $\mathbf{u}$ is taken to be n.

It is now assumed that a nominal trajectory from $\mathbf{x}(0) = \mathbf{c}$ is given. The sequence of states along this trajectory is denoted as $\{\mathbf{x}^{(0)}(k)\}$, where $\mathbf{x}^{(0)}(0) = \mathbf{c}$.

Next, one of the n state variables, say x_{i_1}, is selected. A dynamic programming problem is then solved in which the sequence of all other states, namely $\{x_i^{(0)}(k)\}$, $i \neq i_1$, is held fixed. The problem thus has only one state variable, x_{i_1}. The control vector, on the other hand, still has n components. However, the constraint that the sequence of $(n-1)$ of the states must remain fixed imposes $(n-1)$ equality constraints on the control vector, so that effectively there is only one control variable as well. The performance criterion remains the same as in Eq. (12.2), and the constraints are as in Eq. (12.3). The solution to this optimization problem is a new control sequence, $\{\mathbf{u}^{(1)}(k)\}$, and a new trajectory $\{\mathbf{x}^{(1)}(k)\}$.

Next, a different state variable, say x_{i_2}, is selected and the above optimization procedure repeated. This continues until every state variable has been selected for optimization at least once. Termination occurs when the same control sequence and trajectory are obtained for optimization with respect to any of the state variables.

The advantage of this procedure is that the solution of an n-dimensional dynamic programming problem is reduced to the solution of a sequence of one-dimensional problems. Thus, the computational requirements, both time and storage, increase linearly with n rather than exponentially. The computational savings in high-dimensional problems can be truly astronomical.

The disadvantage of this technique is that convergence to the true optimum cannot be guaranteed in all cases. However, it can be shown that convergence is always monotonic. Also, if the nominal trajectory is sufficiently close to the true optimum trajectory, convergence to this optimum is obtained.

Bellman [Refs. 1, 2] has suggested a number of techniques for improving the likelihood of finding the true optimum, such as starting from a number of different nominal trajectories. In addition, for certain classes of problems convergence to the true optimum can be rigorously proved [Ref. 3].

The same basic ideas are also useful in problems where there are fewer control variables than state variables. The approach is to break up the problem into several sub-problems, where each sub-problem contains fewer than n state variables and one or more control variables. For example, a problem with 12 state variables and 3 control variables might be broken up into three sub-problems, each with 4 state variables and one control variable. It would then be possible to use the four-dimensional program described in Chapter 8 to solve the original 12-dimensional problem.

12.3 ILLUSTRATIVE EXAMPLES

As a first example of this approach, consider the problem with system equation

$$\begin{aligned} x_1(k+1) &= x_1(k) + x_2(k) + u_1(k) \\ x_2(k+1) &= x_2(k) + u_2(k) \end{aligned} \tag{12.5}$$

performance criterion

$$J = \sum_{k=0}^{4} [x_1^2(k) + x_2^2(k) + u_1^2(k) + u_2^2(k)] + 2.5[x_1(5) - 2]^2 + 2.5[x_2(5) - 1]^2 \tag{12.6}$$

constraints

$$\begin{aligned} 0 &\leq x_1 \leq 2 \\ -1 &\leq x_2 \leq 1 \\ -1 &\leq u_1 \leq 1 \\ -1 &\leq u_2 \leq 1 \end{aligned} \tag{12.7}$$

and initial condition

$$\begin{aligned} x_1(0) &= 2 \\ x_2(0) &= 1. \end{aligned} \tag{12.8}$$

The state variables are quantized in uniform increments of $\Delta x_1 = \frac{1}{2}$ and $\Delta x_2 = \frac{1}{4}$ respectively. The control variables are not quantized, but are allowed to take on any admissible value such that for any given quantized present state the next state is also a quantized state. This eliminates the need for interpolation.

The nominal trajectory is specified in Table 12.1.

Optimization is first performed with x_2 as the state variable. The sequence of states x_1 is held fixed at the values in Table 12.1. The control variable u_1 is adjusted as a function of x_2 so as to maintain this sequence, i.e.,

$$u_1(k) = x_1^{(0)}(k+1) - x_1^{(0)}(k) - x_2(k). \tag{12.9}$$

Table 12.1. Nominal Trajectory for Example I

k	x_1	x_2	u_1	u_2
0	2	1	−1	−1
1	2	0	−1	0
2	1	0	−1	0
3	0	0	1	0
4	1	0	1	1
5	2	1		

Cost = 18.

The optimization problem thus has system equation

$$x_2(k+1) = x_2(k) + u_2(k) \tag{12.10}$$

performance criterion

$$J = \sum_{k=0}^{4} \{[x_1^{(0)}(k)]^2 + x_2^2(k) + [x_1^{(0)}(k+1) - x_1^{(0)}(k) - x_2(k)]^2 + u_2^2(k)\} + 2.5[x_1^{(0)}(5) - 2]^2 + 2.5[x_2(k) - 1]^2 \tag{12.11}$$

and constraints

$$\begin{aligned} -1 &\le x_2(k) \le 1 \\ -1 &\le u_2(k) \le 1 \\ -1 &\le x_1^{(0)}(k+1) - x_1^{(0)}(k) - x_2(k) \le 1. \end{aligned} \tag{12.12}$$

The third constraint in Eq. (12.12), which appears as a state variable constraint on $x_2(k)$, arises from the control variable constraint on $u_1(k)$ and the restriction that the sequence $\{x_1^{(0)}(k)\}$ is maintained.

The result of this optimization is shown in Fig. 12.1. As before, the optimal control is shown below and to the right of the quantized state, while the minimum cost is above and to the right. In addition, as an aid to understanding these calculations, other information is included in this figure. Quantized states for which the third constraint in Eq. (12.12) is not met are denoted by ×'s. The sequence $\{x_1^{(0)}(k)\}$ is shown along the k axis. The optimal trajectory in the (x_2, k) plane starting from $x_2(0) = 1$ is traced out on the figure.

The complete new trajectory from $x_1(0) = 2$, $x_2(0) = 1$ is listed in Table 12.2. Note that the sequences $\{x_2(k)\}$ and $\{u_2(k)\}$ are significantly changed from the nominal trajectory. Note also that although the sequence $\{x_1(k)\}$ is still the same, the sequence $\{u_1(k)\}$ has been changed. This latter change results from the effects of the variations in $\{x_2(k)\}$ on Eq. (12.9). Finally, note that the cost has been reduced in one iteration from 18 to 16.34.

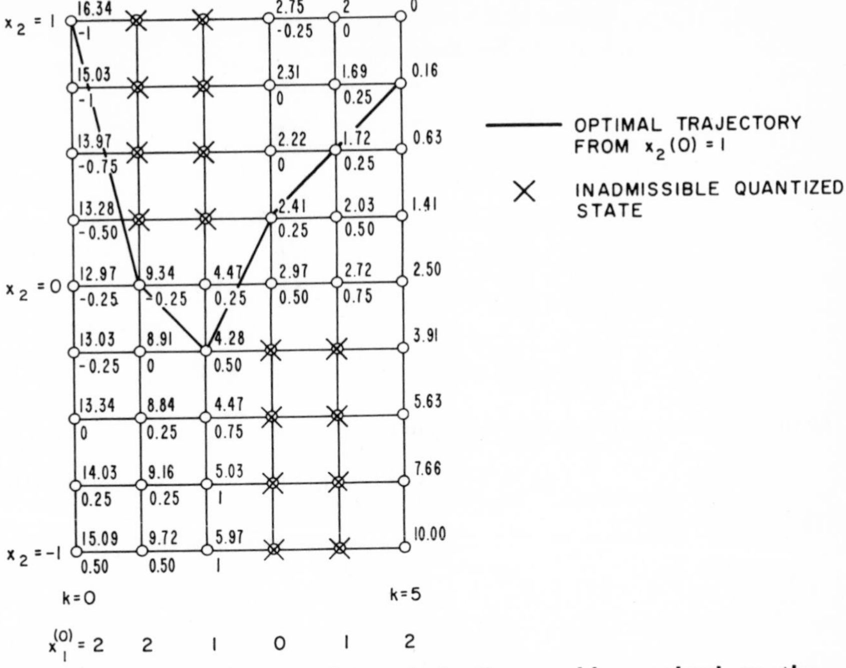

Fig. 12.1. Solution to the optimization problem solved on the first iteration.

On the next iteration the sequence $\{x_2^{(1)}(k)\}$ is held fixed and the optimization is performed with x_1 as the state variable. For this optimization the system equation is

$$x_1(k+1) = x_1(k) + x_2^{(1)}(k) + u_1(k) \tag{12.13}$$

the performance criterion is

$$J = \sum_{k=0}^{4} x_1^2(k) + [x_2^{(1)}(k)]^2 + u_1^2(k) + [u_2^{(1)}(k)]^2 + 2.5[x_1(5) - 2]^2 + 2.5[x_2^{(1)}(5) - 1]^2 \tag{12.14}$$

Table 12.2. Trajectory After One Iteration in Example I

k	x_1	x_2	u_1	u_2
0	2.00	1.00	−1.00	−1.00
1	2.00	0.00	−1.00	−0.25
2	1.00	−0.25	−0.75	0.50
3	0.00	0.25	0.75	0.25
4	1.00	0.50	0.50	0.25
5	2.00	0.75		

Cost = 16.25.

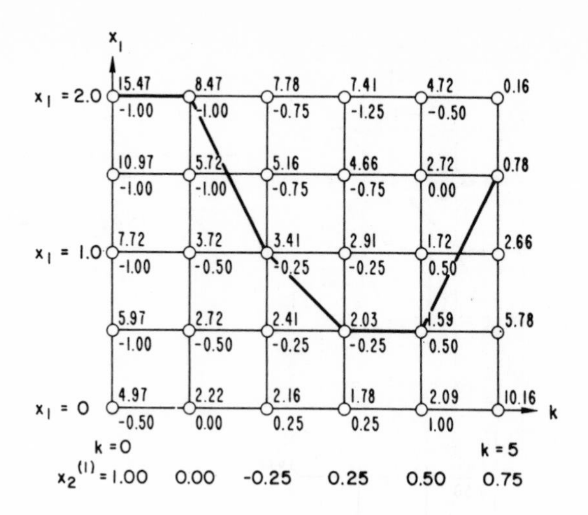

Fig. 12.2. Solution to the optimization problem solved on the second iteration.

and the constraints are

$$0 \leq x_1(k) \leq 2$$
$$-1 \leq u_1(k) \leq 1. \tag{12.15}$$

Because the system equation for $x_2(k)$ does not involve $x_1(k)$, it is true that

$$u_2(k) = x_2^{(1)}(k+1) - x_2^{(1)}(k) = u_2^{(1)}(k). \tag{12.16}$$

Thus, in this case the optimization with respect to $x_1(k)$ does not affect the control $u_2(k)$, and there is no constraint on $x_1(k)$ due to constraints on $u_2(k)$.

The results of this optimization appear in Fig. 12.2. The resulting trajectory from $x_1(0) = 2$, $x_2(0) = 1$ appears in Table 12.3. In this iteration the cost has been reduced from 16.25 to 15.47.

Table 12.3. Trajectory After Two Iterations in Example I

k	x_1	x_2	u_1	u_2
0	2.00	1.00	−1.00	−1.00
1	2.00	0.00	−1.00	−0.25
2	1.00	−0.25	−0.25	0.50
3	0.50	0.25	−0.25	0.25
4	0.50	0.50	0.50	0.25
5	1.50	0.75		

Cost = 15.47.

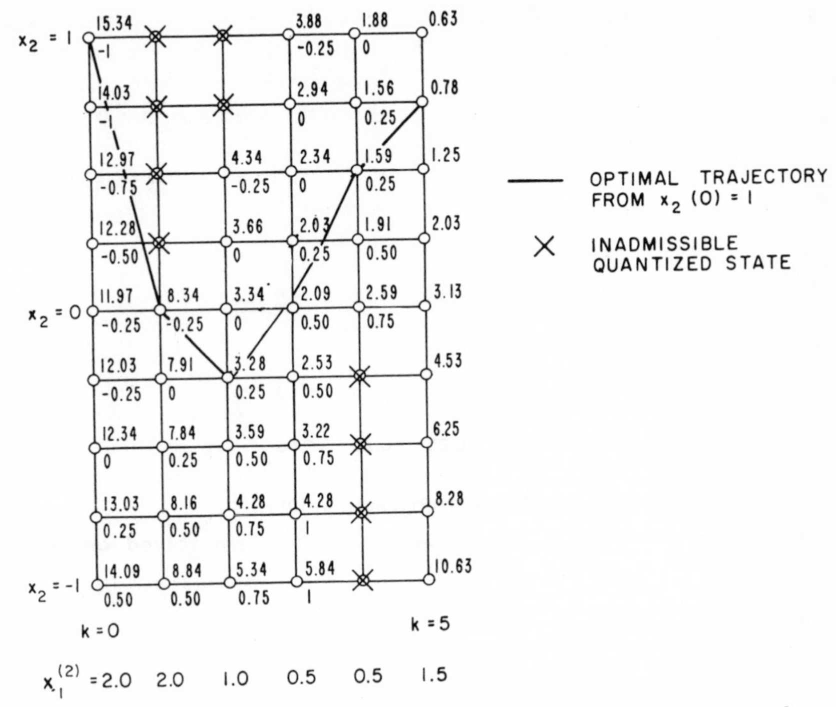

Fig. 12.3. Solution to the optimization problem solved on the third iteration.

In the third iteration the optimization is again with x_2 as a state variable. The problem formulation is as in Eqs. (12.10) through (12.12), with $u_1(k)$ determined by Eq. (12.9). The only difference between this problem and the optimization problem in iteration 1 is that the sequence $\{x_1(k)\}$ comes from Table 12.3 rather than from the original nominal trajectory. The results of the optimization are shown in Fig. 12.3. The new trajectory from $x_1(0) = 2$, $x_2(0) = 1$ appears in Table 12.4.

Table 12.4. Trajectory After Three Iterations in Example 1; Optimal Trajectory

k	x_1	x_2	u_1	u_2
0	2.00	1.00	−1.00	−1.00
1	2.00	0.00	−1.00	−0.25
2	1.00	−0.25	−0.25	0.25
3	0.50	0.00	0.00	0.50
4	0.50	0.50	0.50	0.25
5	1.50	0.75		

Cost = 15.34.

On the fourth iteration the optimization is with x_2 as the state variable. The results of this optimization appear in Fig. 12.4. In this case the same optimal trajectory as on the previous iteration is obtained. Further iterations do not change this trajectory. Thus in three iterations the procedure has converged to the trajectory shown in Table 12.4.

In this case the problem is of sufficiently low dimensionality that an exact solution can be obtained by the conventional dynamic programming

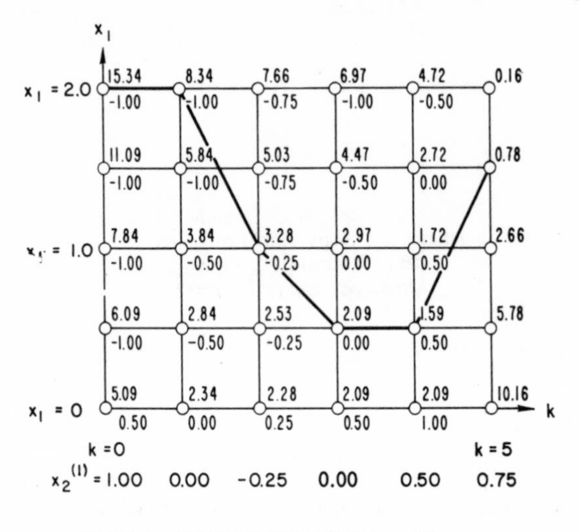

Fig. 12.4. Solution to the optimization problem solved on the fourth iteration.

procedure. When the same increment sizes are used, the same optimal trajectory is obtained. Thus, in this problem convergence to the true optimum solution is obtained.

As another example, consider the problem with system equations

$$\begin{aligned} x_1(k+1) &= 0.625x_1(k) + 0.25x_2(k) + u_1(k) \\ x_2(k+1) &= -0.1875x_1(k) + 0.125x_2(k) + u_2(k) \end{aligned} \tag{12.17}$$

performance criterion

$$J = \sum_{k=0}^{4} [x_1^2(k) + x_2^2(k) + u_1^2(k) + u_2^2(k)] \tag{12.18}$$

constraints

$$\begin{aligned} -5 &\leq x_1 \leq 5 \\ -5 &\leq x_2 \leq 5 \\ -2 &\leq u_1 \leq 2 \\ -2 &\leq u_2 \leq 2 \end{aligned} \tag{12.19}$$

Table 12.5. Nominal Trajectory for Example 2

k	x_1	x_2	u_1	u_2
0	5	5	−0.375	0.3125
1	4	0	0.500	0.7500
2	3	0	0.125	0.5625
3	2	0	−0.250	0.3750
4	1	0	−0.625	0.1875
5	0	0		

Cost = 82.01.

and initial condition

$$x_1(0) = 5$$

$$x_2(0) = 5. \tag{12.20}$$

The state variables are quantized in uniform increments of $\Delta x_1 = \frac{1}{2}$ and $\Delta x_2 = \frac{1}{2}$ respectively. As in the previous example, the control variables are not quantized, but instead are determined such that for any given quantized present state the next is also a quantized state. This eliminates the need for interpolation.

The nominal trajectory was taken to be as in Table 12.5.

After the first iteration, in which x_2 is the state variable, the trajectory is as in Table 12.6. The cost has been reduced from 82.01 to 81.71. The small reduction occurs because most of the cost is on the sequence $\{x_1(k)\}$, which is not allowed to change.

After the second iteration, in which x_1 is the state variable, the trajectory is as in Table 12.7. The cost is now reduced from 81.71 to 62.20, a substantial reduction.

Table 12.6. Trajectory After One Iteration for Example 2

k	x_1	x_2	u_1	u_2
0	5	5.0	−0.375	0.3125
1	4	0.0	0.500	0.2500
2	3	−0.5	0.250	0.1250
3	2	−0.5	−0.125	−0.0625
4	1	−0.5	−0.500	−0.2500
5	0	−0.5		

Cost = 81.71.

Table 12.7. Trajectory After Two Iterations for Example 2

k	x_1	x_2	u_1	u_2
0	5.0	5.0	−1.8750	0.3125
1	2.5	0.0	−1.0625	0.2500
2	0.5	−0.5	−0.1875	0.1250
3	0.0	−0.5	0.1250	−0.0625
4	0.0	−0.5	0.1250	−0.2500
5	0.0	−0.5		

Cost = 62.20.

On the third iteration, where x_2 is the state variable, the trajectory is as in Table 12.8. The cost is reduced from 62.20 to 61.49. Further iterations do not change this trajectory, so convergence is obtained in three iterations. A comparison with the solution generated by the conventional dynamic programming procedure shows that convergence is again to the true optimum.

In further experiments with second-order examples of this type results similar to these two cases were found. In all cases convergence to the true optimum was obtained, even when the nominal trajectory was very far from the true optimum trajectory. Four iterations for convergence was the maximum number observed. The total computer time required for each case was on the order of 20 seconds on a B-5500.

12.4 OPTIMIZATION OF A MULTIPURPOSE FOUR-RESERVOIR SYSTEM

In this section, the optimum operation over 24 hours of a multipurpose four-reservoir system is determined. The reservoir network, which contains

Table 12.8. Trajectory After Three Iterations for Example 2; Optimal Trajectory

k	x_1	x_2	u_1	u_2
0	5.0	5.0	−1.8750	−0.18750
1	2.5	−0.5	−0.9375	0.03125
2	0.5	−0.5	−0.1875	0.15625
3	0.0	0.0	0.0000	0.00000
4	0.0	0.0	0.0000	0.00000
5	.0	0.0		

Cost = 61.49.

both series and parallel connections, is shown in Fig. 12.5. In this optimization, use of water for power generation, irrigation, flood control and recreation is considered. Interaction of the short-term optimization with longer-term operating policies is also taken into account.

The state variables for the problem are taken to be amounts of water contained in each reservoir. The amount of water in the ith reservoir is denoted as x_i, $i = 1, 2, 3, 4$, where each x_i is expressed in normalized units.

On the basis of potential use of the reservoir for recreation purposes, a minimum water level for each reservoir is specified; the amount of water

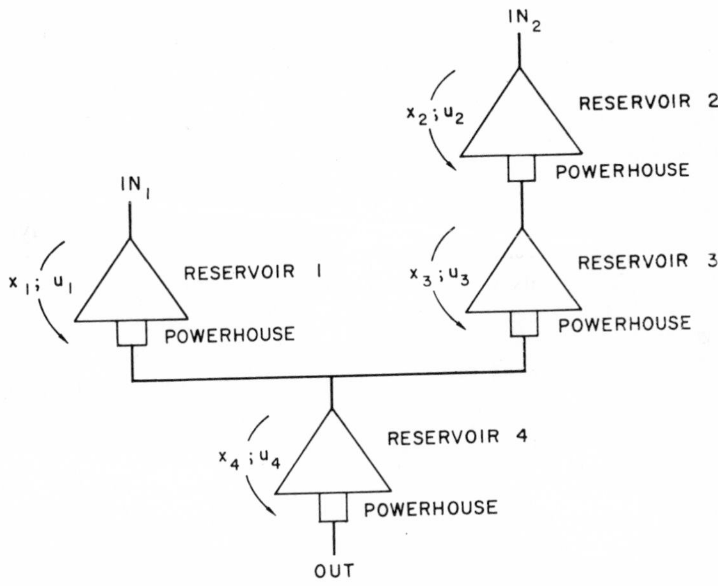

Fig. 12.5. Network configuration of four-reservoir problem.

needed to achieve this level is arbitrarily set as $x_i = 0$, and a constraint is imposed that the amount of water in each reservoir cannot drop below this value.

On the basis of flood control considerations, a maximum water level for each reservoir is established. The amount of water needed to raise the level from the minimum to the maximum value is then expressed in terms of the normalized units, and a constraint is imposed that each x_i cannot exceed this level. The particular constraints considered in this example are expressed as

$$\begin{aligned} 0 \le x_1 \le 10 \\ 0 \le x_2 \le 10 \\ 0 \le x_3 \le 10 \\ 0 \le x_4 \le 15. \end{aligned} \tag{12.21}$$

The control variables are taken to be the flows out of the reservoirs. These quantities are also expressed in the same normalized units; the variables $u_i(k)$, $i = 1, 2, \ldots, 4$, specify the amount of water released from the ith reservoir over the kth time interval. In this example each time interval is two hours.

For each reservoir a maximum flow is determined by the capacity of the power generators, and a minimum flow is determined by considering the use of the downstream flow for navigation, conservation, and municipal and industrial water supplies. For this example the constraints were

$$\begin{aligned} 0 &\le u_1 \le 3 \\ 0 &\le u_2 \le 4 \\ 0 &\le u_3 \le 4 \\ 0 &\le u_4 \le 7 \end{aligned} \tag{12.22}$$

The system equations express how the water flows between the reservoirs. They are

$$\begin{aligned} x_1(k+1) &= x_1(k) - u_1(k) + IN_1 \\ x_2(k+1) &= x_2(k) - u_2(k) + IN_2 \\ x_3(k+1) &= x_3(k) - u_3(k) + u_2(k) \\ x_4(k+1) &= x_4(k) - u_4(k) + u_3(k) + u_1(k) \qquad (k = 0, 1, \ldots, 11). \end{aligned} \tag{12.23}$$

The inflows IN_1 and IN_2 are assumed constant over the day as

$$\begin{aligned} IN_1 &= 2 \\ IN_2 &= 3. \end{aligned} \tag{12.24}$$

The performance criterion considers the use of water for both power generation and irrigation. It is assumed that there is a power generation station at each reservoir outflow. The benefit from the flow over a given two-hour period is assumed to be a linear function of the flow, i.e., the benefit from a flow out of reservoir i at time k is $c_i(k)u_i(k)$. The function $c_i(k)$ is based on the power curve in Ref. 4. The values of $c_i(k)$ are shifted in k with respect to each other to account for the transport delay of water between reservoirs. This delay is four hours from Reservoir 1 to Reservoir 4, four hours from Reservoir 2 to Reservoir 3, and two hours from Reservoir 3 to Reservoir 4. The values of $c_i(k)$, $i = 1, 2, 3, 4$ are shown in Table 12.9.

Irrigation benefits are considered only for the outflow from reservoir 4. The benefit is again linear with flow—i.e., the benefit from flow $u_4(k)$ is $c_5(k)u_4(k)$. The function $c_5(k)$ is shown in Table 12.9.

The benefit function also includes a terminal cost for failing to reach a specified level for each reservoir at the end of the day. This function accounts for the long-term policy of filling or emptying the reservoir during a

Table 12.9. Time Functions Used in Calculating Benefit

k	$c_1(k)$	$c_2(k)$	$c_3(k)$	$c_4(k)$	$c_5(k)$
0	1.1	1.4	1.0	1.0	1.6
1	1.0	1.1	1.0	1.2	1.7
2	1.0	1.0	1.2	1.8	1.8
3	1.2	1.0	1.8	2.5	1.9
4	1.8	1.2	2.5	2.2	2.0
5	2.5	1.8	2.2	2.0	2.0
6	2.2	2.5	2.0	1.8	2.0
7	2.0	2.2	1.8	2.2	1.9
8	1.8	2.0	2.2	1.8	1.8
9	2.2	1.8	1.8	1.4	1.7
10	1.8	2.2	1.4	1.1	1.6
11	1.4	1.8	1.1	1.0	1.5

particular season. This function assesses a heavy penalty for having less than the specified amount of water at the end of the day, but gives no credit for having more than this amount. The particular function used was

$$\psi_i[x_i(12), m_i] = \begin{Bmatrix} -40[x_i(12) - m_i]^2, & x_i(12) \leq m_i \\ 0, & \text{otherwise} \end{Bmatrix}, \tag{12.25}$$

where m_i = desired level of reservoir i at end of the day ($k = 12$).

The initial state was taken to be

$$\begin{aligned} x_1(0) &= 5 \\ x_2(0) &= 5 \\ x_3(0) &= 5 \\ x_4(0) &= 5. \end{aligned} \tag{12.26}$$

The desired final state was

$$\begin{aligned} m_1 &= 5 \\ m_2 &= 5 \\ m_3 &= 5 \\ m_4 &= 7. \end{aligned} \tag{12.27}$$

The system dynamic equations are as in Eqs. (37) and (38). The constraints are expressed in Eqs. (35) and (36). The performance criterion is

$$J = \sum_{k=0}^{11} \sum_{i=1}^{4} c_i(k)u_i(k) + \sum_{k=0}^{11} c_5(k)u_4(k) + \sum_{i=1}^{4} \psi_i[x_i(12), m_i] \tag{12.28}$$

where $c_i(k)$, $i = 1, 2, \ldots, 5$ is specified in Table 12.9, $\psi_i[x_i(12), m_i]$ is shown in Eq. (12.25) and m_i, $i = 1, 2, 3, 4$, are given in Eq. (12.27).

Table 12.10. Initial Policy

k	$x_1(k)$	$x_2(k)$	$x_3(k)$	$x_4(k)$	$u_1(k)$	$u_2(k)$	$u_3(k)$	$u_4(k)$
0	5	5	5	5	2	3	3	5
1	5	5	5	5	2	3	3	5
2	5	5	5	5	2	3	3	5
3	5	5	5	5	2	3	3	5
4	5	5	5	5	2	3	3	5
5	5	5	5	5	2	3	3	5
6	5	5	5	5	2	3	3	5
7	5	5	5	5	2	3	3	5
8	5	5	5	5	2	3	3	5
9	5	5	5	5	2	3	3	5
10	5	5	5	5	2	3	3	5
11	5	5	5	5	2	3	3	3
12	5	5	5	7				

Total Benefit = 362.5

The initial policy chosen is shown in Table 12.10. Basically, this policy consists of setting outflow equal to inflow at every time period, so that the water level in each reservoir remains constant. The only exception to this policy occurs at the end of the day, when the terminal cost function is taken account.

The final results of applying a particular successive approximations algorithm [Ref. 4] to this problem are shown in Table 12.11. The improvement in total benefit was from 362.5 to 401.3 units. Note that the optimum policy

Table 12.11. Optimum Policy

k	$x_1(k)$	$x_2(k)$	$x_3(k)$	$x_4(k)$	$u_1(k)$	$u_2(k)$	$u_3(k)$	$u_4(k)$
0	5	5	5	5	1	4	0	0
1	6	4	8	7	0	1	0	2
2	8	5	10	5	0	2	4	7
3	10	7	8	1	2	0	4	7
4	10	10	4	0	3	3	4	7
5	9	10	3	0	3	4	4	7
6	8	9	3	0	3	4	4	7
7	7	8	3	0	3	4	4	7
8	6	7	3	0	3	4	4	7
9	5	6	3	0	3	4	4	7
10	4	5	3	0	3	4	4	0
11	3	4	3	7	0	2	0	0
12	5	5	5	7				

Total Benefit = 401.3

is very different from the initial policy. Nine iterations were required to obtain convergence; these calculations required approximately 30 seconds on the B-5500. Because of the linearity of the performance criterion, it was possible to compute the true optimum solution by linear programming; for this problem the successive approximations solution was the same as the true optimum solution.

For several other cases substantially the same results as quoted here were obtained. The extension to more complex performance criteria and constraints presents little difficulty. Stochastic inflows can be explicitly taken into account.

12.5 PROCEDURE FOR OBTAINING THE NOMINAL TRAJECTORY

A characteristic of the method of successive approximations is that a nominal trajectory is required. As remarked in Section 12.2, if this nominal is not sufficiently close to the true optimum solution, there is a possibility of convergence to a local optimum rather than the true optimum. Some methods for obtaining a nominal trajectory were discussed in Sec. 6.6. In this section another method, which has been used in several applications to obtain a nominal very close to the true optimum, is discussed. This approach is closely related to the method of successive approximation, and for this reason it is presented here.

This procedure for obtaining a nominal trajectory will be presented only for a special class of problems. However, many practical cases, including water resource problems of the type discussed in the previous section and the airline scheduling problem discussed in the next section, fit this formulation. Extensions to other classes of problems can be made, but a general algorithm has not yet been developed.

Within the general problem formulation of Eqs. (12.1–4), the class of problems to which this method applies is distinguished by a small amount of coupling between the various state and control variables. For the case where there are as many control variables as state variables, the system equations must have the form

$$x_i(k+1) = g_i[x_i(k), \ldots, x_n(k), u_i(k), \ldots, u_n(k), k] \qquad (i = 1, 2, \ldots, n) \tag{12.29}$$

i.e., the system equation for the ith state variable depends only on state and control variables with an index higher than i. In addition, the single-stage cost function must be separable into the form

$$l[\mathbf{x}(k), \mathbf{u}(k), k] = \sum_{i=1}^{n} l_i[x_i(k), u_i(k), k] \tag{12.30}$$

i.e., the sum of terms, each of which involves only one state and/or control variable. Finally, the constraints must have the form

$$x_i \in X_i(k)$$
$$u_i \in U_i(x_i, k) \qquad (i = 1, 2, \ldots, n). \tag{12.31}$$

The initial condition for the problem is some specified $\mathbf{x}(0)$.

The condition that is usually most difficult to satisfy in practice is Eq. (12.29), the form of the system equations. Sometimes, a transformation of the state variables to another coordinate system and/or a re-ordering of variables will yield the desired relations.

If the system equations instead had the form

$$x_i(k+1) = g_i[x_i(k), u_i(k), k] \qquad (i = 1, 2, \ldots, n) \tag{12.32}$$

then the problem could immediately be separated into n one-dimensional problems each with system equation as in Eq. (12.32), performance criterion

$$\sum_{k=0}^{K} l_i[x_i(k), u_i(k), k],$$

constraints as in Eq. (12.31), and initial condition $x_i(0)$. The complete optimal control vector would have as components the $\hat{u}_i(x_i, k)$ determined from the individual optimizations, and the total minimum cost would be the sum of the individual minimum costs.

Although this separation is not directly possible, a good approximation to the true optimum can be found by solving the following sequence of optimization problems: The first problem solved has system equation

$$x_n(k+1) = g_n[x_n(k), u_n(k), k] \tag{12.33}$$

performance criterion

$$J = \sum_{k=0}^{K} l_n[x_n(k), u_n(k), k] \tag{12.34}$$

constraints

$$x_n \in X_i(k)$$
$$u_n \in U_i(x_n, k) \tag{12.35}$$

and initial condition $x_n(0)$. The resulting optimal trajectory and control sequence are denoted as $\{\hat{x}_n(k)\}$ and $\{\hat{u}_n(k)\}$ respectively. The next optimization problem solved has system equation

$$x_{n-1}(k+1) = g_{n-1}[x_{n-1}(k)\hat{x}_n(k), u_{n-1}(k), \hat{u}_n(k), k] \tag{12.36}$$

performance criterion

$$J = \sum_{k=0}^{K} l_{n-1}[x_{n-1}(k), u_{n-1}(k), k] \tag{12.37}$$

constraints

$$x_{n-1} \in X_{n-1}(k)$$
$$u_{n-1} \in U_{n-1}(x_{n-1}, k) \tag{12.38}$$

and initial condition $x_{n-1}(0)$. The resulting optimal trajectory and control sequence are denoted as $\{\hat{x}_{n-1}(k)\}$ and $\{\hat{u}_{n-1}(k)\}$ respectively. The remainder of the sequence of problems is defined as above in order of decreasing n, where in the ith problem the functions $x_j(k)$ and $u_j(k)$, $j > i$, are taken from the sequences $\{\hat{x}_j(k)\}$ and $\{\hat{u}_j(k)\}$, respectively.

The control sequence $\hat{\mathbf{u}}(\mathbf{x}, k)$, which has components $\hat{u}_i(k), i = 1, 2, \ldots, n$, is not necessarily the optimum control sequence, but it generally is not far from the optimum. Thus, the corresponding trajectory is a good nominal trajectory for use in the successive approximations algorithm. In addition, as will be shown in the next section, this policy may actually be the optimum policy, or it may be so close to optimum that further iterations are not warranted; thus, this technique is also a good method for obtaining near-optimum policies. Like the successive approximations technique, this method can be generalized to systems having fewer control than state variables.

12.6 A SUCCESSIVE APPROXIMATIONS PROCEDURE FOR THE AIRLINE SCHEDULING PROBLEM

The airline scheduling problem has already been formulated in Sec. 10.3. In this section a computational algorithm for this problem based on the procedure of the previous section is described. The procedure has been used in applications involving as many as 100 aircraft, and it is readily extendable to larger problems.

To review, the airline scheduling problem can be formulated as follows: Given:

(i) A specification of the airline system that includes for each origin-destination pair the flight time.
(ii) A set of scheduled flights, for each of which the origin, destination, departure time and net benefit are specified.
(iii) A set of allowable ferries (unscheduled empty flights) and their net costs.
(iv) A fleet of available aircraft.
(v) A set of constraints on how the aircraft can be operated.

Find: The assignment of aircraft to scheduled flights and ferries such that the total benefit of scheduled flights minus the cost of ferries is maximized, subject to the constraints.

In Sec. 10.3 the net benefit of a scheduled flight was computed on the basis of passenger revenue and direct operating cost, while the cost of a ferry was taken to be the direct operating cost. Many other types of benefit and cost functions can be used without changing the method.

Several constraints that can be applied to the system were mentioned in Sec. 10.3. These include restrictions on the initial and final locations of the various aircraft, on the availability of a given aircraft because of unscheduled maintenance or lack of a qualified crew, on access to a given airport because of weather, and on the number of hours a given aircraft can fly before scheduled maintenance must be performed. Constraints such as these are of primary importance in short-term optimization problems, over time periods of 24 hours or less. In longer-term applications, such as schedule planning, somewhat different constraints may be more appropriate. For example, the initial and final locations of each aircraft may not be specified, but instead the number of aircraft in each airport at the start and end of the optimization may be specified. Alternatively, there may be no constraint at all on initial and final aircraft location. In the area of total hours of flying time, the constraint may merely be that no aircraft is allowed to fly more than a given number of hours per day. Often, in planning applications, ferries are not allowed. All of these modifications can very easily be taken into account.

The procedure of the previous section can be applied to this problem. As a result, a sequence of one-dimensional optimizations are performed. After each such optimization the flights taken are deleted from the schedule available to the remaining aircraft. The exact methods for implementing constraints vary with the particular application. For example, if the initial and final location of the aircraft being considered are both fixed, this constraint is applied directly. On the other hand, if the number of aircraft in each airport at the start and end of the optimization are specified, then all initial and final locations for which these numbers have not yet been exceeded are allowed. It is also possible to mix these constraints in a single problem; for example, a few aircraft may have very stringent constraints, including specified initial and final locations, while for the remainder of the fleet the constraint is only on the number of aircraft at the start and end of the optimization. This is typically the case in planning operations over a 24-hour period.

In the one-dimensional optimizations a very important computational simplification takes place—the number of discrete states is equal to the number of airports, and no intermediate states of the type discussed in Sec. 10.3 are necessary. Scheduled flights and ferries appear as transitions from one discrete state to another, where the time of the transition is the flight time. Typical transitions are shown pictorially in Fig. 12.6. This type of transition makes the consideration of non-integral flight times very straightforward.

It has been shown that optimum performance can be obtained by considering transitions only when there is either a scheduled flight available or an aircraft has just arrived at an airport on a scheduled flight (the latter case is the only circumstance under which ferries need to be considered). Thus, even though the transition times and departure times are allowed to vary continuously, only a finite number of flights need to be considered in a particular one-dimensional optimization. These types of transitions are quite similar to those in certain state increment dynamic programming procedures.

The computational requirements of this procedure are very reasonable, and applications involving up to 100 airplanes and 25 airports have been solved in a few minutes on a medium-speed computer.* Because the individual optimization problems involve only one state variable, which is quantized only to as many discrete levels as there are airports, there is virtually no high-speed storage problem. Computing time increases linearly with the number of aircraft and slightly faster than linearly with the number of airports.

A number of examples have been solved with this procedure. In the few simple cases where the true optimum could be computed by other methods, such as the procedure of Sec. 10.3, it was found that the procedure obtained

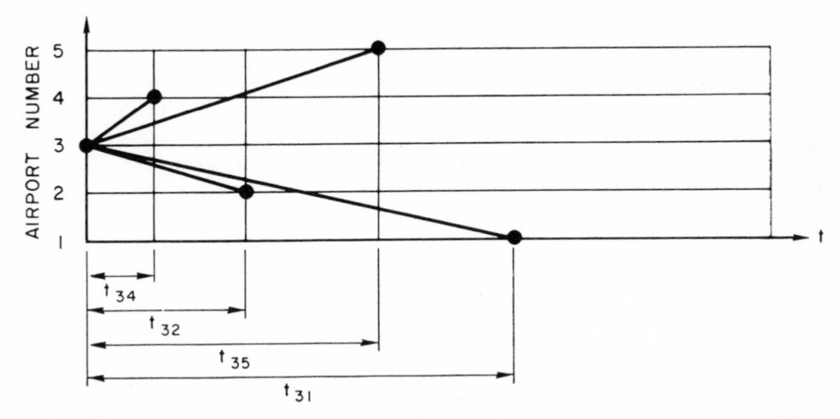

Fig. 12.6. A typical set of transitions from a given airport in the one-dimensional optimization.

this solution. Although it is conceptually possible to construct situations in which the procedure would fail to find the true optimum, such cases are easy to detect and correct by appropriate modifications of the algorithm. Work is presently being carried out in developing a complete successive approximations procedure for iterating on the solution generated by this algorithm. From the experience gained so far in this area, it appears that a relatively small number of iterations would be required to obtain convergence and that in most cases the convergence would be to the true optimum.

A representative schedule planning problem is shown in Fig. 12.7 and Tables 12.12 to 12.15. The objective in this study was to determine an operating policy such that a minimum number of aircraft, all of the same type, was required to meet a given schedule, and that for this number of aircraft net benefit was maximized.

The airline system in the study contains 10 airports. These airports are shown in Fig. 12.7; airport pairs between which there are scheduled flights are

* B-5500.

connected by lines. The schedule was designed so that 60 aircraft could serve all scheduled flights without any ferries. The complete schedule of 305 flights is shown in Table 12.12. Each column in the table contains entries expressing net benefit, in arbitrary units, of the scheduled flights between a given origin-destination pair. The entry for a particular flight appears in the row

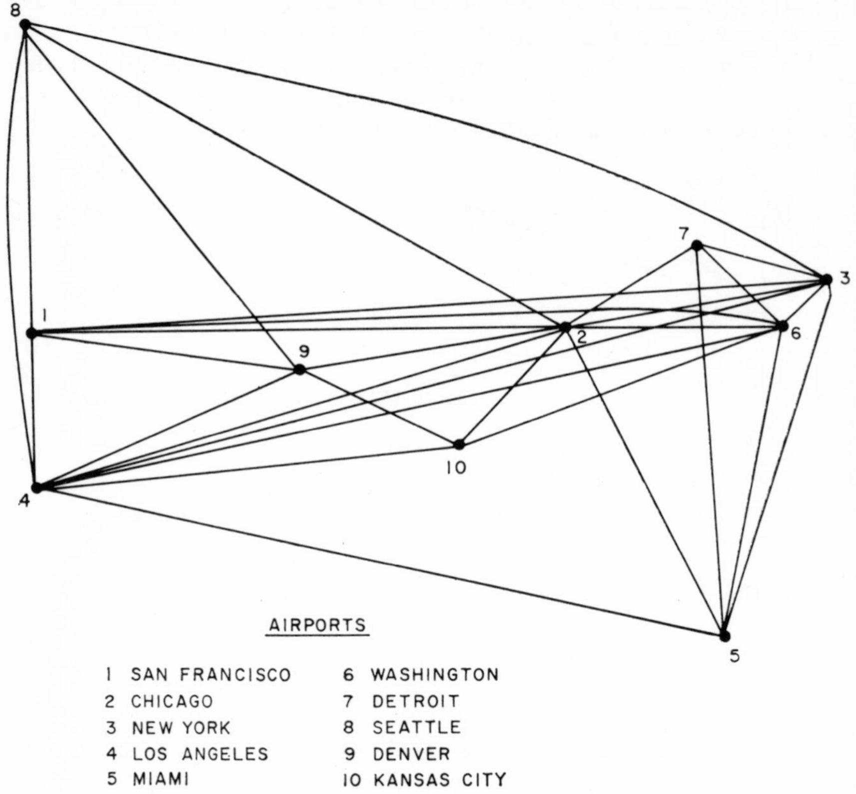

Fig. 12.7. System map. (Allowed origin-destination pairs are indicated by a line between the two airports).

corresponding to the time of departure. The flight times, in hours, are shown in Table 12.13. Because ferries are allowed between all airport pairs, this table contains an entry for every possible combination. The cost of a ferry, in the same arbitrary units as the benefit, is taken to be 10 units for each hour of flying time. There are no constraints on the initial or final locations of the aircraft. However, each aircraft is constrained to fly at most 20 hours during the optimization period (24 hours).

The algorithm described in this section was applied to this problem. The computing time required was 3 minutes on the B-5500.

Table 12.12. Scheduled Flights and Corresponding Benefits (All flights for a given origin-destination pair are listed in the same column. Origin-destination appears at the head of the column. For each scheduled flight the benefit is listed in the row corresponding to the departure time.)

O/D	*San Francisco*						*Chicago*									*New York*				
Time of departure	1/2	1/3	1/4	1/6	1/8	1/9	2/1	2/3	2/4	2/5	2/6	2/7	2/8	2/9	2/10	3/1	3/2	3/4	3/5	3/6
0	100		30	150					100	80		30		55	30		55	150	80	30
1		150									55					150				
2							100						100							
3								55												
4																	55			
5			30												30					
6	100		30		55	55			100	80	55		100	55		150	55	150		30
7			30	150			100	55											80	30
8	100	150	30						100			30					55			
9			30								55			55		150	55	150		30
10			30		55		100	55		80		30								30
11	100		30						100											
12		150	30	150		55		55						55		150	55			30
13			30								55				30			150	80	
14			30				100	55					100							30
15			30														55		80	
16		150	30		55			55		80	55	30		55						
17	100		30					55									55			30
18		150	30	150	55	55	100	55					100			150		150		30
19			30						100		55			55			55		80	
20	100		30				100													30
21			30					55			55						55			
22			30									30								
23			30																	

Table 12.12 Scheduled Flights and Corresponding Benefits (Continued)

Time of departure \ O/D	New York		Los Angeles								Miami					Washington				
	3/7	3/8	4/1	4/2	4/3	4/5	4/6	4/8	4/9	4/10	5/2	5/3	5/4	5/6	5/7	6/1	6/2	6/3	6/4	6/5
0				100	150			80				80		80						
1			30							80	80					150		30		80
2	30																			
3																				
4									55										150	
5			30												80		55			
6	30		30		150	125	150					80	125	80				30		
7	30	150	30	100						80							55			
8			30					80			80							30	150	
9			30											80		150				
10			30	100			150	80	55								55	30		80
11			30												80					
12			30	100	150							80				150	55	30	150	
13			30				150				80		125							
14	30		30						55						80					80
15			30		150												55	30		
16			30	100														30		
17			30			125						80					55	30	150	
18	30	150	30		150		150			80	80			80		150		30		
19			30	100								80								
20	30		30								80									
21			30														55	30		80
22			30																	
23			30																	

Table 12.12 Scheduled Flights and Corresponding Benefits (Continued)

Time of departure \ O/D	Washington		Detroit				Seattle					Denver					Kansas City			
	6/7	6/10	7/2	7/3	7/5	7/6	8/1	8/2	8/3	8/4	8/9	9/1	9/2	9/4	9/8	9/10	10/2	10/4	10/6	10/9
0								100		80										
1		80	30														30	80		
2	30												55							
3																				
4					80															30
5						30				80	80			55						
6			30	30				100							80	30		80		30
7	30								150			55					30		80	
8			30	30	80		55						55	55						
9						30					80						30			
10							55												80	30
11				30									55							
12	30					30		100							80	30	30			
13		80		30						80		55								
14			30				55						55							
15														55				80		
16					80															
17	30			30							80		55			30				
18		80						100	150	80	80	55							80	30
19			30												80					
20	30						55						55	55						
21																30				
22				30													30			
23																				

Table 12.13 Flight Times and Costs of Ferries (The flight time in hours is the first number, and the cost of the ferry in arbitrary units is the second number.)

Destination \ Origin	1	2	3	4	5	6	7	8	9	10
1		4, −40	6, −60	1, −10	6, −60	6, −60	5, −50	2, −20	2, −20	3, −30
2	4, −40		2, −20	4, −40	3, −30	2, −20	1, −10	4, −40	2, −20	1, −10
3	6, −60	2, −20		6, −60	3, −30	1, −10	1, −10	6, −60	4, −40	3, −30
4	1, −10	4, −40	6, −60		5, −50	6, −60	5, −50	3, −30	2, −20	3, −30
5	6, −60	3, −30	3, −30	5, −50		3, −30	3, −30	7, −70	4, −40	3, −30
6	6, −60	2, −20	1, −10	6, −60	3, −30		1, −10	6, −60	4, −40	3, −30
7	5, −50	1, −10	1, −10	5, −50	3, −30	1, −10		4, −40	3, −30	2, −20
8	2, −20	4, −40	6, −60	3, −30	7, −70	6, −60	4, −40		3, −30	4, −40
9	2, −20	2, −20	4, −40	2, −20	4, −40	4, −40	3, −30	3, −30		1, −10
10	3, −30	1, −10	3, −30	3, −30	3, −30	3, −30	2, −20	4, −40	1, −10	

Table 12.14 Solution to the One-Dimensional Optimization Problem for the First Aircraft
(The entry at a given airport and time denotes the maximum profit, the airport from which the optimal flight just came, and the accumulated number of flight hours. Underscored entries specify the optimal flight sequence.)

Time ↓ / *Airport number* →	1	2	3	4	5	6	7	8	9	10
0	0. 0. 0	0. 0. 0	0. 0. 0	0. 0. 0	0. 0. 0	0. 0. 0	0. 0. 0	0. 0. 0	0. 0. 0	0. 0. 0
1	0. 1. 0	0. 2. 0	0. 3. 0	30. 1. 1	0. 5. 0	30. 3. 1	30. 2. 1	0. 8. 0	0. 9. 0	30. 2. 1
2	60. 4. 2	60.10. 2	60. 6. 2	30. 4. 1	0. 5. 0	30. 6. 1	30. 7. 1	0. 8. 0	55. 2. 2	30.10. 1
3	60. 1. 2	60. 2. 2	80. 5. 3	80. 8. 3	80. 2. 3	80. 5. 3	90. 3. 3	80. 4. 3	55. 9. 2	50. 2. 3
4	70. 4. 4	110. 9. 4	80. 7. 4	110.10. 4	110. 6. 4	80. 7. 4	90. 7. 3	80. 8. 3	55. 9. 2	110. 6. 4
5	100. 4. 5	110. 2. 4	115. 2. 4	110. 5. 4	110. 5. 4	80. 7. 4	100. 2. 5	80. 8. 3	140.10. 5	110.10. 4
6	160. 2. 6	135. 3. 6	150. 4. 6	150. 3. 6	110. 5. 4	150. 1. 6	105. 3. 6	160. 2. 6	165. 4. 6	140. 2. 5
7	180. 4. 7	135. 7. 6	180. 6. 7	195. 9. 7	170. 7. 6	180. 3. 7	180. 3. 7	160. 8. 6	170.10. 6	195. 9. 7
8	225. 4. 8	225.10. 8	180. 3. 7	210. 1. 8	170. 5. 6	210. 3. 8	210. 6. 8	215. 1. 8	215. 1. 8	195.10. 7
9	240. 4. 9	240. 7. 9	240. 7. 9	255. 1. 9	215. 2. 9	210. 6. 8	255. 2. 9	245. 9. 9	215. 9. 8	215. 2. 9
10	285. 4.10	270. 9.10	280. 2.10	270. 1.10	260. 3.10	285. 7.10	255. 7. 9	245. 8. 9	215. 9. 8	275. 4.10
11	300. 4.11	295. 3.11	315. 6.11	315. 1.11	290. 7.11	310. 3.11	300. 2.11	290. 4.11	305.10.11	275.10.10
12	345. 4.12	340. 6.12	330. 7.12	330. 1.12	290. 5.11	310. 6.11	305. 3.22	340. 1.12	335. 8.12	295. 9.12
13	360. 4.13	360. 9.13	340. 6.12	375. 1.13	365. 6.13	360. 3.13	340. 6.12	350. 4.13	335. 9.12	365. 9.13
14	405. 4.14	385. 3.14	395. 2.14	390. 1.14	365. 5.13	360. 6.13	370. 5.14	350. 8.13	400. 1.14	390. 2.14
15	420. 4.15	400. 7.15	395. 3.14	435. 1.15	365. 5.13	425. 3.15	425. 3.15	415. 9.15	400. 9.14	390. 9.15
16	465. 4.16	455. 9.16	455. 6.16	450. 1.16	420. 3.15	425. 6.15	425. 7.15	415. 8.15	445. 4.16	440. 6.16
17	480. 4.17	480. 6.17	455. 6.16	495. 1.17	440. 6.16	445. 3.17	485. 2.17	415. 8.15	445. 9.16	445. 2.17
18	525. 4.18	480. 2.17	515. 7.18	510. 1.18	475. 3.17	510. 2.18	485. 7.17	520. 1.18	510. 2.18	475. 9.17
19	540. 4.19	510. 3.18	540. 6.19	555. 1.19	535. 2.10	545. 3.19	545. 3.19	520. 8.18	510. 9.18	500. 9.19
20	585. 4.20	575. 7.20	540. 3.19	570. 1.20	535. 5.19	545. 6.19	545. 7.10	580. 1.20	580. 1.20	500. 9.19
21	585. 1.20	575. 2.20	540. 3.19	570. 4.20	535. 5.19	570. 3.20	575. 6.20	580. 8.20	580. 9.20	500.10.19
22	585. 1.20	575. 2.20	540. 3.19	570. 4.20	535. 5.19	570. 6.20	575. 7.20	580. 8.20	580. 9.20	500.10.19
23	585. 1.20	575. 2.20	540. 3.19	570. 4.20	535. 5.19	570. 6.20	575. 7.20	580. 8.20	580. 9.20	500.10.19
24	585. 1.20	575. 2.20	540. 3.19	570. 4.20	535. 5.19	570. 6.20	575. 7.20	580. 8.20	580. 9.20	500.10.19

Table 12.15* Optimum Schedules for the 47 Aircraft

Aircraft 1

Hour	*Org*	*Dest*	*Profit*	*Hrs*
0	3	6	30	1
1	6	10	80	4
4	10	9	30	5
5	9	4	55	7
7	4	1	30	8
8	1	4	30	9
9	4	1	30	10
10	1	4	30	11
11	4	1	30	12
12	1	4	30	13
13	4	1	30	14
14	1	4	30	15
15	4	1	30	16
16	1	4	30	17
17	4	1	30	18
18	1	4	30	19
19	4	1	30	20

Total profit = 585

Aircraft 2

Hour	*Org*	*Dest*	*Profit*	*Hrs*
0	2	10	30	1
1	10	4	80	4
4	4	9	55	6
6	9	10	30	7
7	10	2	30	8
8	2	7	30	9
9	7	6	30	10
10	6	2	55	12
12	2	9	55	14
14	9	2	55	16
16	2	7	30	17
17	7	3	30	18
18	3	7	30	19
19	7	2	30	20

Total profit = 570

Aircraft 3

Hour	*Org*	*Dest*	*Profit*	*Hrs*
0	3	4	150	6
6	4	1	30	7
7	1	4	30	8
8	4	1	30	9
9	1	4	30	10
10	4	1	30	11
11	1	4	30	12
12	4	1	30	13
13	1	4	30	14
14	4	1	30	15
15	1	4	30	16
16	4	1	30	17
17	1	4	30	18
18	4	1	30	19
19	1	4	30	20

Total profit = 570

Aircraft 4

Hour	*Org*	*Dest*	*Profit*	*Hrs*
0	2	7	30	1
1	7	2	30	2
2	2	1	100	6
6	1	4	30	7
7	4	10	80	10
10	10	9	30	11
11	9	2	55	13
13	2	6	55	15
15	6	2	55	17
18	2	3	55	19
20	3	6	30	20

Total profit = 550

LEGEND*

Hour = Time of Departure
Org = Origin
Dest = Destination
Profit = Net Benefit of a Scheduled Flight or Cost of a Ferry
Hrs = Accumulated Number of Hours Flown

Table 12.15 Optimum Schedules for the 47 Aircraft (Continued)

Aircraft 5

Hour	*Org*	*Dest*	*Profit*	*Hrs*
0	4	3	150	6
6	3	6	30	7
7	6	7	30	8
8	7	3	30	9
9	3	6	30	10
10	6	5	80	13
13	5	2	80	16
16	2	3	55	18
18	3	6	30	19
20	6	7	30	20

Total profit = 545

Aircraft 6

Hour	*Org*	*Dest*	*Profit*	*Hrs*
0	5	3	80	3
4	3	2	55	5
6	2	9	55	7
8	9	4	55	9
10	4	6	150	15
16	6	3	30	16
17	3	6	30	17
18	6	3	30	18
19	3	2	55	20

Total profit = 540

Aircraft 7

Hour	*Org*	*Dest*	*Profit*	*Hrs*
0	1	6	150	6
6	6	3	30	7
7	3	7	30	8
8	7	5	80	11
11	5	7	80	14
14	7	2	30	15
16	2	9	55	17
18	9	1	55	19
20	1	4	30	20

Total profit = 540

Aircraft 8

Hour	*Org*	*Dest*	*Profit*	*Hrs*
0	3	2	55	2
2	2	8	100	6
6	8	2	100	10
10	2	7	30	11
11	7	3	30	12
12	3	2	55	14
14	2	3	55	16
17	3	2	55	18
19	2	6	55	20

Total profit = 535

Aircraft 9

Hour	*Org*	*Dest*	*Profit*	*Hrs*
0	2	4	100	4
5	4	1	30	5
6	1	9	55	7
8	9	2	55	9
10	2	3	55	11
12	3	6	30	12
14	6	5	80	15
17	5	3	80	18
20	3	7	30	19
22	7	3	30	20

Total profit = 545

Aircraft 10

Hour	*Org*	*Dest*	*Profit*	*Hrs*
0	2	9	55	2
2	9	2	55	4
6	2	5	80	7
9	5	6	80	10
12	6	7	30	11
13	7	3	30	12
15	3	5	80	15
18	5	2	80	18
21	2	3	55	20

Total profit = 545

Table 12.15 Optimum Schedules for the 47 Aircraft (*Continued*)

Aircraft 11

Hour	*Org*	*Dest*	*Profit*	*Hrs*
1	1	3	150	6
7	3	6	30	7
8	6	4	150	13
14	4	9	55	15
17	9	10	30	16
18	10	6	80	19
21	6	3	30	20

Total profit = 525

Aircraft 14

Hour	*Org*	*Dest*	*Profit*	*Hrs*
1	10	2	30	1
3	2	3	55	3
6	3	2	55	5
8	2	4	100	9
12	4	2	100	13
16	2	4	100	17
20	4	1	30	18
21	1	4	30	19
23	4	1	30	20

Total profit = 530

Aircraft 12

Hour	*Org*	*Dest*	*Profit*	*Hrs*
0	5	6	80	3
4	6	4	150	9
10	4	8	80	12
13	8	4	80	15
16	4	2	100	19
22	2	7	30	20

Total profit = 520

Aircraft 15

Hour	*Org*	*Dest*	*Profit*	*Hrs*
0	8	4	80	3
4	4	1	−10	4
5	1	4	30	5
6	4	6	150	11
12	6	3	30	12
13	3	5	80	15
18	5	6	80	18
21	6	2	55	20

Total profit = 495

Aircraft 13

Hour	*Org*	*Dest*	*Profit*	*Hrs*
0	1	4	30	1
1	4	10	80	4
4	10	2	−10	5
5	2	10	30	6
6	10	9	30	7
7	9	1	55	9
10	1	8	55	11
12	8	2	100	15
16	2	6	55	17
18	6	10	80	20

Total profit = 505

Aircraft 16

Hour	*Org*	*Dest*	*Profit*	*Hrs*
1	6	5	80	3
5	5	7	80	6
8	7	2	30	7
9	2	6	55	9
12	6	2	55	11
14	2	1	100	15
18	1	9	55	17
20	9	4	55	19
22	4	1	30	20

Total profit = 540

Table 12.15 Optimum Schedules for the 47 Aircraft (Continued)

Aircraft 17

Hour	Org	Dest	Profit	Hrs
0	1	2	100	4
6	2	6	55	6
8	6	3	30	7
10	3	6	30	8
12	6	1	150	14
18	1	3	150	20

Total profit = 515

Aircraft 18

Hour	Org	Dest	Profit	Hrs
0	4	2	100	4
6	2	4	100	8
10	4	2	100	12
14	2	8	100	16
18	8	4	80	19
21	4	1	30	20

Total profit = 510

Aircraft 19

Hour	Org	Dest	Profit	Hrs
1	6	1	150	6
8	1	2	100	10
12	2	3	55	12
14	3	7	30	13
16	7	5	80	16
20	5	2	80	19

Total profit = 495

Aircraft 20

Hour	Org	Dest	Profit	Hrs
0	8	2	100	4
6	2	8	100	8
10	8	1	55	10
12	1	9	55	12
14	9	10	−10	13
15	10	4	80	16
18	4	10	80	19
22	10	2	30	20

Total profit = 490

Aircraft 21

Hour	Org	Dest	Profit	Hrs
1	2	6	55	2
5	6	2	55	4
7	2	3	55	6
9	3	2	55	8
11	2	4	100	12
15	4	3	150	18
21	3	2	55	20

Total profit = 525

Aircraft 22

Hour	Org	Dest	Profit	Hrs
0	2	5	80	3
6	5	6	80	6
9	6	1	150	12
16	1	8	55	14
18	8	3	150	20

Total profit = 515

Table 12.15 Optimum Schedules for the 47 Aircraft (*Continued*)

Aircraft 23

Hour	*Org*	*Dest*	*Profit*	*Hrs*
1	3	1	150	6
7	1	6	150	12
13	6	10	80	15
18	10	9	30	16
19	9	8	80	19

Total profit = 490

Aircraft 24

Hour	*Org*	*Dest*	*Profit*	*Hrs*
0	4	8	80	3
7	8	3	150	9
13	3	4	150	15
19	4	2	100	19

Total profit = 480

Aircraft 25

Hour	*Org*	*Dest*	*Profit*	*Hrs*
0	3	5	80	3
6	5	3	80	6
9	3	6	−10	7
10	6	3	30	8
12	3	1	150	14
18	1	6	150	20

Total profit = 480

Aircraft 26

Hour	*Org*	*Dest*	*Profit*	*Hrs*
1	5	2	80	3
5	2	7	−10	4
6	7	2	30	5
7	2	1	100	9
11	1	2	100	13
16	2	5	80	16
19	5	3	80	19

Total profit = 460

Aircraft 27

Hour	*Org*	*Dest*	*Profit*	*Hrs*
1	6	3	30	1
2	3	7	30	2
6	7	3	30	3
8	3	2	55	5
10	2	5	80	8
13	5	4	125	13
18	4	3	150	19

Total profit = 500

Aircraft 28

Hour	*Org*	*Dest*	*Profit*	*Hrs*
2	6	7	30	1
5	7	3	−10	2
6	3	1	150	8
12	1	6	150	14
18	6	1	150	20

Total profit = 470

Aircraft 29

Hour	*Org*	*Dest*	*Profit*	*Hrs*
6	3	4	150	6
12	4	3	150	12
18	3	1	150	18

Total profit = 450

Aircraft 30

Hour	*Org*	*Dest*	*Profit*	*Hrs*
1	4	1	30	1
5	1	4	−10	2
6	4	5	125	7
12	5	3	80	10
15	3	2	55	12
17	2	3	55	14
19	3	5	80	17

Total profit = 415

Table 12.15 Optimum Schedules for the 47 Aircraft (Continued)

Aircraft 31

Hour	*Org*	*Dest*	*Profit*	*Hrs*
5	8	9	80	3
9	9	10	30	4
10	10	6	80	7
13	6	3	−10	8
14	3	6	30	9
15	6	3	30	10
16	3	6	−10	11
17	6	3	30	12
18	3	4	150	18

Total profit = 410

Aircraft 32

Hour	*Org*	*Dest*	*Profit*	*Hrs*
4	7	5	80	3
8	5	2	80	6
11	2	10	−10	7
12	10	2	30	8
13	2	10	30	9
14	10	9	−10	10
15	9	4	55	12
18	4	6	150	18

Total profit = 405

Aircraft 33

Hour	*Org*	*Dest*	*Profit*	*Hrs*
6	9	8	80	3
9	8	9	80	6
12	9	8	80	9
17	8	9	80	12
20	9	2	55	14

Total profit = 375

Aircraft 34

Hour	*Org*	*Dest*	*Profit*	*Hrs*
5	7	6	30	1
6	6	3	−10	2
7	3	8	150	8
14	8	1	55	10
17	1	2	100	14
21	2	6	55	16

Total profit = 380

Aircraft 35

Hour	*Org*	*Dest*	*Profit*	*Hrs*
6	5	4	125	5
11	4	1	−10	6
12	1	3	100	12
18	3	8	150	18

Total profit = 365

Aircraft 36

Hour	*Org*	*Dest*	*Profit*	*Hrs*
6	1	2	100	4
10	2	1	100	8
16	1	3	150	14

Total profit = 350

Aircraft 37

Hour	*Org*	*Dest*	*Profit*	*Hrs*
6	10	4	80	3
10	4	9	55	5
13	9	1	55	7
18	1	8	55	9
20	8	1	55	11
23	1	4	30	12

Total profit = 330

Table 12.15 Optimum Schedules for the 47 Aircraft (*Continued*)

Aircraft 38

Hour	*Org*	*Dest*	*Profit*	*Hrs*
6	1	8	55	2
8	8	1	55	4
12	1	4	−10	5
13	4	6	150	11
21	6	5	80	14

Total profit = 330

Aircraft 39

Hour	*Org*	*Dest*	*Profit*	*Hrs*
7	10	6	80	3
12	6	4	150	9
19	4	1	−10	10
20	1	2	100	14

Total profit = 320

Aircraft 40

Hour	*Org*	*Dest*	*Profit*	*Hrs*
6	3	7	30	1
8	7	3	−10	2
9	3	4	150	8
15	4	9	−20	10
17	9	2	55	12
20	2	1	100	16

Total profit = 305

Aircraft 41

Hour	*Org*	*Dest*	*Profit*	*Hrs*
8	1	3	150	6
16	3	6	−10	7
17	6	2	55	9
19	2	4	100	13

Total profit = 295

Aircraft 42

Hour	*Org*	*Dest*	*Profit*	*Hrs*
6	4	3	150	6
16	3	6	−10	7
17	6	4	150	13

Total profit = 290

Aircraft 43

Hour	*Org*	*Dest*	*Profit*	*Hrs*
7	3	5	80	3
14	5	7	80	6
17	7	2	−10	7
18	2	1	100	11
22	1	4	30	12

Total profit = 280

Aircraft 44

Hour	*Org*	*Dest*	*Profit*	*Hrs*
9	3	1	150	6
16	1	4	−10	7
17	4	5	125	12

Total profit = 265

Aircraft 45

Hour	*Org*	*Dest*	*Profit*	*Hrs*
5	8	4	80	3
8	4	8	80	6
18	8	2	100	10

Total profit = 260

Table 12.15 Optimum Schedules for the 47 Aircraft (*Continued*)

Aircraft 46

Hour	*Org*	*Dest*	*Profit*	*Hrs*
7	6	2	55	2
9	2	9	55	4
12	9	10	30	5
17	10	2	−10	6
18	2	8	100	10

Total profit = 230

Aircraft 47

Hour	*Org*	*Dest*	*Profit*	*Hrs*
7	4	2	100	4
11	2	7	−10	5
12	7	6	30	6
17	6	7	30	7
18	7	2	−10	8
19	2	9	55	10
21	9	10	30	11

Total profit = 225

The results of the first one-dimensional optimization problem appear in Table 12.14. Forward dynamic programming was used to generate the table. Entries at each airport and hour specify the maximum net benefit, the optimum airport from which the airplane just came (the present airport if the optimum policy was for the aircraft to be on the ground for the previous one-hour time period), and the accumulated flying hours. The latter entry is necessary for imposing the 20-hour limit on flying hours for each aircraft.

Since there is no constraint on the final location of the aircraft, the terminal state is selected on the basis of maximum net benefit. Examination of the bottom row of the table shows that the maximum value appears in the column corresponding to airport 1 (San Francisco). The scheduling can then be recovered by tracing backwards in time on the basis of the previous airport entries in this table. The resulting sequence for this problem is specified by the circled entries in Table 12.14. Note that when optimum flights are longer than one hour, there are no circled entries in the table for this time period—for example, because the circled entry at airport 10 (Kansas City), time 4, specifies the previous airport as airport 6 (Washington), which is a three-hour flight from airport 10, the next entry in the sequence appears at airport 6, time 1, and there are no circled entries at times 2 and 3. Since there is no constraint on the initial location of this aircraft, the first airport in the sequence can be any of the 10; in this case, it happens to be airport 3 (New York).

In this problem it is found that 47 aircraft suffice to meet all scheduled flights. The schedules for all 47 aircraft are shown in Table 12.15. Thus, as a result of this procedure, 13 less aircraft are required than originally provided. The number of ferries required is 24, and the cost of the ferries is 250 units (corresponding to 25 hours of flying time). The total net benefit, including the cost of ferries, is 20,850 units; the reduction from 60 to 47 aircraft thus entails a cost in ferries of a little more than 1% of the total net benefit.

There is no guarantee that this solution is the absolute optimum. However, there is no obvious way to modify this solution so as to either reduce the number of aircraft required or to increase the net benefit with the 47 aircraft. The aircraft are clearly being used efficiently; 23 of the aircraft are flown for a maximum of 20 hours, only 15 are flown for less than 18 hours, and the minimum number of hours flown is 10. Improvements over this schedule might be obtained by iterations of a successive approximations procedure, but it is doubtful that the increase in net benefit would be significant.

If the departure times of scheduled flights are allowed to be changed, it is possible that a reduction in the number of ferries and/or required aircraft might be obtained. Techniques for doing this have been developed and applied to other examples, in some cases with significant improvements. However, these methods have not been tried here.

It is instructive to note the computational requirements of solving this problem according to the procedure of Sec. 10.3. If intermediate states are defined as in that section, a total of 409 discrete states are obtained. If a simultaneous optimization involving 47 aircraft is attempted, the high-speed memory requirement is $(409)^{47} \approx 10^{80}$, a truly astronomical figure. Computing time and the low-speed memory requirement are also tremendously large. Thus, the computational difficulties of this problem are reduced from totally impossible levels to those of a routine calculation on a medium-size computer.

REFERENCES

1. Bellman, R., *Adaptive Control Processes*, Princeton University Press, Princeton, New Jersey (1961).
2. Bellman, R., and Dreyfus, S., *Applied Dynamic Programming*, Princeton University Press, Princeton, New Jersey (1962).
3. Korsak, A. J., and Larson, R. E., "Convergence Proofs for a Dynamic Programming Successive Approximations Technique," submitted to Fourth IFAC Congress, Warsaw, Poland (June 1969).
4. Larson, R. E., and Keckler, W. G., "Applications of Dynamic Programming to Water Resource Problems," presented at IFAC Haifa Conference on Computer Control of Natural Resources and Public Utilities, Haifa, Israel (September 1967).

AUTHOR INDEX

Page numbers set in italics denote the pages on which the complete literature references are given.

SUBJECT INDEX